The Flexible Electronics Opportunity

Committee on
Best Practice in National Innovation Programs for Flexible Electronics

Board on Science, Technology, and Economic Policy

Policy and Global Affairs

Donald Siegel
Committee Chair

Sujai Shivakumar
Editor

NATIONAL RESEARCH COUNCIL
OF THE NATIONAL ACADEMIES

THE NATIONAL ACADEMIES PRESS
Washington, D.C.
www.nap.edu

THE NATIONAL ACADEMIES PRESS 500 Fifth Street, NW Washington, DC 20001

NOTICE: The project that is the subject of this report was approved by the Governing Board of the National Research Council, whose members are drawn from the councils of the National Academy of Sciences, the National Academy of Engineering, and the Institute of Medicine. The members of the committee responsible for the report were chosen for their special competences and with regard for appropriate balance.

This study was supported by: Contract/Grant No. SB1341-12-CQ-0036/13-020, TO #2, and Contract/Grant No. SB1341-03-C-0032 between the National Academy of Sciences and the National Institute of Standards and Technology and Contract/Grant No. DE-DT0000236, TO #28 (base award DE-AM01-04PI45013), between the National Academy of Sciences and the Department of Energy.

Any opinions, findings, conclusions, or recommendations expressed in this publication are those of the author and do not necessarily reflect the views of the organizations or agencies that provided support for the project.

International Standard Book Number-13: 978-0-309-30591-4
International Standard Book Number-10: 0-309-30591-8
Library of Congress Control Number: 2014952749

Additional copies of this report are available for sale from the National Academies Press, 500 Fifth Street, NW, Keck 360, Washington, DC 20001; (800) 624-6242 or (202) 334-3313; http://www.nap.edu/.

Printed in the United States of America

THE NATIONAL ACADEMIES

Advisers to the Nation on Science, Engineering, and Medicine

The **National Academy of Sciences** is a private, nonprofit, self-perpetuating society of distinguished scholars engaged in scientific and engineering research, dedicated to the furtherance of science and technology and to their use for the general welfare. Upon the authority of the charter granted to it by the Congress in 1863, the Academy has a mandate that requires it to advise the federal government on scientific and technical matters. Dr. Ralph J. Cicerone is president of the National Academy of Sciences.

The **National Academy of Engineering** was established in 1964, under the charter of the National Academy of Sciences, as a parallel organization of outstanding engineers. It is autonomous in its administration and in the selection of its members, sharing with the National Academy of Sciences the responsibility for advising the federal government. The National Academy of Engineering also sponsors engineering programs aimed at meeting national needs, encourages education and research, and recognizes the superior achievements of engineers. Dr. C. D. Mote, Jr., is president of the National Academy of Engineering.

The **Institute of Medicine** was established in 1970 by the National Academy of Sciences to secure the services of eminent members of appropriate professions in the examination of policy matters pertaining to the health of the public. The Institute acts under the responsibility given to the National Academy of Sciences by its congressional charter to be an adviser to the federal government and, upon its own initiative, to identify issues of medical care, research, and education. Dr. Victor J. Dzau is president of the Institute of Medicine.

The **National Research Council** was organized by the National Academy of Sciences in 1916 to associate the broad community of science and technology with the Academy's purposes of furthering knowledge and advising the federal government. Functioning in accordance with general policies determined by the Academy, the Council has become the principal operating agency of both the National Academy of Sciences and the National Academy of Engineering in providing services to the government, the public, and the scientific and engineering communities. The Council is administered jointly by both Academies and the Institute of Medicine. Dr. Ralph J. Cicerone and Dr. C. D. Mote, Jr., are chair and vice chair, respectively, of the National Research Council.

www.national-academies.org

Committee on
Best Practice in National Innovation Programs for Flexible Electronics

Donald Siegel, *Chair*
Dean and Professor
School of Business
University at Albany, SUNY

A. Michael Andrews, II*
Vice President for Research and Engineering and Chief Technology Officer (*retired*)
L-3 Communications Corporation

Byron C. Clayton
Vice President
Nortech

Nick Colaneri
Director, Flexible Display Center
Arizona State University

Stephen R. Forrest (NAE)
Professor
Departments of EECS, Physics and Materials Science & Engineering
University of Michigan

Russell Gaudiana
Vice President, Research
Konarka Technologies, Inc.
(*Member: 7/30/2010-3/8/2013*)

Mary L. Good (NAE)
Dean Emeritus, Donaghey College of Engineering and Information Technology
Special Advisor to the Chancellor for Economic Development
University of Arkansas at Little Rock

Frank Jeffrey
Chief Executive Officer
Power Films
(*Member: 7/30/2010-7/14/2011*)

James Turner
Senior Counsel
Association of Public and Land-grant Universities

Project Staff

Sujai J. Shivakumar
Study Director

McAlister T. Clabaugh
Program Officer

David E. Dierksheide
Program Officer

Karolina E. Konarzewska
Program Coordinator

Thomas R. Howell
Consultant

Charles W. Wessner
Program Director
(through March 2014)

---

*Co-chair through September 25, 2013.

For the National Research Council (NRC), this project was overseen by the Board on Science, Technology, and Economic Policy (STEP), a standing board of the NRC established by the National Academies of Sciences and Engineering and the Institute of Medicine in 1991. The mandate of the Board on Science, Technology, and Economic Policy is to advise federal, state, and local governments and inform the public about economic and related public policies to promote the creation, diffusion, and application of new scientific and technical knowledge to enhance the productivity and competitiveness of the U.S. economy and foster economic prosperity for all Americans. The STEP Board and its committees marshal research and the expertise of scholars, industrial managers, investors, and former public officials in a wide range of policy areas that affect the speed and direction of scientific and technological change and their contributions to the growth of the U.S. and global economies. Results are communicated through reports, conferences, workshops, briefings, and electronic media subject to the procedures of the National Academies to ensure their authoritativeness, independence, and objectivity. The members of the STEP Board* and the NRC staff are listed below:

*As of September 2014.

David T. Morgenthaler
Founder
Morgenthaler Ventures

Luis M. Proenza
President Emeritus
University of Akron

Kathryn L. Shaw
Ernest C. Arbuckle Professor of Economics
Graduate School of Business
Stanford University

Laura D'Andrea Tyson
Professor of Business Administration and Economics
Director, Institute for Business & Social Impact
Haas School of Business
University of California, Berkeley

Harold R. Varian
Chief Economist
Google Inc.

Jay Walker
Chairman
Patent Properties, Inc.

STEP Staff

Stephen A. Merrill
Director
(through March 2014)

Paul T. Beaton
Program Officer

McAlister T. Clabaugh
Program Officer

Aqila A. Coulthurst
Program Coordinator

Sujai J. Shivakumar
Senior Program Officer

David E. Dierksheide
Program Officer

Karolina E. Konarzewska
Program Coordinator

Preface

Flexible electronics refers to technologies that enable flexibility in the manufacturing process as well as flexibility as a characteristic of the final product. Features such as unconventional forms and ease of manufacturability provide important advantages for flexible electronics over conventional electronics built on rigid substrates. Today, examples of flexible electronics technologies are found in flexible flat-panel displays, medical image sensors, photovoltaic sheets, and electronic paper.

Some industry experts predict that the global market for flexible electronics will experience a double-digit growth rate, reaching $250 billion by 2025.[1] However, most experts believe that the United States is not adequately poised to capitalize on this opportunity. A recent study commissioned by the National Science Foundation and the Office of Naval Research concluded that "the relatively low prevalence of actual manufacturing and advanced systems research and development in the United States has led to an incomplete hybrid flexible electronics R&D scenario for this country."[2] Furthermore, the report observed that "manufacturing is moving to regions of the world that provide greater investment and commitment to product development. It then becomes questionable as to whether this approach is a healthy one and can be sustained in the long term."[3]

---

[1] TMR, *Flexible Electronics Market—Global Industry Size, Share, Trends, Analysis and Forecasts 2012–2018* (2013).

[2] Ananth Dodabalapur et al., *European Research and Development in Hybrid Flexible Electronics* (Baltimore: WTEC, July 2010).

[3] Ibid.

THIS STUDY

Responding to a congressional request, a committee of the National Research Council's (NRC's) Board on Science, Technology, and Economic Policy (STEP) has reviewed foreign and domestic innovation programs and their potential to advance the production of flexible electronics technologies. It has sought to understand their structure, focus, funding, and likely impact and to determine what appropriate steps the United States might consider to develop a robust flexible electronics industry. (See Box P-1.)

THE STEP BOARD'S RESEARCH ON INNOVATION AND COMPETITIVENESS

Since 1991, the National Research Council, under the auspices of the Board on Science, Technology, and Economic Policy, has undertaken a program of activities to improve policy makers' understandings of the interconnections of science, technology, and economic policy and their importance for the American economy and its international competitive position. The board's activities have corresponded with increased policy recognition of the importance of knowledge

BOX P-1
Statement of Task

An ad hoc committee will examine and compare selected innovation programs both foreign and domestic, and their potential to advance the production of flexible electronics technology in the United States. The analysis, carried out under the direction of the committee, will include a review of the goals, concept, structure, operation, funding levels, and evaluation of foreign programs similar to major U.S. programs, e.g., innovation awards, S&T parks, and consortia. To assess these programs, the committee will convene a series of meetings to gather data from responsible officials and program managers, and encourage a systematic dissemination of information and analysis as a means of better understanding the transition of flexible electronics research into products and to identify specific recommendations to improve and to develop U.S. programs.

Specifically, the committee will examine the role of research consortia around the world to advance flexible electronics technology, comparing their structure, focus, funding, and likely impact, and determining what appropriate steps the United States might consider to develop the industry in the United States. This review will include the potential of the industry, the possible contributions of a consortium, and other measures contributing to the development of the industry in the United States. The committee will undertake workshops to carry out this analysis, prepare a workshop summary capturing the tacit knowledge expressed, commission additional analyses, and develop findings and recommendations for inclusion in the committee's final consensus report.

and technology to economic growth. New economic growth theory emphasizes the role of technology creation, which is believed to be characterized by significant growth externalities.[4] In addition, many economists have recognized the limitations of traditional trade theory, particularly with respect to the reality of imperfect international competition. Public–private partnerships are increasingly recognized for their contributions to the commercialization of state and national investments in research and development. Such partnerships help address the challenges associated with the transition of research into products ready for the marketplace.[5]

One important element of NRC analysis has concerned the growth and impact of foreign technology programs. U.S. competitors have launched substantial programs to support new technologies, small firm development, innovative production at large companies, and consortia among large and small firms to strengthen national and regional positions in sectors they consider strategic for the development of their economies. Some foreign governments have chosen to provide public support to research and the commercialization of that research to overcome the market imperfections apparent in their national innovation systems.[6] They believe that the rising costs and risks associated with new potentially high-payoff technologies, and the growing global dispersal of technical expertise, underscore the need for national research and development programs to support new and existing high-technology firms within their borders.[7]

ACKNOWLEDGMENTS

On behalf of the National Academies, we express our appreciation and recognition for the insights, experiences, and perspectives made available by the many participants of workshops and site visits held over the course of this study. We would particularly like to recognize Thomas Howell for his research and preparation of a draft of this report, and David Dierksheide and Karolina Konarzewska of the NRC staff for their assistance in preparing this report for publication.

[4] National Research Council, *Enhancing Productivity Growth in the Information Age*, eds. D. W. Jorgenson and C. Wessner (Washington, DC: The National Academies Press, 2007).

[5] National Research Council, *Government-Industry Partnerships for the Development of New Technologies: Summary Report*, ed. C. Wessner (Washington, DC: The National Academies Press, 2003).

[6] For a review of leading national programs to support applied research and commercialization of new products, see National Research Council, *21st Century Manufacturing: The Role of the MEP Program*, eds. P. Shapira and C. Wessner (Washington, DC: The National Academies Press, 2013), Appendix A.

[7] For a review of the challenges and opportunities faced by the United States in the face of unprecedented global competition for developing, commercializing, and manufacturing the next generation of technologies, see National Research Council, *Rising to the Challenge, U.S. Innovation Policy for the Global Economy,* eds. C. Wessner and A. Wolff (Washington, DC: The National Academies Press, 2012).

ACKNOWLEDGMENT OF REVIEWERS

This report has been reviewed in draft form by individuals chosen for their diverse perspectives and technical expertise, in accordance with procedures approved by the National Academies' Report Review Committee. The purpose of this independent review is to provide candid and critical comments that will assist the institution in making its published report as sound as possible and to ensure that the report meets institutional standards for objectivity, evidence, and responsiveness to the study charge. The review comments and draft manuscript remain confidential to protect the integrity of the process.

We wish to thank the following individuals for their review of this report: Peter Boer, Tiger Scientific Inc.; Michael Ettenberg, Dolce Technologies; John Ettlie, Rochester Institute of Technology; Antonio Facchetti, Polyera Corporation; Pradeep Fulay, West Virginia University; Tobin Marks, Northwestern University; Robert Pfahl, International Electronics Manufacturing Initiative, Inc. (retired); Gregory Tassey, University of Washington; Nicholas Vonortas, The George Washington University; and Thomas Weller, University of South Florida.

Although the reviewers listed above have provided many constructive comments and suggestions, they were not asked to endorse the conclusions or recommendations, nor did they see the final draft of the report before its release. The review of this report was overseen by Elsa Garmire, Dartmouth College. Appointed by the National Academies, she was responsible for making certain that an independent examination of this report was carried out in accordance with institutional procedures and that all review comments were carefully considered. Responsibility for the final content of this report rests entirely with the authoring committee and the institution.

Donald Siegel Sujai Shivakumar

Contents

EXECUTIVE SUMMARY 1
Key Findings, 1
Key Recommendations, 2

1 FLEXIBLE ELECTRONICS AND THE MANUFACTURING CHALLENGE 3
A Renewed Focus on Manufacturing, 4
From Basic to Applied Research, 11
The Foreign Competitive Challenge, 18
The Potential for Applied Research Partnerships, 19
Learning from the Sematech Experience, 21
Overview of this Report, 38

2 ASSESSING THE POTENTIAL OF FLEXIBLE ELECTRONICS 41
Advantages of Flexible Electronics, 41
Potential Applications, 45
Market Growth, 48
Challenges, 52

3 GOVERNMENT PROMOTIONAL EFFORTS 59
Government Funding Levels, 61
Cooperative Research Centers, 64
Promotion of Innovation Clusters, 67

4 THE EMERGING COMPETITIVE LANDSCAPE 71
The East Asian Edge in Displays, 73
Initial U.S. and European Advantage in Materials, 80
Military Applications—U.S. Edge, 81
U.S./European Leadership in Organic Photovoltaics, 82
The Flow of Technology to Asia, 83
Perspectives from Recent SWOT Analyses, 89

5 EUROPEAN INITIATIVES 97
European Union Programs, 97
United Kingdom, 105
Netherlands/Belgium, 124
Germany, 133
Finland, 155

6 EAST ASIA 161
South Korea, 161
China, 182
Taiwan, 200
Japan, 217

7 UNITED STATES 235
Federal Support for Flexible Electronics, 236
FlexTech Alliance, 244
ASU Flexible Electronics and Display Center, 247
University Research, 255
Key Companies, 260

8 FINDINGS AND RECOMMENDATIONS 273
Findings, 273
Recommendations, 279

BIBLIOGRAPHY 283

Executive Summary

In response to a congressional request, the National Research Council's Board on Science, Technology, and Economic Policy has examined and compared selected innovation programs, both foreign and domestic, and their potential to advance the production of flexible electronics technology. To augment their extensive knowledge of the industry, the committee convened two workshops to draw on the perspectives of industry leaders, academics, and senior government officials, conducted several site visits, and commissioned research on policies and programs that have been implemented in East Asia, Europe, and the United States.

The analysis summarized in this volume includes a review of the role of research consortia around the world to advance flexible electronics technology. Based on an in-depth understanding of the structure, focus, funding, and likely impact of these programs, the committee has identified several recommendations to develop a robust flexible electronics industry in the United States.

KEY FINDINGS

Flexible electronics technologies involve circuits that can bend and stretch, enabling significant versatility in applications and the prospect of low-cost manufacturing processes. They represent an important technological advance, in terms of their performance characteristics and potential range of applications, including medical care, packaging, lighting and signage, consumer electronics, and alternative energy (especially solar energy.) What these technologies have in common is a dependence on efficient manufacturing that currently requires improved processes, tooling, and materials, as well as ongoing research. (Finding A)

Seeking to capture the global market in flexible electronics, major U.S. competitors in East Asia and Europe have launched targeted, large-scale programs with significant government funding to develop these new technologies, refine them, and ultimately manufacture them within their national borders. National and regional investment undertaken by our foreign competitors is significantly larger than comparable U.S. investment and more weighted toward later-stage applied research and development. (Finding E)

Significant U.S. expansion in the market for flexible electronics is not likely to occur in the absence of mechanisms to address investment risk, the sharing of intellectual property, and the diverse technology requirements associated with developing and manufacturing flexible electronics technologies. Linking industry, university, and government is a proven means to galvanize industry and promote cooperation in applied research and development. (Finding G)

Accordingly, the committee finds that collaboration among industry, universities, and government offers the best prospect for achieving sufficient levels of investment and the acceleration of new technology development that is required to develop a vibrant flexible electronics industry. (Finding H)

KEY RECOMMENDATIONS

Consistent with the challenges and opportunities outlined in the findings above, the committee recommends the following:

A. The United States should increase funding of basic research related to flexible electronics and augment support for university-based consortia to develop prototypes, manufacturing processes, and products in close collaboration with contributing industrial partners.
B. Consortia, bringing together industry, universities, and various levels of government, should be used as a means of fostering precompetitive applied research in flexible electronics.
C. The United States should establish and support a network of user facilities dedicated to flexible electronics.
D. Where possible, federal efforts to support the growth of competitive flexible electronics industries should leverage state and regional developmental efforts, with the objective of establishing co-located local supply chains and capturing the associated cluster synergies.
E. Agency mission needs should help drive demand for flexible electronics technologies, while lowering costs, improving capabilities, and contributing to the development of a skilled workforce.

1

Flexible Electronics and the Manufacturing Challenge

The global financial crisis of 2007-2008 has forced a sharper focus on the manufacturing sector in the United States and other advanced economies.[1] The bursting of financial bubbles and the spectacular collapse of major financial institutions has dispelled the once widely held view that the decline of U.S.-based manufacturing can be shrugged off in light of the advent of a dynamic and innovative financial services sector.[2] The recession that followed this crisis hit U.S.-based manufacturing hard—between December 2007 and December 2009, it lost 2 million jobs, or 17 percent of the total workforce, and manufacturing employment fell to its lowest level since March 1941.[3] By mid-2013, roughly 12 million Americans worked in manufacturing, down from a peak of about 20 million in the

[1] A Korean industry analyst observed in 2012 that "the global financial crisis made even advanced economies emphasize the importance of manufacturing. Korea must foster high-value-added manufacturing industries." Chae Seung-byung, senior analyst, Samsung Research Institute, "Manufacturing Sector at Its Worst in Three Years," *The Korea Herald Online*, December 9, 2012.

[2] "Not so long ago, it was fashionable in Whitehall and Westminster to claim that [Britain's] relative industrial decline didn't matter. Then came the banking crisis, which showed what happened when you bet the lot on the City and the bet didn't come off." Aditya Chakrabortty, "David Cameron's Talk About Reviving British Industry is Nonsense," *The Guardian*, November 1, 2010, see James Dyson, *Ingenious Britain: Making the UK the Leading High Tech Exporter in Europe*, March 2010.

[3] Megan M. Barker, "Manufacturing Employment Hard Hit During the 2007-09 Recession," *Monthly Labor Review*, April 2011, p. 33. U.S. manufacturing output peaked in December 2007, bottoming out 19 percent below the 2007 average in December 2009. By early 2013, 5 years later, U.S. manufacturing output was still 4 percent below the 2007 average. Federal Reserve data reproduced in SCDigest Editorial Staff, "Supply Chain News: How Is US Manufacturing Doing Five Years After the Great Recession?" *Supply Chain Digest*, April 10, 2013.

late 1970s.[4] Surveying these developments, the President's Council of Advisors on Science and Technology (PCAST) warned in 2011 that the U.S. competitive position in manufacturing was declining relative to other high-wage countries such as Japan and Germany, and the United States was losing leadership in technology-intensive industries employing high-skilled workers.[5]

A RENEWED FOCUS ON MANUFACTURING

At both the federal and state levels, a significant aspect of the policy response to the "Great Recession" has been initiatives to bolster the U.S. manufacturing sector to reverse its long-term decline, foster onshore manufacturing operations, create new domestic jobs, and exploit the U.S. science and research base. In 2011, the Administration established the Advanced Manufacturing Partnership (AMP) to coordinate investments in manufacturing by governments, universities, and industry.[6] In 2012, the Administration announced that it would create 15 manufacturing research institutes around the country to strengthen U.S. manufacturing infrastructure, the National Network for Manufacturing Innovation (NNMI).[7] A number of new state and regional initiatives feature significant investments in local manufacturing innovation, infrastructure, and workforce development.[8]

U.S. Advantages in Manufacturing

Notwithstanding the erosion of many industries, the United States may be able once again to position itself to compete internationally in manufacturing.

[4] See Figure 2 in Martin Neil Baily and Barry P. Bosworth, "US Manufacturing: Understanding Its Past and Its Potential Future," *The Journal of Economic Perspectives* 28, no. 1 (2014): 3–25. See also, Pamela M. Prah, "Has U.S. Manufacturing's Comeback Stalled?" *USA Today*, July 30, 2013.

[5] PCAST, *Report to the President on Ensuring American Leadership in Advanced Manufacturing* (June 2011), i.

[6] AMP is being coordinated through the National Institute of Standards and Technology (NIST). The President has defined the object of AMP as leveraging existing programs in manufacturing, augmented by $500 million in investments. "President Obama Launches Advanced Manufacturing Partnership," White House Press Release, June 24, 2011.

[7] The NNMI features collaboration between the private sector, universities, and government organizations at the federal, state, and local levels. The first research institute, the National Additive Manufacturing Institute, will be located in Youngstown, Ohio, and will specialize in additive manufacturing, a potentially paradigm-shifting manufacturing process based on 3-D printing. The process uses 3-D software to control machines that project various materials including resins, metal powders, and polymers to "print" 3-D in layers, with little if any waste of materials. "Obama Names Youngstown as Model for the New Tech," *Youngstown Vindicator,* February 13, 2013; "Obama Push for 3-D Hub to Turn 'Rust Belt' City Into 'Tech City,'" *Columbus Examiner*, August 17, 2012.

[8] See National Research Council, *Best Practices in State and Regional Innovation Initiatives* (Washington, DC: The National Academies Press, 2013), 78–83, 115–116, 131–140, 149–161, 174–182. For a detailed case study of initiatives in New York State to promote innovation and manufacturing in semiconductors, see National Research Council, *New York's Nanotechnology Model*, rapporteur C. Wessner (Washington, DC: The National Academies Press, 2013).

The United States has the world's foremost system of research universities, which have a long tradition of working with industry to promote innovation in manufacturing.[9] For many generations the United States has been a leader in the introduction and improvement of mechanized production methods, and U.S. manufacturers have responded to the recession, in part, through intensified investments in automated production.[10] The advent of new techniques for domestic extraction of natural gas and oil is lowering the energy costs of U.S.-based manufacturers relative to their foreign competitors.[11]

A 2012 survey by Massachusetts Institute of Technology found that a number of factors were combining to favor "re-shoring" of manufacturing in the United States, citing U.S. advances in automation and manufacturing techniques, declines in domestic natural gas prices, and the advent of additive manufacturing and nanotechnology.[12] A widely cited 2013 survey by the Boston Consulting Group (BCG) of 200 U.S. executives of large manufacturers found that 21 percent were either already relocating production to the United States or planning to do so within the next 2 years, and another 33 percent were considering or would do so in the near future, which represents sharp increases from 2012's results of 10 and 27 percent, respectively.[13] Based on these trends, the BCG forecasted that U.S. manufacturing would return $70 billion to $115 billion of export business to the United States by 2020.[14] An international survey of global executives by

[9] National Research Council, *State and Regional Innovation Initiatives*, 49–59; N. Rosenberg and R.R. Nelson, "American Universities and Technical Advance in Industry," *Research Policy* 23 (1994): 323–348.

[10] "Making a Comeback—Bluffs Manufacturer's Revival in the Wake of the Recession Echoes Nationally," *Omaha World-Herald*, June 21, 2013; "More Businesses are Returning to Made-in-USA Mentality," *Spokane Spokesman-Review*, May 26, 2013. In 1854, an English machine tool maker, Joseph Whitworth, who had been appointed as a British commissioner to the New York Industrial Exhibition, presented a report to Parliament based on site visits to American manufacturers. Profoundly impressed with American methods and the workforce, he stated: "It is this [comparative labor scarcity] and this eager resort to machinery whenever it can be applied, to which, under the guidance of superior education and intelligence, the remarkable prosperity of the United States is mainly due. . . . The workmen hail with satisfaction all mechanical improvements, the importance and value of which, as releasing them from the drudgery of unskilled labor, they are enabled by education to understand and appreciate." "English Views of American Manufacturers: Special Report of Mr. Joseph Whitworth," *Greenough's American Polytechnic Journal of Science, Mechanic Arts and Engineering* 3–4 (1854): 344–345.

[11] Price Waterhouse Cooper, "Shale Gas: A Renaissance in U.S. Manufacturing? (2011), Accessed on April 14, 2014, <*http.www.pwc.com/en_US/us/industrial-products/assets/pwc-shale-gas-as-us-manufacturing-renaissance.pdf*>.

[12] MIT, "U.S. Re-Shoring: A Turning Point," *MIT Forum for Supply Chain Innovation and Supply Chain Digest* (Cambridge, MA: MIT, 2012); TD Economics "Onshoring, and the Rebirth of American Manufacturing," October 15, 2012.

[13] "U.S. Manufacturers 'Relocating' from China," *Financial Times*, September 23, 2013.

[14] "Overseas Jobs are Coming Home—S.C. Business," Columbia, SC, *The State,* September 8, 2013.

the consultancy Stanton Chase cited concern about product and process quality as a key motivating factor:

> [A] significant geographical separation of research & development and production is not always ideal . . . [I]t is increasingly seen as beneficial to have part of production close to R&D.[15]

Emblematic of this trend, in 2013 Motorola, then owned by Google, indicated that its new Moto-X smartphone would be manufactured at a Flextronics facility in Fort Worth, Texas—the only smartphone to date to be made in the United States.[16] The same trends are observable in Europe. A 2012 survey of German manufacturers found that the share of German companies that offshored a part of production declined between 2003 and 2009, and a November 2013 survey found that in the preceding 12 months, 11 percent of UK companies brought production back home while only 4 percent moved production offshore.[17]

During the past half century, U.S. industry has performed particularly well in sectors where U.S. firms were the first movers in entirely new technologies, such as microelectronics, software and computer systems, and biotechnology. In a number of other technology fields in which U.S. researchers made cutting-edge discoveries, however, most if not all of the manufacturing activity and manufacturing jobs have come to be located outside of the United States.[18] "Flexible electronics" is an emerging technology field with revolutionary potential that represents an opportunity to establish another new high-technology industry that would be a major source of domestic manufacturing income and jobs and that would address national needs in areas such as national security, energy, and sustainable growth.

---

[15] Dieter Hagmann of Stanton Chase in "Reshoring Is an Issue for Europe Too," *Finanz & Wirtschaft,* October 23, 2013.

[16] "More Companies Making Commitments to Made in the USA," *Bucks County Courier Times,* June 30, 2013. A 2013 study by the MIT Forum for Supply Chain Innovation cited a survey of companies that found that one-third of them were considering "reshoring" offshore manufacturing into the United States and that 15 percent were "definitively" planning to do so (ibid). On January 2014, Google announced that it would sell its Motorola business to Lenovo of China for about $2.9 billion. At this writing, the impact of this transaction, if any, on Motorola's manufacturing arrangement with Flextronics is unclear. Lenovo issued a statement in January 2014 that "there are now no plans to change Motorola's approach to manufacturing." "What Lenovo's deal could mean for American manufacturing," *Engadget*, January 2014.

[17] "Offshoring and Reshoring Trends: European Data," *The Operations Room,* September 27, 2013; Bernhard Dachs et al., *The Offshoring of Production Activities in European Manufacturing* (Austrian Institute of Technology and Fraunhofer Institute for Systems and Innovation Research ISI, December 2012); "Reshoring Trend in the UK," *FDI—Foreign Direct Investment (UK),* February 1, 2014; "The Mark of the Makers Seems to Have Got Going," *Coventry Telegraph,* February 10, 2014.

[18] U.S.-invented "lost technologies" produced mainly abroad include lithium-ion batteries, oxide ceramics, wafer steppers, solar cells, interactive electronic games, laptop computers and many types of industrial robots. PCAST, *Ensuring American Leadership in Advanced Manufacturing*, 4–5.

The Opportunity in Flexible Electronics

"Flexible electronics" refers to electronic devices that can be bent, folded, stretched, or conformed regardless of their material composition without losing functionality.[19] Although conventional electronics such as integrated circuits or solar cells are typically built on thick inflexible substrates, "flexible electronics—built on substrates like plastic or metallic foil—can be folded, wrapped, rolled, and twisted with negligible effect on its electronic function."[20] (See Box 1-1.)

Flexible electronics have multiple potential applications that include bendable and reliable displays for smartphones that can be rolled up or folded; stretchable electronics "skin" and other medical devices that can monitor health; "smart fabrics" that can modify their characteristics in response to external stimuli; and flexible photovoltaic and lighting systems that can be adapted to irregular or curved surfaces. With respect to national defense, flexible electronics can support a variety of applications for equipment that is rugged, lightweight, waterproof, and versatile, including displays, batteries, communications, and physiological monitoring systems that may be embedded in soldiers' uniforms. To the extent such devices can be produced through additive processes such as printing, utilizing roll-to-roll methods, another benefit will be a dramatic reduction in the production cost of many kinds of electronic devices, a significant development given the escalation in cost of semiconductor fabs to multiple billions of dollars. The energy savings and biodegradability likely to be associated with most flexible electronics technologies will make major contributions to sustainability.[21] Although estimates vary as to the eventual market for flexible electronics products, the range for potential annual revenues by the early 2020s is pegged by informed observers at $75 billion to $190 billion.[22]

The United States is in a strong position to compete successfully in the global flexible electronics industry. The U.S. system of academic research universities remains the world's best and enjoys a strong track record with respect to commercializing research.[23] U.S. universities and government organizations are engaged in a broad range of research topics directly relevant to flexible electronics products and associated production processes. The U.S. microelectronics

---

[19] Closely related fields are defined by materials composition ("organic electronics," or electronic devices composed of organic as opposed to inorganic matter) and production process ("printed electronics," or electronic devices manufactured through printing technologies). NorTech FlexMatters, "Developing a Roadmap for Northeast Ohio's Flexible Electronics Sector."

[20] Muhammad A. Alam and Satish Kumar, "Definition of Flexible Electronics," ed. Bharat Bhushan, *Encyclopedia of Nanotechnology* (Dordrecht, Springer Reference, 2014).

[21] Chemical Sciences and Society Summit, *Organic Electronics for a Better Tomorrow: Innovation, Accessibility, Sustainability*, September 2012.

[22] Adam Page and Smithers Pira, "Market for Organic and Printed Electronics," in OE-A, *Organic and Printed Electronics* (2013), 34; IDTechEx, "Printed, Flexible and Organic Electronics Sees 15.3% CAGR," *Printed Electronics World,* May 15, 2013.

[23] Richard C. Atkinson and William A. Blanpied, "Research Universities: Core of the U.S. Science and Technology Systems," *Technology and Society* 30 (2008): 30–48.

BOX 1-1
Flexible Electronics at the Frontier

According to a recent report by the National Academy of Engineering, "One of the frontier goals in electronics research is to transform conventional fabrication processes to meet the demands of soft, pliant, and often easily damaged surfaces. Research in new materials and patterning technologies has enabled flexible electronics that push the boundaries of how electronics are made and used toward the possibility of incorporating electronic control and power sources into any object.

Unlike conventional silicon electronics that are limited to rigid wafers, flexible electronic devices have been demonstrated on plastics, paper, fibers, and even biological tissues. These flexible devices enable a wide range of applications, in fields ranging from energy sustainability to smart sensor networks to bioelectronics. Some specific examples are energy-efficient, stretchable lighting, lightweight photovoltaics, smart-sensing wallpaper, and dissolvable electronic implants.

To make flexible electronics that are compatible with delicate surfaces, low-temperature processing is required. This need has led to the development of materials such as organic conductors and semiconductors as well as advanced solution-based techniques that enable low-temperature processing. Thus flexible electronics not only enable novel applications but also promote the use of alternative manufacturing technologies, such as roll-to-roll printing for electronics."[a]

[a]Yueh-Lin (Lynn) Loo and Tse Nga (Tina) Ng, "Flexible Electronics," in National Academy of Engineering, *Frontiers of Engineering: Reports on Leading-Edge Engineering from the 2013 Symposium* (Washington, DC: The National Academies Press, 2014).

industry is the world leader not only in technology but also in manufacturing of the most advanced devices. Furthermore, the printing industry, which has relevant technologies and know-how with respect to manufacturing processes, is well established in the United States, although buffeted by the decline of book and newspaper readership.[24] U.S.-based companies are global leaders in the manufacturing of a number of key materials with applications in flexible electronics including Corning (bendable glass), Universal Display Corporation (phosphorescent emitter materials), and DuPont Teijin (films with electronics applications). The U.S. Army and Air Force and other elements of the defense establishment have been supporting the development of flexible displays for military and dual-use applications for more than a decade, and other federal

[24] IBIS World, *Out of Print: The Industry Struggles as Printed Media Lose Consumers to Web*, August 2012; "Gravure Makes Inroads in Printed Electronics," *Printed Electronics Now*, February 2011; "Printed Electronics: Fexo and Screen," *Printed Electronics Now*, February 2009; "A New Industry Shapes the Future of Printing," *Printed Electronics Now*, December 2008.

agencies have provided R&D support. [25] Promising startups spun off from U.S. research universities are commercializing academic discoveries such as Cambrios (silver nanowires with touchscreen applications at MIT); MC10 (stretchable electronics at the University of Illinois), and Imprint Energy (bendable batteries at the University of California at Berkeley).[26] U.S. companies have also pioneered the development of organic photovoltaics.

Key Challenges

Despite these strengths, the United States faces major challenges in establishing a strong manufacturing base in this emerging field. Most R&D in flexible electronics conducted by major U.S. firms has been precompetitive. The technological hurdles to commercial applications in the form of new products and processes have proven sufficiently daunting worldwide that they have confounded numerous optimistic forecasts of industry growth.[27] Repeated delays in commercialization of flexible electronics technologies attributable to materials and manufacturing challenges have affected large established electronics firms as well as startups and have occurred in all of the geographic regions examined in this study. These difficulties underscore the reality that widespread commercialization will require further advances in efficient manufacturing technology, materials, equipment, and processes.

Asian manufacturers have developed such a complete dominance in the area of conventional electronic displays that they are poised to dominate flexible displays with mainstream commercial applications in products such as smartphones, TVs, and e-readers.[28] To date, a number of innovative U.S. companies developing

---

[25] The 2010 World Technology Evaluation Center, Inc. (WTEC) study group that reported on European research programs in flexible electronics reported that "principal strengths of U.S. efforts, as perceived by European scientists [include] strong support from Federal agencies such as NSF, ONR, Department of Defense, Department of Energy, etc." WTEC, *European Research and Development in Hybrid Flexible Electronics* (July 2010), xvi. Mariana Mazzucato, a professor of economics at the University of Sussex, observes in a 2013 book that "despite the perception of the U.S. as the epitome of private sector–led wealth creation, in reality it is the State that has been engaged on a massive scale in entrepreneurial risk taking to spur innovation." She details four highly successful U.S. institutions including the Defense Advanced Research Projects Agency and the Small Business Innovation Research program. Mariana Mazzucato, *The Entrepreneurial State: Debunking Public vs. Private Sector Myths* (London, New York, and Delhi: Anthem Press, 2013), 72–86.

[26] "A new battery that could revolutionize wearables," *Gigaom*, January 8, 2013; "MC10 Reshaping Electronics," *Flexible Substrates*, September 2012; "Honey I shrunk the technology—NU making a big difference in the world of nanotech," *Chicago Sun-Times*, August 13, 2007.

[27] "Printed Electronics—Many New Directions," *Printed Electronics World,* February 21, 2011; Lawrence Gasorian, "Notable Developments in Flexible Glass," *Flexible Substrate,* January 2013; Nick Colaneri, "Manufacturing Flexible Displays: The Challenges of Handling Plastic," *Solid State Technology,* May 1, 2013.

[28] HSBC Global Research, *Flexible Display: Fantastic Plastic—A Shape Shifting Game Changer,* April 2013.

new flexible electronics technologies have chosen to manufacture their designs in Asia or have been acquired outright by Asian firms.

Although the U.S. government is investing in R&D relevant to flexible electronics through a number of institutional channels, and several states are making significant commitments to R&D, cumulative government financial support in the United States falls short of government funding in this field in Europe and East Asia.

Perhaps most significantly, while the United States has developed an onshore supply chain to support flexible electronics manufacturing, no large U.S.-based champion has yet emerged that is prepared to engage in large-scale commercial manufacturing of products that integrate the various flexible and printed electronics technologies being developed by the U.S. research base. Some industry experts reckon that if only market forces act, then a champion never will emerge, particularly in the displays field, reflecting the demonstrated risks associated with investments in this sector.[29]

The Role of Public Investments

A 2011 study by PCAST identified flexible electronics as one of four "early-stage technologies that have transformative potential" that may likely not develop if left to market forces ("market failures").[30] Flexible electronics is a "General Purpose Technology" (GPT), that is, a technology capable of having a major impact in a broad number of sectors in manners similar to those that occurred with

---

[29] DuPont's Steven C. Freilich offers a perspective from the history of the display industry that underscores the market deterrents that inhibit investment in promising but uncertain new technology areas. "In 2005, four major technologies were jostling for market share as the size of display panels grew ever larger: the traditional cathode ray tube (CRT); plasma displays, that was pushing the CRT for dominance in the mid-sized displays; liquid crystal displays (LCDs), which dominated the high definition hand-held market and had suddenly solved problems in larger dimensions; and rear projection, which had dominated the larger sizes. Plasma was quickly pushed by LCDs out of the mid-sized market, which then pressured the rear projection displays, leaving the market to just two technologies instead of four. The investment of every one of these technologies amounts to billions of dollars per fabrication unit, and yet companies must be prepared to shift quickly to keep up with evolving technologies, consumer tastes and price changes. You may think that [the display competition] is over now, but up in the corner [of this chart] you find organic light emitting diodes, and in a couple of years we're going to see this shift happening all over." Steven C. Freilich, "DuPont Reflections on Photovoltaics," in National Research Council, *The Future of Photovoltaic Manufacturing in the United States* (Washington, DC: The National Academies Press, 2011), 68.

[30] PCAST observed with respect to flexible electronics that the R&D "needed to advance this technology is costly and complex, requiring facilities for prototyping and pilot-scale manufacture. Yet, the benefits are broadly applicable across industries and cannot be fully captured." The other three sectors identified by PCAST were nanoscale carbon materials, next-generation optoelectronics, and nanotechnology-enabled medical diagnostic devices and therapeutics. PCAST, *Report to the President,* 19–20.

respect to the steam engine, microelectronics, and the Internet.[31] However, as the academic analyses of GPTs have concluded, the private sector will not necessarily make the investments necessary to develop such technologies. Professor Vernon W. Ruttan, a development economist at the University of Minnesota, observed in 2006:

> A second issue is whether the private sector can be relied upon as a source of major new general purpose technology development. The quick answer is that it cannot! Each of the general purpose technologies that I have reviewed has required at least several decades of public support to reach the threshold of military and commercial viability. Decision makers in the private sector seldom have access to the patient capital implied by a time horizon measured in decades rather than years.[32]

A recent study by Mariana Mazzucato of the University of Sussex cites considerable academic research and empirical evidence to the effect that large-scale and long-term government investment "has been the engine behind almost every GPT in the last century," including the U.S. system of mass production, aviation technology, nuclear power, space technology, information technology, and the Internet.[33]

FROM BASIC TO APPLIED RESEARCH

The question currently confronting the United States is whether industry, universities, and government can work together to realize the opportunity presented by flexible electronics in the face of significant technological hurdles and intense international competition. In particular, the question is whether U.S. research discoveries and technological advances in flexible electronics can be translated into significant onshore manufacturing capability and domestic manufacturing employment. The United States arguably has no peer in basic research in this field. However, the President's National Science and Technology Council has

---

[31] The term was coined by Timothy F. Bresnahan and Manuel Trajtenberg in 1995 and is widely used in analyses of innovation and its effects on economic development. Bresnahan and Trajtenberg, "General Purpose Technologies: Engines of Growth?" (NBER Working Paper No. w4148, 1995). GPTs are identified by three characteristics: (1) they are extremely pervasive and are used as inputs across wide areas of the economy; (2) they continually improve over time to the benefit of their users; and (3) they facilitate innovation. Elhanan Helpman and Manual Trajtenberg, "A Time to Sow and a Time to Reap: Growth Based on General Purpose Technologies" (NBER Working Paper No. 4854, September 1994).

[32] Vernon W. Ruttan, "Is War Necessary for Economic Growth?" Clemons Lecture, Saint Johns University, Collegeville, Minnesota (October 9, 2006), 30.

[33] Mariana Mazzucato, *The Entrepreneurial State*. Among others Mazzucato cites the work of Dr. Vernon W. Ruttan, who demonstrated that military and defense research, development, and procurement "have been major sources of technology development across a broad spectrum of industries that account for an important share of United States industrial production." Ruttan, "Is War Necessary for Economic Growth?"

highlighted shortcomings in the U.S. system of applied research, which is the practical application of science—an area where many U.S. companies have excelled individually, but which remains an area of relative weakness in the overall landscape of U.S. manufacturing innovation.[34]

The Innovation Infrastructure

The challenge facing the United States in innovation is to create an infrastructure that recognizes and capitalizes on the changing nature of research where the boundaries between basic and applied research have become more blurred.[35] Technological advances are now enabling a degree of creativity in applied research of the kind traditionally associated with basic research—a phenomenon that is being exploited systematically in intermediate research organizations such as Europe's Fraunhofer institutes and IMEC and in leading companies such as DuPont, General Electric, Proctor and Gamble, and Intel. Dr. Kristina Johnson, then the U.S. Undersecretary of Energy, underscored this fundamental shift in a 2011 presentation at a National Academies symposium.[36] The traditional view of scientific research, articulated by Vannevar Bush, she observed, was two-dimensional, with knowledge flowing in one direction from basic research to application, development, and commercialization.

> But this model, to the extent it was ever realistic, was changed by the advent of transistors, computers and the Internet—the features of the information age. In the old model, scientists would think about how things work, and engineers

[34] One factor underlying the U.S. position on applied research has been the downsizing and even disappearance in the country of large industrial laboratories, which have historically been centers of applied research and innovation aimed at commercial objectives. See Erich Bloch, "Seizing U.S. Research Strength," *Issues in Science and Technology,* Summer 2003. Corporate laboratories have been affected, among other things, by financial restructuring. "General Electric donated the David Sarnoff Research Laboratory to SRI International, which General Electric acquired in its purchase of RCA. Regional telephone companies formed from the split of AT&T created Bell-core as a separate laboratory, and soon it was sold. Kodak acquired Sterling Drug and then sold. General Motors took over Hughes Aircraft, and DuPont acquired Conoco. General Electric aerospace merged into Martin Marietta, which in turn merged into Lockheed. Kodak transferred research from its central laboratory into operating divisions. Allied Signal, Armstrong World Industries, and W. R. Grace completely eliminated corporate support for central research." Roli Varma, "Changing Research Cultures in U.S. Industry," *Science, Technology, and Human Values,* Autumn 2000, 400.

[35] It is beyond the scope of this study to survey the vast, evolving, and occasionally discordant theoretical literature on innovation and its management, exploitation, and diffusion. However, see generally John E. Ettlie, *Managing Innovation: New Technology, New Products, and New Services in a Global Economy* (Burlington, MA: Elsevier-Butterworth-Heinemann, 2006); Bengt-Ake Lundvall (ed.), *National Systems of Innovation: Toward a Theory of Innovation and Interactive Learning* (London: Anthem Press, 2010); Ruud E. Smits, Stefan Kuhlmann and Philip Shapira (eds.), *The Theory and Practice of Innovation Policy: An International Research Handbook* (Cheltenham and Northampton, MA: Edward Elgar, 2012).

[36] Kristina Johnson, "The U.S. Department of Energy's Perspective," in National Research Council, *Future of Photovoltaics Manufacturing,* 150.

> would make them work. The new availability of knowledge allowed the engineers to invent new things as well. They could design them more intelligently, using mathematical models. Computers brought everyone the same platform and tools, allowing not only engineers but social scientists to become more quantitative and take more analytic approaches to the deployment of technology. This is a fundamental change that has not been clearly recognized.[37]

Although some individual U.S. firms have been able to break down traditional walls between basic and applied research in a highly successful manner (see Box 1-2), as the MIT Task Force on Production and Innovation recently concluded, these firms are somewhat anomalous in the U.S. industrial landscape.[38]

The Missing Middle

But while individual U.S. companies have proven adept at translating scientific knowledge into useful products and processes, in contrast to countries such as Germany, the United States does not have a nationwide infrastructure supporting translational innovation by companies—particularly small companies—broadly and systematically. In Germany, small companies can access sophisticated computer design and simulation tools and platforms, prototyping facilities, precision measuring and test equipment, and a deep bench of expertise at a dense network of Fraunhofer institutes and research centers bridging the space between the research base and industry. Equipment makers can test their machines on pilot lines in a factory environment.[39] Although some of these capabilities are present

---

[37] Ibid. A complementary perspective is provided by Michael Idelchik, GE's Vice President for Advanced Technologies at GE Global Research, who told an MIT task force on innovation that today's advanced industrial R&D involves "the management of concurrent, non-sequential interactions with multiple exchanges between scientists and engineers and manufacturing specialists and with the product passing back and forth between the hands of experts with fundamentally different competencies." He cited the example of GE's intermetallic turbine blades for the new GEnx engine. "The new alloy was patented in 1989, but the complexities of working with this material meant that it took until 1992 to be cast. The casting involved continuous interactions between [specialists in] manufacturing and R&D. There are patents in the blades, but most of the casting and materials depend on trade secrets. . . . The first engine test was in 1993—and this took collaboration between designers, R&D, and manufacturing." Suzanne Berger, *Making in America: From Innovation to Market* (Cambridge, MA, and London: MIT Press, 2013), 58.

[38] The MIT Task Force on Production and Innovation documented a number of highly innovative U.S. companies and industry groups that have scaled up and manufactured new technologies successfully. But it observed that these examples, while "promising," were also "fortuitous." "When we looked in detail at their origins, we found special individuals and companies who were willing to go far beyond their ordinary job descriptions and ways of doing business. The United States abounds in entrepreneurial people but without institutional bases, their efforts often remain isolated and hard to sustain. These positive but chancy interactions are unlikely to produce complementary capabilities on the scale needed to transform regional ecosystems across the United States." Berger, *Making in America*, 211.

[39] See Appendix A2, "Fraunhofer Gesellschaft: The German Model of Applied Research," in National Research Council, *21st Century Manufacturing: The Role of the Manufacturing Extension Partnership* (Washington, DC: The National Academies Press, 2013), 224–284.

BOX 1-2
Intel Corporation—Integrating Basic and Applied Research

One of the most successful microelectronics enterprises of all time, Intel Corporation not only reduced barriers between basic and applied research, but also was created as a spinoff of Fairchild with no central research laboratory whatsoever: *[B]ecause of problems Gordon Moore and Bob Noyce experienced in transferring technology from Fairchild's central research laboratory to their manufacturing lines, they established Intel without a separate R&D laboratory. They did this because competition is driven by time-to-market and technology transfer from a central laboratory is too slow to synchronize with the product cycle. In addition, the exactness of understanding that researchers wish to pursue is too rarely met in semiconductor development where process designers make empirical choices and try them to see if they work.*[a]

Intel continued to operate on the Noyce principle of "minimum information," that is, rather than trying to understand completely a given phenomenon, perhaps resulting in a published scientific article, Intel "tries to get by with as little information as possible. . . . Locating development and manufacturing together allows Intel to explore variations on its existing technologies very efficiently." Semiconductor companies that "made great investments in long-term research at the expense of focus on the next product cycle have lost market share to Intel, a company that has kept the next product cycle in clear focus while annually investing $2 billion in R&D."[b] This approach, which was an important factor in Intel's success, was carried over into Sematech, an industry-government applied research consortium, with the result that the developmental pace of the U.S. semiconductor industry accelerated dramatically.[c]

---

[a]Elias C. Carayannis and James Gover, "The Sematech-Sandia National Laboratories Partnership: A Case Study," *Technovation,* 2002, 586. The MIT Task Force on Production and Innovation, which interviewed numerous innovative companies, reported that "there was general consensus across the senior executives that bringing new ideas through early stages of prototyping, testing, demonstration, and pilot production works most efficiently when it is kept close to key scientists and engineers within the company. . . . Proximity is important not only for control and to avert disaster, but to accelerate them to market and to explore and develop multiple variants (and price points) of a new product." Berger, *Making in America,* 57–58.

[b]Carayannis and Gover, "The Sematech-Sandia National Laboratories Partnership," 586. At the same time, there are limits to the extent to which applied research can be pushed forward on its own without parallel advances in basic science. For example, in a 2011 interview, Dr. Bernard Kippelen, Director of Georgia Tech's Center for Organic Photonics and Electronics, cited significant recent strides achieved in improving the performance of organic thin-film transistors (OTFTs) and the potential for further improvements. He said, "One of the biggest hurdles [to further progress] is the mismatch between the performance level reached by OTFTs and the lack of fundamental understanding of their properties. There is a need for further study as to develop predictive modelling capabilities, especially predictive control of morphology of organic semiconductors." "A Look at Printed Electronics: Printed Electronics Now Interview with Dr. Bernard Kippelen," *Printed Electronics Now,* July 2011.

[c]"Moore's Law" is a rule of thumb articulated by Intel's Gordon Moore that the number of transistors on integrated circuits will double approximately every 2 years, reflecting advances

continued

BOX 1-2 Continued

in microelectronics technology. The existence of Moore's Law and its acceptance by the semiconductor industry gave individual companies a method for benchmarking its own development. With the advent of Sematech, the industry as a whole was able to accelerate the pace of development through collective understandings on specific needs and the timing of their introduction. "What Moore's Law has done is to give the research process a cadence. You try to get from one node, or minimum feature size, to the next as fast as possible. That has served to excite the industry to beat the roadmap, and they have done that. It wouldn't have happened without that expectation or cadence that Moore's Law provides." Larry Sumney, "The Semiconductor Research Corporation (SRC): A Proven Means to Fund Relevant Research," in National Research Council, *Future of Photovoltaic Manufacturing*, 186.

at individual sites in the United States, nothing comparable to the German system as a whole exists.[40]

The Administration's National Network for Manufacturing Innovation is intended to address this gap, closely linking the U.S. research base to industry and the workforce. A recent government "Frequently Asked Questions" relating to NNMI framed the issue as follows:

> Q: Don't we already have many Federal programs that achieve the goals of the NNMI?
>
> A: The simple answer is no. There are no current federal programs that have the required attributes to significantly influence the nation's competitiveness and successfully bridge "the missing middle" in the manner and magnitude proposed for the NNMI. A whole-of-government approach is needed.[41]

---

[40] The MIT Task Force on Production and Innovation observed in 2013 that "The American company stands alone. The German firm has access to a rich and diverse set of external resources . . . to which it contributes, but which it does not have to generate on its own. It hires employees who have been educated in technical schools and universities linked to hands-on experience and credentialing within companies. . . . For a price that seems acceptable even to small and medium businesses, the German firm can get expert advice and use expensive equipment at para-public institutions like the Fraunhofer Institutes." Berger, *Making in America*, 87.

[41] "National Network for Manufacturing Innovation Frequently Asked Questions," accessed June 7, 2013, <*http://manufacturing.gov/docs/nnmi_faa.pdf*>. National Science and Technology Council, *National Network for Manufacturing Innovation: A Preliminary Design* (Washington, DC: Executive Office of the President, January 2013), 2. Under Secretary of Commerce for Standards and Technology Patrick Gallagher stated in 2012 congressional testimony that an interagency group comprised of the Departments of Commerce, Defense, and Energy under the National Science and Technology Council's Committee on Technology had concluded, after study, that "the acceleration of innovation for advanced manufacturing required bridging a number of gaps in the present U.S. innovation system, particularly the gap between R&D activities and, the deployment of technological innovations in domestic manufacturing production." Patrick D. Gallagher, "Assembling the Facts: Examining the Proposed National Network for Manufacturing Innovation," May 31, 2012.

Although it is beyond the scope of this study to access the NNMI, it should be noted that the initiative draws on the example of Germany's Fraunhofer-Gesellschaft, a public network of large-scale institutes of applied industrial research.[42] As this study indicates, at least 10 Fraunhofer institutes are currently deploying major resources, including equipment, prototyping and simulation tools, demonstration lines, and their own deep competencies, to the development of processes and products across a broad spectrum of potential commercial applications of flexible electronics. There is nothing comparable in the United States, notwithstanding the existence of several noteworthy research centers at U.S. universities addressed in Chapter 7. The Fraunhofer's methods have a track record of commercial success, reflected in the competitive prowess and reputation of German engineering and manufactured goods around the world and Germany's relative success in retaining advanced manufacturing jobs at home.[43] Whether or not NNMI replicates the Fraunhofer's level of achievement, and depending on its ultimate areas of technology focus, the NNMI has the potential to help bridge the gap between basic research and commercialization in flexible electronics.

Modeling the Innovation Process

The National Institute of Standards and Technology (NIST) has developed a sophisticated schematic, breaking down the innovation process into basic components and identifying the points in that process where public sector investment can have a positive impact. Analysis by Gregory Tassey, formerly NIST's chief economist, identifies the development of generic technology platforms (enabling technologies that can be shared by everyone) and "infratechnologies" as key areas in which public participation (augmented by private efforts) can make major contributions to innovation. (See Figure 1-1.)

Infratechnologies are tools which include "measurement methods for R&D and production control, technical support for interfaces between components of systems technologies, scientific and engineering databases, techniques such as quality assurance procedures, and test methods for facilitating marketplace transactions of complex technology-based products. They are ubiquitous in technology-based industries and often exert their impacts in the form of industry standards."[44] Infratechnologies support R&D, but they also play a critical role in the later developmental stages of production, system integration, and market development.

---

[42] Congressional Research Service, *The Obama Administration's Proposal to Establish a National Network for Manufacturing Innovation* (January 29, 2014), 4.

[43] See, generally, National Research Council, *21st Century Manufacturing*, 224–284.

[44] Gregory Tassey, "Modeling and Measuring the Economic Roles of Technology Infrastructure" *Economics of Innovation and New Technology* (January 2008), p. 624. See also Gregory Tassey, "Technological Infrastructure and the Role of Government," Working Paper Series, December 19, 1995; Gregory Tassey, *The Technology Imperative* (Cheltenham and Northhampton, MA: Edward Elgar, 2007).

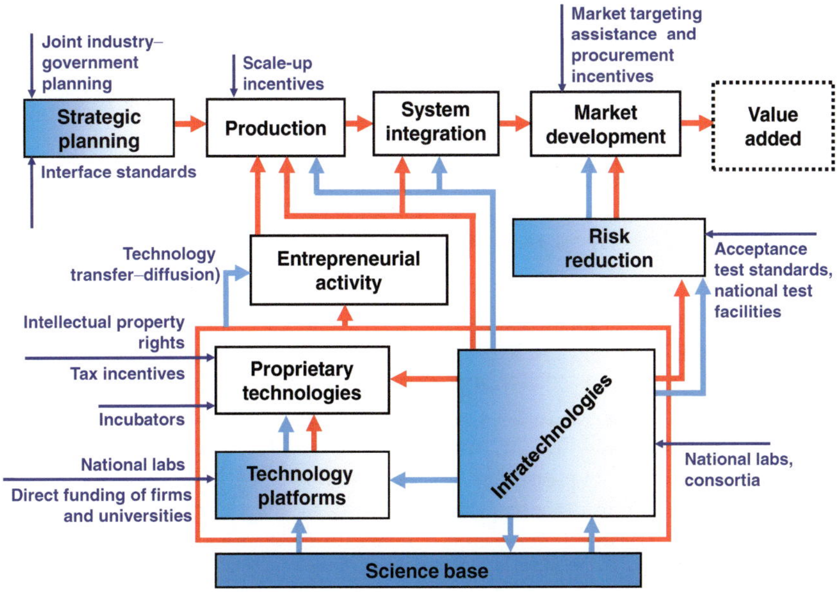

FIGURE 1-1 Targets for science, technology, innovation and diffusion policy.
SOURCE: Figure 6, "Managing the entire technology life cycle: Policy roles in response to market failures" in Gregory Tassey, "Beyond the business cycle: The need for a technology-based growth strategy," Science and Public Policy (2013) 40(3):293-315, by permission of Oxford University Press.

Tassey concludes that infratechnologies mitigate risks associated with commercialization of new products and thus facilitate entrepreneurial investment.[45] NIST Director Patrick Gallagher observed in 2010 that Tassey's model is a useful taxonomy of stages in the R&D process which enables a more efficient application of government support at the stages in the process where it can have the greatest effect.

> [The model] goes from planning and production and goes to value add[ed] and then what you see is the development of technology in these lower boxes, which include entrepreneurial activity, risk reduction technology methodologies, which include standards development and regulators and other things that

[45] Infratechnologies reduce risk, among other ways, by establishing industry norms, standards and benchmarks which ensure that a new product will "fix" or "interface" with other products in a given technology system. The development of such standards by Sematech—and industry-government consortium—with respect to semiconductor manufacturing equipment and materials was critically important to the vendor base which supplied semiconductor manufacturers. See Tassey, "Technological Infrastructure" (1995) op. cit. pp. 11–12; Tassey, "Modeling the Economic Roles of Technology Infrastructure" (2008) op. cit. pp. 620–622.

> lower the risk of introducing a technology into the market. This is where the real technology innovation is occurring. . . . [Y]ou can divide the technologies into several pieces. We tend to think of the technology as the intellectually protected part of it, the IP part of technology. But in fact it's also equally important to have technology infrastructure, manufacturing process technology is vital to productivity increases and, in fact, are essential for this system to work. And very often proprietary technologies are built on underlying generic technology platforms. . . . [F]rom a government side it means that we should really start focusing on these generic technology issues.[46]

The global flexible electronics industry is in its infancy. Scaled-up production for commercial applications exists in only a few areas (e-paper, RFID tags, OLED screens). Most of the government programs surveyed in this study are found in the early stages of development—basic research and the development of generic technology platforms (such as roll-to-roll manufacturing processes) and "infratechnologies." While some government-supported R&D in flexible electronics may result in proprietary technology owned by individual firms (such as the German Fraunhofer's contract research for industry), it is overwhelmingly concentrated on industry-government consortia developing generic technologies that can be used by multiple participating (and in some cases nonparticipating) industry partners. Scale-up incentives for production are far less common than in mature sectors like semiconductors and in some countries may be inhibited by rules (EU state aids requirements) or domestic opposition (United States) in the future. In China, where massive state financial support is being deployed to support scaling up of production of AMOLED displays,[47] the principal state actors are municipal governments such as those of Shanghai, Ordos, Huizhou, and Beijing.[48] Given the local boosterism that is intrinsic to many economies, including our own, the local character of government support for production that is evident in China may prove the norm in many of the countries surveyed in this report.

THE FOREIGN COMPETITIVE CHALLENGE

The international competitive prospects confronting any U.S. effort to establish onshore manufacturing capability in flexible electronics are sobering. East

---

[46] Patrick Gallagher, "Strengthening the Connections: Research Innovation and Economic Growth," University of Pittsburgh, October 7, 2010. Tassey's criticism of neoclassical economics, and the alleged indifference of conventional economic thinking to the offshoring of manufacturing and the importance of industrial technology, has drawn sharp criticism, not so much because his model itself is flawed but because of the importance he places on the role of government in the innovation process. See Claude Barfield, "Commentary on Gregory Tassey's 'Rationales and Mechanisms for Revitalizing U.S. Manufacturing R&D Strategies,' " *Journal of Technology Transfer* (2010), vol. 35, issue 3.

[47] AMOLED (active-matrix organic light-emitting diode) is a display technology for use in mobile devices and televisions.

[48] See Chapter 6.

Asian firms dominate the manufacture of conventional displays and are using their installed manufacturing base in that field to leverage their entry into flexible displays with consumer applications. They are backed by government programs that emphasize applied research in industry and government research institutes, some of which are among the best of their kind in the world. Large Asian industrial groups enjoy not only ample financial resources but also extensive industrial and technological competencies in relevant fields such as microelectronics, optoelectronics, materials science, and printing. Particularly in Korea and Taiwan, company management has proven adept at making bold, rapid moves into risky but promising new fields.

East Asian efforts in flexible electronics are heavily concentrated in the area of displays for consumer use, holding out the prospect that a far broader U.S. approach, based on applications in areas such as medical devices, photovoltaics, lighting, and smart textiles, could be successful. However, a European effort based on just such an approach is already well under way, with a growing emphasis on establishing indigenous manufacturing capability in what in European parlance is termed "OLAE" (organic and large area electronics). Europe enjoys not only a strong fundamental research base but also a formidable infrastructure for applied research in relevant technology areas, which includes Germany's renowned Fraunhofer institutes, a new group of research centers in the United Kingdom, and world-class institutes in the Low Countries (IMEC and its affiliate, the Holst Centre) and Finland (VTT). The European developmental effort is broad in both a geographic and technological sense, is supported by successive layers of government at the community, national, regional, and local levels, and engages companies with a long tradition of collaboration to achieve technological objectives.

THE POTENTIAL FOR APPLIED RESEARCH PARTNERSHIPS

Collaborative applied research efforts involving companies, government, and universities may present advantages in fields where technological and commercial risks are substantial and where no single company, no matter how large and well endowed, can command the resources and full range of technologies needed to successfully commercialize new technologies. There is a growing consensus that research partnerships, whether formal or informal, represent an organizational form that can encourage collaboration by reducing technological risks to participants, reducing time to market, fostering cost savings, and improving appropriability of R&D results. (See Box 1-3.)

Most current U.S. development initiatives in flexible electronics are not connected or mutually supportive. In contrast, the European Union and its Member States have overwhelmingly opted to pursue the development of flexible electronics through a variety of collaborative organizational structures—despite differences in language, nationality, and innovation culture. This is indicative

BOX 1-3
Economic Foundations for Applied Research Partnerships

Recent reviews of the academic and business literature as well as a growing body of empirical evidence affirm that research joint ventures that include firms, universities, and government organizations are effective.[a] In their review of this literature, Combs and Link (2003) argue that "a rich theoretical foundation has emerged upon which one can, after the fact, justify the historic and current policy interest in research partnerships." They find that well-structured research partnerships and the collaborations they create improve cost efficiencies, increase competition in the marketplace, and improve consumer surplus through improved products.[b] In their survey, Caloghirou, Ioannides, and Vonortas (2003) note further that "inter-firm cooperative agreements to create and disseminate technological knowledge are now viewed as veritable competitive mechanisms, right at the strategic core of most companies in high technology industries."

The analytical literature documents a variety of incentives for firms to join a research joint venture that includes R&D cost-sharing, pooling of risks and reducing uncertainties, internalizing knowledge spillovers, ensuring the continuity of R&D efforts through access to finance, complementary resources and skills, and advancing R&D through capturing research synergies and the effective deployment of extant resources. Joint venture participants also find that research joint ventures can improve their strategic flexibility, promote shared technical standards, and increase the market power of the industry.[c] Indeed, Cologhirou and colleagues make the case that "it would not be unreasonable to argue that there is some evidence in support of each of these arguments."

---

[a]See Yannis Caloghirou, Stavros Ioannides, and Nicholas S. Vonortas, "Research Joint Ventures," *Journal of Economic Surveys* 17, no. 4 (2003): 541–570. Kathryn L. Combs and Albert N. Link, "Innovation Policy in Search of an Economic Foundation: The Case of Research Partnerships in the United States," *Technology Analysis and Strategic Management* 15 (2003): 177–187. Albert N. Link and John T. Scott, "Public/Private Partnerships: Stimulating Competition in a Dynamic Market," *International Journal of Industrial Organization,* 19, no. 5 (2001): 763–794. See also Albert N. Link and Donald S. Siegel, *Innovation, Entrepreneurship, and Technological Change* (New York: Oxford University Press, 2007). Stephen Martin and John T. Scott, "The Nature of Innovation Market Failure and the Design of Public Support for Private Innovation," *Research Policy* 29, no. 4–5 (2000): 437–448.

[b]Effective management and organization contribute to the success of research consortia. See Lee G. Branstetter and Mariko Sakakibara, "When Do Research Consortia Work Well and Why? Evidence from Japanese Panel Data" (NBER Working Paper No. 7972, October 2000). See also Sarah M. Greene, Gene Hart, and Edward H. Wagner, "Measuring and Improving Performance in Multicenter Research Consortia," *JNCI Monographs* 2005, no. 35: 26–32.

[c]In spite of the fact that much is known about how to effectively manage consortia in the United States, there remains much unevenness in putting this knowledge into practice. Early studies evaluating the National Science Foundation cooperative R&D centers concluded that R&D consortia are a good idea with payoffs in the range of 3 to 1, similar to the National Labs payback. See Barry Bozeman, "Technology Transfer and Public Policy: A Review of Research and Theory," *Research Policy* 29, no. 4–5 (April 2000): 627–655. Studies that followed, taking a more systematic, often quasi-experimental design approach to evaluation found a greater variation in outcomes. For example, some later studies found that the most important benefits

continued

BOX 1-3 Continued

of university consortia and cooperative R&D centers were indirect. See Eliezer Geisler, "Technology Transfer: Toward Mapping the Field, a Review, and Research Directions," *The Journal of Technology Transfer* 18, no. 3–4 (Summer–Fall 1993): 88–93. Firm benefits were often paid back in human capital: participating firms got to work and recruit the best university technical students as part of project work. There continue to be good examples. See Jonathan A. Morell, "Why Are There Unintended Consequences of Program Action, and What Are the Implications for Doing Evaluation?" *American Journal of Evaluation* 26, no. 4 (December 2005): 444–463.

that they see that the advantages associated with such structures are real and applicable in this sector.[49]

Can a collaborative effort work in the United States? Dr. Malcolm Thompson, former Chairman and CEO of RPO, Inc., a developer of touch screens, outlined the advantages for the United States of a consortium approach in flexible electronics in a 2010 National Academies conference.[50] He pointed out that in light of the fact that addressing the relevant process, materials, and equipment challenges was beyond the ability of even a very large company, a collaboration would reach across the relevant topical areas, generate precompetitive research results capable of broad adoption, and minimize duplication of cost and effort. He summarized the alignment of interests and challenges confronting industry, government, and academia that favored a collective effort. (See Table 1-1.)

LEARNING FROM THE SEMATECH EXPERIENCE

Many industry and government leaders closely associated with the development of flexible electronics technologies advocate the establishment of a "Sematech-style" industry-government manufacturing consortium to facilitate the development of flexible electronics manufacturing competencies and the necessary tools and materials.[51] That fact, and the reality that few other comparable

[49] "After ten years of sustained efforts, Europe has now managed to create a leading position in R&D on OLAE (organic and large area electronics), mainly thanks to R&D collaboration efforts between research institutes, SMEs and large companies established through the support of public funding." The FP7-ICT Coordination Action OPERA and EM Commission DG INFSO Unit G5 "Photonics," *An Overview of OLAE Innovation Clusters and Competence Centres* (September 2011), 4.

[50] Malcolm J. Thompson, "A Consortium in Flexible Electronics for Security, Manufacturing and Growth in the US," in National Research Council, *Flexible Electronics for Security, Manufacturing, and Growth in the United States*, rapporteur S. Shivakumar (Washington, DC: The National Academies Press, 2013).

[51] In 2009, a blue-ribbon panel of experts sponsored by the National Science Foundation and the Office of Naval Research, which conducted a thorough study of European R&D in flexible electronics, recommended that the United States establish a Sematech-like organization for hybrid

TABLE 1-1 Factors Favoring a U.S. Consortium in Flexible Electronics

Government	Industry	Academia
Interests	***Interests***	***Interests***
• Job creation • National security • National competitiveness • Economic growth • Cost of government services	• Profit • Revenue • Market dominance • Product leadership • Intellectual property	• Educating and training students • New innovative R&D • New materials and processes
Issues	***Issues***	***Issues***
• Do not like picking winners and losers • Precompetitive R&D • Each government department has narrow focus	• Cannot exist alone no matter how big the company • Need complete new industry supply chain • Have fewer R&D $ • Few big R&D centers left such as Bell Labs, etc.	• Less R&D funding • Relevance • Needs better connection with industry • Difficulty in accessing fabrication facilities for prototype creation and systems integration

U.S. analogues are readily apparent, suggests that a closer assessment of the relevance of the Sematech experience to the challenges facing the United States in flexible electronics is warranted.

Sematech is widely acknowledged to have played a major role in the U.S. semiconductor industry's market resurgence and recapture of global leadership in the 1990s.[52] Numerous subsequent U.S. and foreign R&D initiatives have used Sematech as a model and point of reference; indeed, the very name "Sematech" has taken on an iconic cast not necessarily coextensive with the actual experience of the Sematech consortium.[53]

---

flexible electronics for precompetitive research, involving multiple companies and universities. Ananth Dodabalapur, "Panel on Flexible Hybrid Electronics," WTEC Workshop: International Assessment of R&D in Flexible Hybrid Electronics (June 30, 2009). The U.S. Army's launch of the Flexible Electronics and Display Center at Arizona State University in 2004 was "based on a scaled-down version of the Sematech model." "Army to Create a 'Sematech'-Like Consortium to Develop U.S. Flexible Display Industry," *Manufacturing & Technology News,* May 2, 2003.

[52] Kenneth Flamm and Qifei Wang, "Sematech Revisited; Assessing Consortium Impacts on Semiconductor R&D," in National Research Council, *Securing the Future: Regional and National Programs to Support the Semiconductor Industry* (Washington, DC: The National Academies Press, 2003).

[53] "Sematech has become a model for how industry and government can work together to restore manufacturing industries—or help jump-start new ones. The National Alliance for Advanced Transportation Battery Cell Manufacture, formed in 2008, was designed on the model of Sematech, for instance. So is the Department of Energy's new SunShot Initiative, which aims to reduce the cost of solar energy by 75 percent by 2020." "Lessons from Sematech," *MIT Technology Review,* July 25, 2011. The National Academies' Steering Committee for Government-Industry Partnerships made a comparable observation in 2003 in assessing the applicability of the Sematech model to an initiative to develop U.S. industrial capability in solid-state lighting. "[F]irms in the semiconductor industry

No Panacea

In fact the "Sematech model" per se does not appear to represent a ready-made solution for the challenges confronting the nascent U.S. flexible electronics industry. Sematech was a consortium embracing large players in an established industry (i.e., IBM, Intel, AT&T, Texas Instruments) pursuing research themes in an agreed technology area (i.e., complementary metal oxide semiconductor or CMOS) pursuant to certain commonly accepted principles (such as Moore's Law). Moreover, Sematech was a national effort arising out of a perceived national emergency affecting U.S. defense readiness in the Cold War.[54] None of these elements is present in flexible electronics.

Moreover, although it is often overlooked, the initial implementation of Sematech itself—despite the factors cited above—proved so difficult that the application of the Sematech model remains a challenge for even well-established industries, to say nothing of newer industries such as flexible electronics.[55] Cultural differences between companies exerted powerful centrifugal force at Sematech, and it is difficult to envision how they would not do so in future flexible electronics consortia of even more diverse members.[56] A number of original Sematech members pulled out when the Sematech research agenda did not match their own objectives.[57] Similar problems, writ large, are likely to confront consortia formed to implement a Sematech-style mission in flexible electronics, particularly given the diversity of potential technology paths.

Applicable Lessons

At the same time, advocates of collaboration in flexible electronics point to the role that intermediary organizations can play in bridging the cultural factors that divide relevant industry segments. They point out that this new field will require industries that have seldom, if ever, worked together, such as printing and

---

appear to have had the advantage of a much clearer research path than that being confronted by the solid-state lighting industry today." National Research Council, *Government-Industry Partnerships for the Development of New Technologies* (Washington, DC: The National Academies Press, 2003), 86n.

[54] Defense Science Board, *Task Force on Semiconductor Dependency,* November 30, 1986.

[55] "The Sematech semiconductor equipment consortium represents another model for successful public–private cooperation in technology development. But it has not been replicated in other industries." Vernon W. Ruttan, "Is War Necessary for Economic Growth?" September 26, 2006, 24.

[56] At Sematech "IBM employees, the 'white shirts,' were put off by the more informal Californians from Advanced Micro Devices. AT&T people were conspicuous for their secretiveness and reluctance to speak up. People from Intel were too blunt." Although these differences were gradually overcome, similar issues will pose a hurdle for prospective consortia. "Sematech Sets Model for Industry Consortium," *Fort Worth Star Telegram,* November 10, 1993. A clash between Sematech's first CEO, Bob Noyce, and COO, Paul Castrucci, arising out of different managing styles, led to the abrupt departure of the latter in 1989. Larry D. Browning and Judy C. Shetler, *Sematech: Saving the U.S. Semiconductor Industry* (College Station, TX: Texas A&M University Press, 2000), 86–99.

[57] Withdrawals included Micron Technology, LSI Logic, and Harris Corporation.

electronics, to cooperate closely in the development of new industrial processes. They believe that R&D collaboration can make this happen, albeit not without difficulty, to facilitate relationships between companies and technological disciplines that "don't necessarily talk to each other."[58]

Despite the many differences that exist between the situation of the U.S. semiconductor industry in the 1980s and the flexible electronics industry of today, a number of aspects of the Sematech experience are relevant to the latter.[59] The federal contribution to Sematech was matched by industry, whose top executives made a strong commitment to the consortium in terms of their own companies' resources and talent. Sematech included not only semiconductor manufacturers, but also equipment and materials suppliers, "who help[ed] set and drive an agenda, [convening] a large network around individual problems and bring[ing] solutions more quickly." Technology roadmaps and development of common manufacturing standards facilitated the development of necessary equipment and materials and reduced time to markets. Sematech also facilitated industry partnerships with universities, NIST, and the Department of Energy's national laboratories.[60]

Although the Sematech model is arguably inapplicable *in toto*, many lessons can be derived from specific operational aspects of the Sematech experience—as well as the broader experience of the U.S. semiconductor industry—which are potentially instructive with respect to the value of collaboration in this new field.

Industry leadership. Most retrospective accounts of Sematech concur that a critical factor in its success was the readiness of the major players in the semiconductor industry to make very substantial, even painful, contributions to the success of the project. Company CEOs and other senior executives served on Sematech's Board of Directors. "Technical advisory groups," comprised of company representatives with specialized expertise, advised on specific thematic projects. Many semiconductor device companies sent their most talented engineers to the consortium, a burden that weighed heavily on smaller firms but an initiative that ensured that the assignees' companies received relevant research results.[61] IBM executives such as Sandy Kane, Obi Oberoi, and Paul Castrucci

[58] Conference call, FlexTech Alliance, November 15, 2013.

[59] One difference is that much of the semiconductor manufacturing was initially located in the United States, whereas many of the companies that are interested in flexible electronics, especially for consumer applications, are not necessarily interested in locating manufacturing in the United States.

[60] Michael Polcari, President and CEO of Sematech, "The Sematech Model: Potential Applications for PV," in National Research Council, *Future of Photovoltaics Manufacturing*, 182.

[61] Sam Harrell, who served as an assignee to Sematech, recalls that "so if you were assigned in a lithography program, you worked in the lithography program. But you were also responsible to make sure that the lithography companies who were members got the information they needed. They got access, whether it's technical meetings or other kinds of opportunities to get together with customer and so on. . . . Most of the [Sematech] employees were not in the technical mainstream. Most of the technical mainstream were assignees." "Oral History of Sam Harrell," November 16, 2004, Austin, Texas (Computer History Museum, 2011). Ted Malanczuk, Vice President of Wafer Fabrication and Technology at National Semiconductor, said in 1988 that the assignment of secondees to Sematech was "not an easy task. National will have to donate up to 25 of its best." LSI Logic, one of two

played key roles in the launch of the consortium, and as Castrucci later recalled, IBM executive (later President) Jack Keuhler "was behind [Sematech] 100 percent. They were putting $50 million a year into it."[62] This industry-led feature of Sematech stands in contrast to the character of most European research consortia, which are government-led projects in which participating companies pursue themes of interest to public authorities.[63]

To date no comparable major corporate U.S. "champions" have emerged in flexible electronics, a fact that represents a significant stumbling block to the realization of a U.S. manufacturing presence. However, a number of world-class U.S. companies are active in precompetitive and upstream research, including Hewlett-Packard, DuPont, Eastman Kodak, Corning, 3M, Universal Display Corporation, and GE. The Palo Alto Research Corporation (PARC), a research arm of Xerox—to which are attributable many of the innovations underpinning the Internet era—is pursuing flexible electronics research themes.[64] Although the critical mass of industry commitment, resources, and personnel to a joint developmental effort has not yet occurred, industry leadership in this field remains possible, particularly with supportive public policies.

Roadmapping. Technology roadmaps are evolving industry plans in areas of emerging technology that identify short-, medium- and longer-term technological objectives for the purpose of forming a rough consensus about technological requirements, the timing of needed solutions, and the future evolution of the technology. By their nature, roadmaps rely on forecasts about the market or consumer preferences, and such predictions can be difficult, especially for new markets. Participants in the symposia convened for this study have emphasized the importance of technology roadmapping in flexible electronics to stimulate the

---

Sematech members with less than $1 billion in annual sales, was reportedly "still wrestling with how to pull at least five key people from its operations to send to Sematech." "Sematech Beckons Brightest Engineers Going to Promised Land," *San Jose Mercury News,* March 14, 1988.

[62] "Oral History of Paul Castrucci," July 18, 2008, Mountain View, California (Computer History Museum, 2008), 24; "Join Hands in the Semiconductor Race," *Los Angeles Times,* June 22, 1989. Charlie Sporck, then head of National Semiconductor and one of the founders of Sematech, observed in an oral history interview that he "went around and found tremendous support including IBM. IBM is not really a big joiner. But that led to the foundation of Sematech." "Interview with Charlie Sporck," February 21, 2000, <*http://silicongenesis.stanford.edu/transcripts/sporck.htm*>. IBM also contributed its extensive knowledge and experience gained from 3 years of operations in processing 200 mm (8 inch) diameter wafers at a time (late 1980s) when most U.S. semiconductor makers had not yet made the transition to 200 mm. AT&T contributed proprietary technology for its static random access memory (SRAM) to serve as a process vehicle as well as key process technology. Robert R. Schaller, *Technological Innovation in the Semiconductor Industry: A Case Study of the International Technology Roadmap for Semiconductors (ITRS).* (Ph.D. dissertation, George Mason University, 2004), 458; Browning and Shetler, *Sematech,* 57–59, 72–74.

[63] Browning and Shetler, *Sematech*; National Research Council, *Government-Industry Partnerships,* 92.

[64] Ross Bringans, Palo Alto Research Center, Inc., "Challenges and Opportunities for the Flexible Electronics Industry," in National Research Council, *Flexible Electronics for Security, Manufacturing, and Growth.*

development of common industry standards, identify technological gaps, and enable equipment and materials suppliers to produce products for which real needs and applications do exist.[65] The relevance of recent experience with roadmapping in semiconductors was acknowledged.

In the semiconductor industry, roadmaps have become essential because semiconductor manufacturing is virtually completely automated, utilizing complex, highly specialized machines and processes developed by multiple vendors that cannot readily be integrated without coordination between players and a clear understanding of what technologies will be required within a given timeframe. Although flexible electronics may ultimately involve manufacturing processes that are simpler relative to those involved in semiconductor fabrication, they will nevertheless also be highly automated, involve multiple technological actors and disciplines, and require close and sophisticated coordination (e.g., roadmapping) in order to succeed.

Technology roadmapping began within individual semiconductor companies in the late 1970s.[66] Collective benchmarking by the industry first occurred in the context of the Defense Department's Very High Speed Integrated Circuit (VHSIC) program, launched in 1980, which was an Army/Navy/Air Force effort to develop semiconductors with military applications exclusively based on substrates made of silicon, the principal technology then in use in the commercial semiconductor industry.[67] The Semiconductor Research Corporation (SRC), a research consortium formed in 1983 by the Semiconductor Industry Association

[65] Dan Gamota, "iNEMI Flexible Electronics Roadmap: From Concept to Product," in National Research Council, *Flexible Electronics for Security, Manufacturing, and Growth.*

[66] Bob Galvin, then the Chairman of Motorola, recalled that beginning in the mid-1970s he spent many hours in the company's semiconductor laboratories, "dropping in on people, finding out what they're working on." He concluded that this front-line knowledge needed to be shared throughout the company in a systematic way. He told his executive team that "I want a roadmap of what we're doing and where we are going," giving them a simple chart (x and y axis) showing what Motorola was developing and resources needed, with a review in 6 months' time. The managers incorporated other factors into the process, such as product life cycles and learning-curve economies, eventually producing a forecasting tool—the "roadmap"—which came to play a central role in Motorola's strategic planning. Interviews and email exchanges with Bob Galvin and Bill Howard, Motorola, cited by Schaller, *Technological Innovation in the Semiconductor Industry*.

[67] Previously defense-related microelectronics R&D efforts had focused on technologies based on substrate materials such as gallium arsenide and germanium, which had important defense-related advantages but limited commercial application. VHSIC established a significant government connection with industry. Larry Sumney, who served as program director of VHSIC, went on to serve as Executive Director of the Semiconductor Research Corporation (SRC), the first R&D consortium in the U.S. semiconductor industry, and later served as acting head of the nascent Sematech consortium. He observed that VHSIC program reviews by participating companies "were a public way for individual firms to gauge or benchmark themselves against each other in a form of open competition. Driven by definitive technology targets (i.e., 'near micron' 1.25 microns for Phase I, and 'sub-micron' 0.5 micron for Phase II, the VHSIC program contained some of the key factors that could be interpreted as a technology roadmap exercise, if not explicitly intended as such." Schaller, *Technological Innovation in the Semiconductor Industry*, 427–428.

to support university-based R&D, originated the first industry-wide goal-setting technology "roadmap" in 1984-1985, a set of 10-year research targets to make possible next-generation dynamic random access memory (DRAM) devices, with the idea "to organize university research through a 'roadmap' versus the traditional 'white space' approach to academic inquiry.[68]

Sematech embraced and facilitated roadmapping from its inception, with planning actually predating the formal startup of the consortium in 1988 in industry-government-academic workshops convened in 1987 and 1988.[69] Paul Castrucci, an IBM executive who served as Sematech's first chief operating officer, took part in the workshops and described in a subsequent oral history interview how they quickly broke down boundaries between people representing different organizations.[70] He took the Sematech job based on his experience in the workshops:

---

[68] Ibid., 442–443. Larry Sumney observed in 2011 that SRC brings relevant sectors together with government in a manufacturing research ecosystem. The sectors lead by jointly identifying the most urgent R&D needs at the precompetitive level, and government provides incentives by co-funding research: "Given the diversity of participants, this ecosystem can be distributed but very coordinated. We see a flow of related ideas and technologies moving in both directions between industry and academia, with government playing a major role. Larry Sumney, "Semiconductor Research Corporation," in National Research Council, *Future of Photovoltaic Manufacturing,* 188. An industry participant who took part in the SRC exercise later recalled that "from the very beginning there was general enthusiasm on talking about forecasting the future," and the discussion was so productive that "consensus on the future was easily reached." Schaller, *Technological Innovation in the Semiconductor Industry*, 444, citing telephone interview with Jim Daughton, Honeywell, August 4, 2000.

[69] Sematech's startup team conducted an intense series of industry-government planning workshops in 1987 and 1988, which produced the "Black Book," a roadmap for achieving technological goals together with resources and business initiatives needed to achieve those goals. The Black Book set forth a 5-year plan for pushing technology levels ahead (Phase I, 0.8 micron design rules; Phase II, 0.5 micron; Phase III, 0.35 micron). The planning workshops primarily involved engineers and scientists from companies, the government, and SRC organized into subsets of 10-15 people according to specific technology topics with some overlap of related workshops. The workshops included "many of the industry's best people" and among other things produced the "first real database of key technologists that 'fanned and penetrated the armor' of the semiconductor community." Schaller, *Technological Innovation in the Semiconductor Industry*, 451–458.

[70] At a 1988 Sematech workshop, "They had 60 people there, different backgrounds, national laboratories, device, people (tooling) people, and universities. And the idea of this two and a half day workshop was to define the 0.25 micron process tools and ground rules. Give the ground rules and the process to Motorola guys, have them design an SRAM in two and a half days. I say 'this I got to see." I couldn't get the Army of Engineers to do that in two and a half months. So they broke them into groups of 20 each, that mix I talked about. It was run by a TI guy, I can't remember his name, but they were worrying about the materials. Lou Parillo from Motorola had the job of the processing tools, I was in that group. And [Obi] Oberoi from IBM had the back end of the line [a mixture of] 20 people in each of those groups, mixture. So I'm in the process group, 20 of us. Lou's up front with a flipchart. Nobody's talking. We've all been told, 'don't talk when you go out.' . . . I said, 'Come on, this is not proprietary, let's get going.' So people started opening up. And Lou got down to maybe step five and then he said, 'I really don't know how to do this next step.' The guy over in the corner says 'Well I know how to do it. Da, da, da, da." After 8 hours, we had a complete process description and all the tools. It was a combined intelligence of those 20 people, collaboration." "Oral History of Paul Castrucci," 23–24.

> The secret of success in the future is going to be collaboration. But there are a lot of people who don't understand that. But Sematech was based on collaboration. So I says 'I'll take that job, John. You know, I think I can do something with it.' I saw what happened in that workshop.[71]

The Sematech workshops were instrumental in giving direction and focus to a vast and comprehensive investment program by the U.S. semiconductor industry and the U.S. government. The roadmapping process made it possible to identify certain potential "show stopping" technological problems in advance and to enable redeployment of resources to address those problems in a timely manner.[72] Sam Harrell, one of the founders of Sematech, later recalled,

> Those were working sessions which drove to some conclusion about the needs and requirement of the industry and what [were the] most likely alternatives to meet those needs and requirements. Those were very powerful interactions that had never been able to happen before. . . . Sematech's [proposed] $100 million from the government and $100 million from industry was peanuts compared to what the industry spends on its own balance sheets. Suppliers alone spend $1.4 billion a year on RD&T [research, development, and testing]. The member companies spend $6 to $7 billion a year on RDT&E [evaluation] in a comparable basis. What the strategic workshop road maps did was to set in motion a bunch of focusing activities of $8 or $9 billion worth of effort, not just $200 million worth of effort.[73]

The emerging field of flexible electronics is already characterized by numerous roadmaps generated by industry groups, consulting organizations, and regional development agencies. However, no universally accepted roadmap yet exists for flexible electronics, and it is not clear that a single comprehensive roadmap is feasible or appropriate over the near term, given the multiplicity of potential manufacturing processes, substrates and conductive materials, encapsulation techniques, and end products. The semiconductor roadmaps of the 1980s and 1990s involved certain technological uncertainties, such as the future of lithography, but all reflected silicon-based CMOS technology and many generally accepted

---

[71] Ibid., 24.

[72] Intel's Paulo Gargini, who participated in semiconductor industry roadmapping in the 1990s, recalls that "in the 1997 roadmap for the first time, for instance, I had the idea that by the middle of the next decade, 2004/5/6, we had to introduce high-k metal gate. And at the end of the meeting, especially the university people were terrorized, because they thought they were not going to think about it until 2010. At that point, this was in '97 . . . in the next six years you have to be ready, so and it was really a terror for all of them all of a sudden, something that was really a low key project, all of a sudden was becoming very important. The benefit of it was that indeed, fortunately for all of us, many of the universities began working on it. . . ." "Oral History of Paolo Gargini," July 27, 2011, Mountain View, California (Computer History Museum, 2011).

[73] Cited in Browning and Shetler, *Sematech*, 42. Beginning in 1998, the U.S. semiconductor industry expanded the roadmap process to include foreign participants, and the evolution of global semiconductor technology is now assessed pursuant to the International Technology Roadmap for Semiconductors (ITRS).

BOX 1-4
International Electronics Manufacturing Initiative (iNEMI) Roadmaps

The International Electronics Manufacturing Initiative has been roadmapping the needs of the electronics industry since 1994. The iNEMI roadmap is modeled on the International Technology Roadmap for Semiconductors (ITRS) and includes 21 individual roadmap chapters in its most recent biannual edition. Five external organizations including the ITRS contribute to the chapters. Dan Gamota, a member of the WTEC task force that reported on European developmental efforts in flexible electronics in 2010, chairs the Printed and Organic Electronics Technical Working Group, which has been developing, publishing, and updating the iNEMI Roadmap for Printed and Organic Electronics in consultation with numerous relevant public and private entities since 2006. Dr. Gamota indicates that the iNEMI roadmap is intended to stimulate the development of industry standards and to identify "gaps and needs" in the industry supply chains, including potential "showstoppers."

manufacturing processes. Despite that fact, the establishment of generally accepted roadmaps was a challenging exercise.[74]

In a recent discussion of the applicability of semiconductor-style roadmapping to the photovoltaic manufacturing industry, it was pointed out that the diversity of photovoltaics (PV) manufacturing themes in areas such as deposition methods, cell architectures, and materials (polycrystalline versus monocrystalline silicon) meant that a single semiconductor-like equipment technology roadmap "would not be appropriate for PV."[75] Full pursuit of the opportunities in this emerging field is likely to require multiple roadmaps, some of them quite divergent in technological terms with respect to materials and processes.[76] The role

[74] Intel's Paolo Gargini, a participant in early semiconductor industry roadmapping exercises, recalls that the process was "dominated by many people who were highly theoretical, and I understood, after a while, that the motivation was to submit proposals to the government for funding, and one of the problems that came back from the government, the national labs, university and industry, would go and propose programs to the government, with completely different roadmaps. . . . "Oral History of Paolo Gargini."

[75] Doug Rose, Sunpower, "Observations on Building a PV Roadmap Model," in National Research Council, *Future of Photovoltaic Manufacturing,* 198–201.

[76] Dr. Steven Freilich of DuPont, speaking in 2011 of the opportunities presented by thin-film photovoltaic devices on flexible substrates, cited the experience of semiconductor roadmapping in the area of lithography as a precedent for "controlling a fast-moving technology and fast-moving markets." The semiconductor roadmap explored five "technologies of note" between 2000 and 2003, one of which was dropped. The roadmap "was able to lay out the objectives for each technology, performance goals, milestones and timing. This is important to a materials supplier . . . because it clearly shows when a research program is not performing well." Steven C. Freilich, "DuPont Reflections on Photovoltaics," in National Research Council, *Future of Photovoltaic Manufacturing*, 69.

played by consortia in roadmapping in the semiconductor context, although not perfectly analogous, remains instructive.[77]

Supply chain integration. A rationale for one or more flexible electronics consortia is to "align the interests of end product manufacturers to tools and materials suppliers."[78] A critically important conclusion of Sematech's early planning workshops was a unanimous view that "the chipmakers' biggest headaches stemmed not only from technological problems but more from difficulties with suppliers" with a need to synchronize the "cadence" of development of tools, materials, and process technology by suppliers needed to support the device makers' advances in design.[79]

Semiconductor device makers complained that U.S.-based equipment and materials vendors lacked ability to recognize and address quality problems associated with their products, while the suppliers faulted the device firms for nonstandardized requirements, unwillingness to share relevant data, and an over-emphasis on squeezing the lowest price, rather than the best quality, out of suppliers. Intel's Gordon Moore, who participated in Sematech during its early years, recalls that Sematech "worked with these [supplier] companies to develop reliable tools, to teach them total quality control, and to help them understand the needs of the industry and the increasing sophistication of the manufacturing process."[80] Suppliers' equipment was put through rigorous endurance testing, a process dubbed

---

[77] Mark Pinto of Applied Materials cited the problems associated with establishing an industry chain in an emerging sector at the National Academies 2011 symposium on photovoltaics. He observed that in the solar industry it was hard to form collaborations "because manufacturing and technical development were still competitive." There were "some three dozen solar start-ups in Silicon Valley developing nearly the same number of [copper indium gallium arsenide] processes, all of them competing. Believe me, I know, because they all want equipment, and it's hard to find." However, he pointed out, even in such situations, "there are areas in which collaboration is useful," such as the modeling necessary to explore optics and electronics. "You can plug in whatever band structure you want for whatever elements and do the model from first principles. This helps to figure out what's going on and how to design these structures. Every company should benefit from that." National Research Council, *Future of Photovoltaic Manufacturing*, 117.

[78] Sam Harrell is former President of SEMI-Sematech, a consortium formed by U.S. tool and materials suppliers to support collaboration with Sematech. "Oral History with Sam Harrell." In 1989, Kristopher Lee, Program Manager for Sematech's tool development program, estimated that U.S. semiconductor production lines could only be used for 25 to 40 percent of a given work week. "Most of the time, they are shut down, because one or more machines requires repair or maintenance." "Sematech Racing the Clock in Effort to Regain US Lead in Chipmaking," *Dallas Morning News,* July 22, 1989.

[79] Ibid.

[80] Moore observes that "much of the important work required to improve manufacturing equipment did not have to be done individually but could be done by the consortium centrally. The consortium developed a cost-of-ownership model for manufacturing tools that described the problems in detail." Gordon Moore, "The Sematech Contribution," in National Research Council, *Securing the Future,* 99–100. Sematech developed a program called "Partnering for Total Quality" (PFTQ) that benchmarked U.S. suppliers' performance against global trends (which showed that "U.S. equipment makers were almost looking at going out of business.") and, jointly with suppliers, developed protocols for PFTQ partner training. Browning and Shetler, *Sematech,* 135.

"Iron Man."[81] Conversely, suppliers successfully demanded that device makers develop and share nonproprietary standards.[82] Sematech roadmaps were modified to include "the intricate coordination between [Sematech's quality program] with equipment and material suppliers."[83] One supplier company executive, Scott Kulicke, who was instrumental in setting up SEMI-Sematech, a parallel organization of suppliers that interfaced with Sematech, observed in 2004 that

> [w]hen all was said and done, the part of Sematech that really worked was in the roadmap, and in the ability of using the roadmap and the Sematech infrastructure to focus the equipment companies on a few stated objectives so we did a better job of bringing it out on time and to spec. And that allowed the semiconductor guys to regain product parity in a rapidly evolving, technologically evolving environment.[84]

In flexible electronics, contrasted with the semiconductor industry of the 1980s, no established group of major device makers exists to demand cost savings and improved quality from upstream suppliers. Rather, a flexible electronics food chain has emerged in the United States that—for the moment at least—primarily serves offshore device makers, but which constitutes a viable infrastructure to support potential device manufacturing in the United States. The toxic character of relations between U.S. semiconductor device makers and their vendors in the 1980s likely would have foreclosed the revival of semiconductor manufacturing in the United States had it not been addressed in a foursquare manner, primarily through the Sematech initiative. In flexible electronics, the U.S. competitive position could probably not survive a comparable protracted period of culture clash between whatever indigenous device producers may emerge ultimately, on the one hand, and suppliers, on the other. Consortia, and other forms of collaboration, may play a role comparable to that of Sematech in mitigating conflict and forging

---

[81] Intel's Paolo Gargini recalls that prior to Sematech's formation, Intel "[B]uilt two adjacent chambers in which we would put a Japanese equipment in one chamber, and we will put a US equivalent in the next chamber, and then we will compel the CEO of the company to watch his equipment fail after two hours, while the Japanese equipment had been going for a week." After Sematech was formed, Intel "essentially … gave Sematech the whole procedure (of testing the endurance of the equipment) that we had developed here in [Intel] and Sematech created the name Iron Man (i.e. endurance test) that was essentially, the equipment was going to be put through the Iron Man until it broke down and then there would be the list of defects (failures) and the supplier was supposed to fix it (the equipment failures), and so forth, so the major benefit of Sematech was really to help not the IC companies, but really, to help the equipment company come back to speed." "Oral History of Paolo Gargini," 27.

[82] Dan Hutcheson, President of the consultancy VLSI Research, observed in 1993 that prior to Sematech "the semiconductor industry looked at the equipment industry like used-car salesmen. The way it usually worked was that a chip maker would take two equipment vendors and pit them against each other. So they'd get the prices down, but there was not incentive to produce reliable equipment." "Sematech Sets Model for Industry Consortium," *Fort Worth Star-Telegram,* November 10, 1993.

[83] Browning and Shetler, *Sematech,* 135.

[84] "Oral History of C. Scott Kulicke," September 22, 2004 (Computer History Museum, 2004).

strong and lasting ties within the industry supply chain. One of the most urgent challenges will be the development of standards.

Development of common standards. The development of widely accepted industry standards is a necessary step in the evolution of a new technology-intensive sector. Standards enable industry participants and customers to understand, measure, and compare various materials, devices, and processes. The existence of common standards is also necessary to enable the interoperability of equipment and systems developed by different companies in complex manufacturing operations. In most cases the development of workable and generally acknowledged industry standards requires extensive collaboration. NIST plays a key role in the development of industry standards, but in the United States such standards "are not handed down by the government, but produced through a collaborative process."[85] Cooperative research organizations may not be essential in all cases to such collaboration, but the historical record illustrates the positive role they can play.

For example, at the time Sematech was formed the semiconductor fabrication process utilized company-specific computer-integrated manufacturing (CIM) systems that "evolved over many years to support the particular needs of an individual company or factory," a reality that resulted in "high support costs and long delays to add new functionality . . . low customer satisfaction and low development/support staff morale."[86] In 1994, Sematech and Texas Instruments eleased a jointly developed protocol, CIM Framework 1.0, which allowed software applications to interoperate with each other in the same manner that Microsoft Windows–compatible applications interface in computers using the Windows operating system. This innovation enabled manufacturers to avoid installation of multiple proprietary CIM systems that were not necessarily compatible with other suppliers' tools. "The use of cooperative equipment standards in an otherwise competitive environment reduced duplication and enhanced manufacturing productivity."[87] A retrospective assessment by NIST, which participated in the effort pursuant to a cooperative research and development agreement (CRADA), observed that Sematech played a key role in developing an industry consensus supporting the CIM framework.[88]

---

[85] Comments of Keat Rochford, Acting Director, Electronics and Electrical Engineering Laboratory, National Institute of Standards, "Measurement and Standards: The Role of NIST," in National Research Council, *Future of Photovoltaic Manufacturing*, 163.

[86] Sematech (CIM) *Framework Specification Version 2.0.*

[87] Browning and Shetler, *Sematech*, 176.

[88] S.L. Stewart and James A. St. Pierre, NIST, "Experiences with a Manufacturing Framework," *Business Object Design and Implementation* (1997), 138. "To achieve Sematech's objectives, it is not sufficient to produce a specification, even one of technical excellence. There must also be widespread agreement among both the suppliers and users of manufacturing software that applications should be based on the specification. This consensus is a necessary step in the road to adoption and success. Sematech has already involved the users' groups from its member companies in the process of developing the specification. It is also contacting independent suppliers and providing orientation

Like semiconductors, flexible electronics production will be highly automated and require the development of industry standards for equipment, software, and materials. Printed electronics manufacturers have complained, for example, that materials received from suppliers are of inconsistent quality, while suppliers counter that users need to specify "a few broad categories of standard material quality, as the semiconductor industry did early on and [that] would allow suppliers to focus on their limited development funds on controlling the key specifications to the necessary degree for multiple customers."[89]

Given the diversity of flexible electronics technologies, it is unlikely that a single public or private organization can establish industry-wide standards and, in fact, standards setting has moved forward in some fields already.[90] However, standards setting may evolve in this industry as NIST indicates. As in other sectors, the progress will be industry-driven, and consortia are likely to play a constructive leadership role.[91]

It is far from inevitable that the formation of industry standards for flexible electronics will await consensus among U.S.-based firms. In 2009, Germany's research ministry BMBF sponsored a collaboration, MaDriX, by local firms to develop standards for the printed electronics industry. This effort involved €15 million, €8 million of which was provided by BMBF. The effort included a printed electronics manufacturer, PolyIC, materials suppliers BASF, Evonik Industries, and ELANTAS Beck and Siemens, which provided the necessary inspection systems. "Through MaDriX, material parameters will be determined and an unvarying test environment for new materials introduced. By doing so, companies can develop new materials quickly and more efficiently. Standardizing test conditions facilitates use of a statistical measurement system developed by Jacobs University Bremen and Siemens to enable companies to compare results.[92]

Leveraging the federal laboratories. A number of federal laboratories are currently engaged in research within or relevant to the field of flexible electronics,

---

and training about the CIM Framework in scheduled classes and public conferences. We believe that this process of awareness, involvement, and training is absolutely essential to the success of a CIM Framework, and we recommend that it be continued and expanded to the limits of the resource available" (ibid).

[89] "Printed Electronics Suppliers Look to Learnings from Electronics World to Help Scale Volume Production," *Printed Electronics Now,* January 2010. The consultancy IDTechEx observed in 2011 that in flexible electronics, "for most devices, there are no globally approved test standards, so while one company may claim exceptionally high mobility for an organic semiconductor, the trade-off may be lifetime and using a very high voltage. Similarly, for lighting and displays, there needs to be test standards to fairly compare different materials and devices," IDTechEx, *Printed, Organic & Flexible Electronics: Forecasts, Players & Opportunities 2011-2021* (2011), 232.

[90] The National Renewable Energy Laboratory in Denver has developed standards for evaluating and comparing materials and devices in the area of organic photovotaics and has been globally acknowledged as the independent test center for these technologies. IDTechEx, *Printed, Organic, and Flexible Electronics,* 232–233.

[91] NIST, Global Standards Information, <*http://gsi.nist.gov/global/index.cfm/L1-5/L2-44/A-165*>.

[92] "MaDriX Sets New Standards for the PE Industry," *Printed Electronics Now,* September 2009.

including NIST and the Ames, Lawrence Berkeley, and Oak Ridge National Laboratories.[93] NIST has been working to address "basic unknowns" associated with flexible electronics materials and nanoscale structures "through sophisticated measuring instruments that range from x-rays and synchrotrons to acoustic surface measurement and scanning tunneling microscopes . . . to help the technology developers better understand the materials science that underlie manufacturing behaviors."[94] The federal laboratories have a broad statutory mandate to transfer technology to industry, including the conclusion of CRADAs.[95] However, as Carol Battershell of the Department of Energy (DOE) noted in 2011, a "perceived weakness [exists] in the commercialization activities of the national laboratories."[96]

The federal laboratories are engaged in a number of initiatives designed to accelerate transfer of technology to industry, with implications for flexible electronics.[97] The Sematech experience offers a case study in how a consortium can lay the groundwork for productive future collaborations between individual firms and the labs. A variety of cultural and institutional factors had historically inhibited working relationships between the semiconductor industry and the federal laboratories, notwithstanding the extraordinary resources and relevant competencies available at some federal facilities.[98] Sematech played a significant

[93] See Chapter 6, "Lawrence Berkeley National Laboratory Develops Micro-Activator that Flexes Under Laser Light," *Flexible Substrate,* January 2013; "ORNL Develops Carbon Nanotube Conductive Coatings for Flexible Electronics," *Flexible Substrate,* September 2011.

[94] Eric K. Lin, NIST, "Advancing Technology Through Measurement Science at NIST," in National Research Council, *Future of Photovoltaic Manufacturing*, 112.

[95] Stevenson-Wydler Technology Innovation Act of 1980 (P.L. 96-480); Bayh-Dole Act of 1980 (P.L. 96-517); Federal Technology Transfer Act of 1986 (P.L. 99-502); National Competitiveness Technology Transfer Act of 1989 (P.L. 101-189); National Technology Transfer and Advancement Act of 1995 (P.L. 104-113); Technology Transfer Commercialization Act of 2000 (P.L. 106-404); America COMPETES Act (P.L. 110-69).

[96] Caroll Battershell, Senior Advisor for Commercialization and Development, Energy Efficiency and Renewable Energy, U.S. Department of Energy, "Bringing Department of Energy Innovations to Market," in National Research Council, *Future of Photovoltaic Manufacturing,* 155. See also NIST, *Federal Laboratory Transfer Fiscal Year 2010,* August 2012; General Accountability Office, *Technology Transfer: Clearer Priorities and Greater Use of Innovative Approaches Could Increase the Effectiveness of Technology Transfer at Department of Energy Laboratories,* June 2009.

[97] DOE operates a Process Development and Integration Laboratory that supports manufacturing-scale R&D and enables DOE National Laboratories to partner with companies to develop commercially relevant manufacturing processes. DOE's Technology Commercialization Fund facilitates the transfer of technologies from the laboratories to companies in a manner designed to bridge the "valley of death" between research and later stage funding. The Entrepreneur in Residence program enables selected entrepreneurs to spend a year in National Laboratories "mining any knowledge that is available," with a commitment to look preferentially at the labs' technologies for commercialization. John Lushetsky, "DOE Solar Energy Technologies Program: Accelerating the U.S. Solar Industry," and Carol Battershell, "Bringing Department of Energy Innovations to Market," in National Research Council, *Future of Photovoltaic Manufacturing*, 153–157.

[98] Sandia National Laboratory in New Mexico, which was DOE's leading research laboratory for semiconductors, held responsibility for all electronics and microelectronics systems in the U.S. nuclear arsenal. Sandia had an on-site semiconductor design, manufacture, and failure analysis capa-

role as "matchmaker" between the industry and federal laboratories, establishing operations at Sandia and Oak Ridge National Laboratories, and NIST.[99] Collaboration with federal laboratories proved particularly valuable for industry-leading device makers and smaller equipment makers.[100]

Improving performance. Sematech's collaborative approach to manufacturing challenges has enhanced the members' productivity and efficiency in numerous ways. A system of "blind benchmarking" enables companies to compare their performance metrics with those of other companies.[101] "Equipment productivity teams are joint efforts by members to identify common problems with a tool, work together with the tool supplier, and share information about how to make a given tool perform more efficiently."[102] There is no reason why closely comparable, if not identical, practices could not be applied in industry collaborations in flexible electronics.

---

bility. In collaboration with Sematech, "Sandia's technical capabilities easily met and often exceeded Sematech's needs." But the industry regarded the federal laboratories as insensitive to commercial time scales and other imperatives, and the latter saw the industry as inexperienced in pursuing public return from research. Companies viewed federal laboratories as variants on university research laboratories without tenure pressure, whose researchers "are more concerned with pushing the frontiers of science than meeting the demands of making timely analysis and recommendations for the engineering of manufacturing equipment." Elias Carayannis and James Glover, "The Sematech-Sandia National Laboratories Partnership—A Case Study," *Technovation*, 22 (2002): 588.

[99] In 1989, Sematech and Sandia concluded an agreement to create the Semiconductor Equipment Technology Center (SETEC) as a vehicle for concentrating Sandia's resources on the development of tool designs and methodologies and equipment performance and reliability. This collaboration led to a $110.7 million CRADA that afforded the U.S. industry access to "the use of first-rate facilities and a complete range of science-based expertise upon which to call when faced with real industry problems." In 1992, Sematech and Sandia concluded a CRADA to establish the Contamination-Free Manufacturing Research Center to solve defect problems associated with larger die sizes and smaller feature sizes required by next-generation semiconductor devices. In 1993, Sematech entered into a CRADA with NIST, with DOE contributing $53.6 million and Sematech $49.4 million, to support semiconductor manufacturing R&D in areas of core competence including materials analysis, equipment modeling and design, and equipment and software reliability. Browning and Shetler, *Sematech,* 173–174; John McGrayer, *SETEC/Semiconductor Manufacturing Technologies Program: 1999 Annual and Final Report* (Sandia Report SAND2000-2845).

[100] McBayer, *SETEC,* 38; Carayannis and Gover, "Sematech-Sandia," 588–589.

[101] Blind benchmarking involves members' sharing 50 performance metrics on a nondisclosed basis. Each company recognizes its own data but not that of other companies. One member, seeing the lower cost of electricity paid by other members, used the data to demonstrate to its power company that its rates were not competitive and secured a reduction. Blind benchmarking can also serve as an early performance wake-up call. "When you see that your 10 percent improvement still leaves you behind by 50 or 60 percent, you realize you have to do something different." Michael Polcari, "The Sematech Model: Potential Applications for PV," in National Research Council, *Future of Photovoltaic Manufacturing*, 183.

[102] In one case, a Sematech member company found that most of the power used by the tool was consumed by the pumps. The research team identified which pumps could be idled at which times, saving substantial energy consumption. The member company then worked with the tool suppliers to adjust the idle modes for maximum efficiency. The information was made available to all Sematech members. Ibid., 183.

Building human capital. Byron Clayton of NorTech, a technology-based economic development intermediary that has established a flexible electronics cluster in northeast Ohio, emphasized at the symposia and meetings convened for this study that training a workforce to support flexible electronics manufacturing is "hugely important." He indicated that local universities and junior colleges need to know in advance the types of skill sets that will be required so that they can implement appropriate curricula. Failure to address workforce concerns will result in local manufacturers "looking elsewhere."[103]

Although the renaissance of the U.S. semiconductor industry is most commonly associated with Sematech, the education and training role of SRC, which was formed to support and coordinate relevant university-based research, was also important. Dr. John E. Kelly, who heads IBM's global research efforts, observes that SRC was "instrumental in sustaining the essential pipeline of skills. If you decide you're going to compete on innovation, versus lowest cost only, you have to have the skills to innovate." At the time SRC was formed in 1982, fewer than 100 students and faculty were conducting silicon-based semiconductor R&D in the United States; by 2008, that figure had grown to 2,226, with a research output "larger in some dimensions than that of some of the largest corporations in the industry." SRC "put thousands of highly qualified students into the industry."[104] SRC's research agenda, which directs government and industry funds to university-based research teams, focuses on the space between "blue sky" basic research and early product development in all parts of the semiconductor value chain. All projects are governed by research contracts with milestones worked out with the principal investigator; intellectual property (IP) is shared, with the universities owning the IP and SRC members allowed royalty-free, non-exclusive access. SRC "make[s] sure there's no blocking IP."[105]

Sematech-style industry collaborations and public–private partnerships are sometimes characterized as "industrial policy," with a variety of negative and counterproductive connotations. This study indicates that such industrial policies are being implemented across Europe and East Asia in flexible electronics, including even in countries with robust Manchestrian traditions, like Britain, and ferocious intra-industry rivalries that have historically inhibited collaboration, like Korea. At least in the European case, the merits of public–private

[103] National Research Council, *Flexible Electronics for Security, Manufacturing, and Growth in the United States: Summary of a Symposium* (Washington, DC: The National Academies Press, 2013).

[104] John E. Kelly, "Collaboration for Success in Semiconductors," in National Research Council, *Future of Photovoltaic Manufacturing*, 73. As of 2011 SRC had invested more than $1.3 billion contributed by members and government, supported 7,500 graduate students through 3,000 research contracts, 1,700 faculty, and 241 universities. SRC support resulted in more than 43,000 technical documents, 326 patents, 579 software tools, and work on 2,315 research tasks or projects. Larry Sumney, "Semiconductor Research Corporation," in National Research Council, *Future of Photovoltaic Manufacturing*, 184–185.

[105] Sumney, "Semiconductor Research Corporation," in National Research Council, *Future of Photovoltaics Manufacturing,* 185.

BOX 1-5
The FlexTech Alliance

The FlexTech Alliance, based in San Jose, California, is a consortium that promotes the supply chain for electronic displays and printed, flexible electronics. Like its organizational predecessor, the U.S. Display Consortium, it encourages collaboration among companies, academia, and research organizations, and it funds R&D projects in the display supply chain. The scale of FlexTech is extremely modest by current international standards, with combined funding support from the Army and Air Force Research Laboratories of $6 million over three years, with matching industry cost-sharing. This is sufficient to fund one or two tool development projects, several project demonstrators, and some materials and device R&D annually.

FlexTech representatives (including company executives) have emphasized the importance of fostering increased collaboration among hitherto disparate but well-developed industry sectors in the United States: electronics, printing, and packaging, among others. They believe that collaboration could be manifested in a variety of ways, as it is in Europe—the formation of industry clusters, joint university-industry R&D efforts, supply chain partnerships, and specialized consortia. "What you create as a whole is greater than the sum of the parts."[a]

[a]Conference call with FlexTech on November 15, 2013. See Chapter 7 for an extended description of the FlexTech Alliance.

collaborations have been validated by the eminent U.S. scientists and engineers who comprised the 2010 World Technology Evaluation Center, Inc. (WTEC) mission. Under such circumstances, and viewed against the concededly one-off experience of Sematech, the question presented does not appear to be why the United States should adopt a public–private collaborative approach in this new field, buy why it should not do so. The WTEC panel observed that in Europe,

> There is close industry-university-innovation center cooperation in precompetitive research. . . . In addition to substantial technical challenges, the cost of research and development in the flexible electronics area is perceived as a significant risk to individual companies, many of which have core expertise in only a subset of required competencies (e.g., circuit/systems design, coating, printing, and materials). European innovation centers mitigate the financial risk by sharing costs among multiple commercial enterprises and by leveraging substantial government funding at both the national and European Union (EU) levels. On the technical front, the centers foster a highly synergistic and interdisciplinary environment in which the complementary expertise of industrial, government, and academic scientists is combined to achieve new systems design goals (e.g., ultra-low-power

systems in foil), enhanced device performance, broader materials choices, and practical, low-cost manufacturing approaches.[106]

OVERVIEW OF THIS REPORT

Examining and comparing selected innovation programs both foreign and domestic, and their potential to advance the production of flexible electronics technology in the United States, an ad hoc committee of the National Academies reviewed the goals, concept, structure, operation, funding levels, and evaluation of foreign programs similar to major U.S. programs.[107] Specifically, the committee examined the role of research consortia around the world to advance flexible electronics technologies, comparing their structure, focus, funding, and likely impact, in order to determine what appropriate steps the United States might consider to the develop the industry in the United States.[108]

This volume surveys flexible electronics developmental efforts under way in Europe, East Asia, and North America with a focus on government support. Chapter 2 summarizes the enormous promise of flexible electronics as well as the technological obstacles to realizing that promise. Chapter 3 presents an overview of government promotional efforts in the three regions surveyed. Chapter 4 examines the beginnings of a competitive landscape in an industry where most of the technology is not yet mature in a commercial sense, with a highlight on expert SWOT (strengths, weaknesses, opportunities, threats) analyses of relevant regions and countries. Chapter 5 surveys European promotional efforts in flexible electronics, with a focus on programs being implemented by the European Union and by national and regional governments in the United Kingdom, the Netherlands, Belgium, Germany, and Finland. Chapter 6 presents an overview of East Asian promotional efforts featuring South Korea, Taiwan, Japan, and China. Chapter 7 summarizes federal and state initiatives in North America. This volume concludes with the committee's consensus findings and recommendations on the potential of the U.S. flexible electronics industry, the possible contributions of intermediate institutions to facilitate applied research and manufacturing, and other measures to support the development of the flexible electronics industry in the United States.

---

[106] WTEC Panel Report on European Research and Development in Hybrid Flexible Electronics (July 2010), xvi, 7. The WTEC panel members were Dr. Ananth Dodabalapur (Chair), Ashley H. Priddy Centennial Professor in Engineering at the University of Texas at Austin; Dr. Ana Claudia Aria, Manager of the Printed Electronic Devices at PARC Inc. (formerly XEROX-PARC); Dr. C. Daniel Frisbie, Professor of Chemical Engineering and Material Science at the University of Minnesota; Mr. Daniel Gamota, co-founder and President of Printovate Inc., a developer of large area electronics in lighting, renewable energy, and point-of-care diagnostics; Dr. Tobin J. Marks, Professor of Chemistry, Material Science, and Engineering at Northwestern University; Dr. Colin E. C. Wood, Research Professor at Texas State University at San Marcos, and Dr. Grant Lewison, an expert on bibliometrics and a Senior Research Fellow at University College, London.

[107] For a list of the members of this committee, see the Front Matter of this report.

[108] See the Preface for the complete text of the Committee's Statement of Task.

BOX 1-6
A Review of Sources and Their Limitations

The committee's review of programs and policies to support national flexible electronics industries has drawn from a variety of sources, including

- Presentations by U.S., Korean, Taiwanese, and German participants at a National Academies symposium convened by the committee in 2010 and presentations by U.S. and Dutch industry representatives in a workshop convened by the committee in 2011.
- Site visits by the committee to the Flexible Display Center at Arizona State University and to flexible electronics firms research centers in the FlexMatters cluster developing in northeast Ohio.
- A 2010 assessment by the WTEC study of European R&D programs in flexible electronics.
- Reports by government and para-public organizations such as the European Commission, VTT (Finland), the European Union's (EU's) OPERA project, the UK House of Commons Innovation, Universities, Science, and Skills Committee, the German National Academy of Science and Engineering (acatech), NorTech (Ohio), the UK Plastic Electronics Leadership Group, PolyMap (EU), and Taiwan's ITRI.
- Reports of U.S., EU, and UK industry associations.
- Company annual reports and regulatory filings.
- Consultant studies, for example, HSBC's *Fantastic Plastic* and IDTechEx's *Printed, Organic & Flexible Electronics Forecasts, Players & Opportunities* (2011).
- Websites and annual reports of each of the organizations known to be involved in flexible electronics R&D, such as individual Fraunhofer institutes, Finland's VTT, the Holst Centre in the Netherlands, ITRI in Taiwan, and various research organizations in Korea and Japan.
- European databases such as CORDIS and the UK's EPSRC database.
- National Academies' studies, for example, *21st Century Manufacturing: The Role of the MEP Program* (2013), which includes an extended appendix on leading manufacturing support programs in Asia and Europe.
- Informed trade press, including publications such as *Flexible Substrate*, *Plastic Electronics*, *The Emitter*, *Printed Electronics Now*, *Nanowerk*, and *Printed Electronics World*.
- Media reports from East Asia and Europe including economic, technical, and mainstream publications such as *Chosun Ilbo*, *JoongAng Daily*, *Taiwan Economic News*, the *South China Morning Post*, and *China Daily* and government press agencies such as Yonhap (S. Korea), Xinhua (China), Central News Agency (Taiwan), and Jiji (Japan).
- Relevant academic articles and books.

These sources are cited in full when referred to in the report.

continued

BOX 1-6 Continued

The committee has, within the scope of its budget, made every effort to draw on publicly available information about programs to support the flexible electronics industry. However, information available about these national programs is limited.

- Existing academic work with respect to government support for flexible electronics is very limited. Although there are numerous technical scientific articles about research findings, these typically address scientific arcana and rarely contain information relevant to the committee's task.
- Published consultants' studies mostly describe what individual companies are doing and what applications could emerge in flexible electronics. Although useful in this regard, they shed little light on government policy measures.
- Commercial data are limited and consist largely of forecasts—inherently subjective—because this is a relatively new industry.
- Articles, studies, or government white papers about relevant programs in East Asian countries are often not made public and are rarely published in English.

2

Assessing the Potential of Flexible Electronics

"Flexible electronics" refers to electronic devices that can be bent, rolled, or folded without losing functionality. The term is largely coextensive, albeit not entirely synonymous with terms such as "plastic electronics" (commonly used in the United Kingdom [UK]), "organic electronics," "OLAE" (organic and large area electronics, a term used in the European Union [EU]), and "printed electronics" and "printed intelligence" (Finland).[1]

ADVANTAGES OF FLEXIBLE ELECTRONICS

Flexible electronics is an emerging technology offering "completely new product concepts combined with low production costs, low energy consumption and environmentally friendly materials and processes."[2] It is a technology field with revolutionary potential. A blue-ribbon panel of U.S. scientists conducting a 2010 study of flexible electronics programs in Europe under the auspices of the World Technology Evaluation Center, Inc. (WTEC) observed that

> [o]rganic/polymeric and inorganic flexible devices integrated in intrinsic and hybridized systems represent a highly promising interdisciplinary area of technology that will provide greatly increased functionality and potential to meet future challenges of scalability, flexibility, low power consumption, light weight,

[1] Although flexible electronics products commonly incorporate "organic" (carbon-based) materials, some of them also contain metals and metal oxides. Although most flexible electronics products are likely to be produced through roll-to-roll printing processes, some are currently being fabricated on conventional liquid crystal display and microelectronics production lines.

[2] Valtion Teknillinen Tutkimuskeskus (VTT), *Promoting Entrepreneurship in Organic and Large Area Electronics in Europe* (2011), 4.

> and reduced cost. . . . Application areas impacted by flexible electronics include energy (e.g., photovoltaic energy conversion systems and energy-efficient lighting), consumer electronics (e.g., portable flexible displays, sensors, and actuator), healthcare (e.g., low cost personal health monitoring systems), communications (e.g., radio-frequency identification systems), and national defense (e.g., networked sensing, intelligent and autonomous systems, and enhancement of individual warfighter capabilities). . . . There is excellent promise in all of these applications for reducing costs through manufacturing processes that utilize printing and lithography methodologies and through the combining of multiple functionalities.[3]

The dramatic potential of flexible electronics technology is widely acknowledged in the global scientific community.[4]

Functionality

Flexible electronics devices will be able to perform functions that conventional electronic devices cannot, including bending, rolling, folding, and stretching, and may ultimately be more durable. According to one account, an e-paper electronic sign in Japan survived the 2011 earthquake and tsunami and continued to "display emergency contact and note information long after other powered-displays fell dark."[5] Flexible electronics devices have enormous potential for use on and in the human body, not only because they enable stretchability, flexibility, and mechanical softness that cannot be achieved with silicon-based technologies, but also because organic electronics devices are more compatible with biological systems than silicon-based alternatives.[6] Printed electronics devices will be able to incorporate multiple functions in a single device that are impossible for silicon devices, such as batteries, microphones and speakers, displays, and solar cells.[7]

Organic light-emitting diodes (OLEDs) are solid-state devices that emit light and can be mounted on a variety of flexible substrates for applications in displays and lighting. OLEDs have advantages over conventional light-emitting technologies that are leading to their deployment in an increasing number of applications. In the lighting field, conventional light-emitting diodes, the alternative

---

[3] WTEC Panel Report, *European Research and Development on Hybrid Flexible Electronics,* July 2010, v, xv. The study was commissioned by the National Science Foundation and the Office of Naval Research. WTEC is a nonprofit research organization originally spun off from Loyola University, Maryland, in 2001. It is the leading organization in the United States conducting international technology assessments through expert review. Its panelists typically are leading scientists in their fields.

[4] See Robert H. Reuss, Babu R. Chalamala, et al., "Macroelectronics: Perspectives on Technology Applications," *Proceedings of the IEEE,* July 2005.

[5] "The Future of Stand-Alone E-Readers," *Flexible Substrate,* November 2012.

[6] Chemical Sciences and Society Summit, *Organic Electronics for a Better Tomorrow: Innovation, Accessibility, Sustainability,* September 2012, 14.

[7] IDTechEx, *Printed, Organic & Flexible Electronics: Forecasts, Players & Opportunities 2011-2021* (2011), 30.

form of solid-state lighting, are small, intensely bright, and operate at high temperatures, whereas OLEDs can be viewed directly, do not require the diffusers used to offset the brightness of LEDs, and operate at lower temperatures, arguably making them more suitable for a variety of lighting applications. With respect to displays, OLEDs offer numerous advantages relative to conventional liquid crystal displays (LCDs), including lower manufacturing and materials cost, ability to operate without a backlight (OLED circuits are light-emissive), lower energy consumption, lighter weight, and higher contrast ratios that enable sharper images.

More Efficient Manufacturing

The prospect that complex electronic devices can be fabricated through a comparatively simple printing process utilizing roll-to-roll (R2R) technology "represents a disruptive manufacturing technology [that] will make it possible to economically generate high value-added technology products at meters-per-minute rates on plastic film, paper or foil, achieving feature dimensions as small as 10 nanometers over areas encompassing billions of identical devices."[8] Most importantly, given the staggering investment costs associated with fabrication of silicon-based microelectronic devices, the prospect that comparable devices could be fabricated with comparatively inexpensive machines and processes intrigues policy makers, scientists, and established companies and entrepreneurs.[9] (See Table 2-1.) Kent Displays, for example, an Ohio-based startup making flexible displays for consumer applications, is producing flexible consumer e-writing products using a roll-to-roll process that has minimized investment and production costs. Although a traditional liquid crystal display manufacturing plant requires investments of a billion dollars or more, the Kent Displays facility was built with capital investments of several million dollars.[10]

---

[8] Jeffrey D. Morse, *Nanofabrication Technologies for Roll-to-Roll Processing: Report from the NIST-NNN Workshop* (September 27-28, 2011), 3. Most printing methods available at present are adaptable to flexible electronics production, including inkjet, flexo, offset, and gravure. Printing is an additive process, that is, material is only deposited where it is required, and often there are only two stages between a bare substrate and a functioning layer of circuits on a substrate, a printing stage and a curing stage. By contrast, traditional photolithography, a subtractive process, involves multiple steps, materials, and machines to produce a single functioning layer on a bare substrate and consumes materials not needed in finished devices. "IDTechEx Releases Printed and Thin Film Transistors and Memory Report," *Flexible Substrate,* February 2013.

[9] Global Foundries' Fab 8 in Malta, New York, a 300 mm wafer fabrication facility with an eventual capacity of 60,000 wafer starts per month, reportedly involved investments of $8.5 billion. Ajit Monocha, "Keynote Address," National Research Council symposium, "New York's Nanotechnology Model: Building the Innovation Economy," Troy, New York, April 3, 2013.

[10] Albert Green, Kent Displays, "Roll to Roll Manufacturing of Flexible Displays," in National Research Council, *Building the Ohio Innovation Economy: Summary of a Symposium* (Washington, DC: The National Academies Press, 2013), 123-124. The NRC Committee on Best Practice in National Innovation Programs in Flexible Electronics visited Kent Displays, among other firms in the emerging northeast Ohio flexible electronics cluster, on June 4, 2013.

TABLE 2-1 Comparison of Electronics Processing Techniques

Conventional	Printed
Subtractive batch process (sheet deposition with photolithographic and etched layers)	• Additive continuous process • Printed material only in necessary areas
Controlled vacuum environment	Ambient temperature and pressure conditions
Fixed, long production runs of identical products	Flexible short production runs
High equipment, materials costs	Lower equipment and materials cost

SOURCE: Zella King and Cathy Curling, "Plastic Electronics: Analysis of Competence Matrices for UK and Germany," *UKDL Newsletter,* Winter 2008/2009.

Potential Eco-friendliness

Organic electronic products are expected to lead to more energy-efficient displays and other electronic devices, and inexpensive and highly versatile organic photovoltaics will enhance the ability of society to expand the use of renewable forms of energy not reliant on fossil fuels.[11] A recent white paper released by the annual Chemical Sciences and Society Summit, which convenes "some of the best minds in chemical research from around the world," observed that

> [a]s chemists continue to study and improve their understanding of the electronic behavior of organic materials, engineers will be able to build devices that last longer and that are recyclable or perhaps even biodegradable.[12]

Although this prediction may or not be borne out by events, the adoption of organic electronics is likely to reduce E-waste and other environmental problems, such as the current reliance on rare-earth metals such as indium, the extraction of which sometimes results in environmental degradation.[13]

[11] CS3, *Organic Electronics for a Better Tomorrow,* 16.

[12] CS3, *Organic Electronics for a Better Tomorrow,* 16.

[13] The term "E-waste" embraces discarded information technology and consumer electronics hardware such as mobile phones, computers, printers, televisions, monitors, and other electronic equipment that contains carcinogenic and other toxic substances. The European Union alone generates an estimated 9 million tons of e-waste annually. E-waste contains toxic lead, mercury, cadmium, and other heavy metals, which can cause neurological damage. While the United States and the EU have rules that require recycling and prohibiting export of e-waste to developing countries, these rules are commonly dodged by "itchy-fingered traders, who effectively dodge recycling costs, and illegally export broken goods to developing countries." "Illegal Trash Trade: E-waste Smuggling Contaminates Developing Countries," *Amman News*, August 6, 2013; "Lust for Upgrades Builds a Mountain of e-waste," *Sydney Morning Herald*, April 16, 2007; "Toxic Waste Shipped by UK Firm Still Lying at Port," *Kenya Daily Nation Online*, August 5, 2013.

POTENTIAL APPLICATIONS

Flexible electronics have been cited in connection with an extraordinarily broad range of product applications where bendable and stretchable characteristics offer value.

- ***Photovoltaics.*** Photovoltaic panels that conform to curved or otherwise irregular surfaces offer numerous advantages over the heavy and rigid panels that account for most of today's photovoltaic electricity generation. "Building-integrated photovoltaics" (BIPV) could allow photovoltaic laminates to be applied directly to building surfaces or incorporated in building products such as shingles and siding.[14] Dr. Harry Zervos of the consultancy IDTechEx observes that "it will be easy to go to the local DIY shop and buy rolled or folded solar panels and fit them in the average family car to take them home. There will be no need to strengthen the home roof before installing such panels."[15] Current organic photovoltaics (PV) technology has reached conversion efficiency of 10-12 percent, and some researchers believe rates of 15-20 percent will be reached.[16]
- ***Lighting.*** OLEDs can be mounted on curved and bendable surfaces to provide lighting with versatility and energy efficiency superior to conventional forms of lighting. Although OLED lighting has not been widely introduced commercially, "limited release prototypes and commercial products have become available to demonstrate potential and allow interested users to try out OLED technology."[17]
- ***Displays.*** Flexible displays will be developed that are nonbreakable, waterproof, rugged, and capable of being rolled up or folded for convenience. One much-discussed potential application is the ability to transform smartphones into much better computing platforms by adding a convenient larger display that can be rolled up or folded when not in use.[18] Another application is flexible, roll-to-roll produced e-paper, which has already enjoyed some success on a trial basis when used to produce shelf tags for supermarkets. That said, "the flexible display market has not developed commercially as quickly as had been hoped, partially due to industry restructuring and competition from tablet computers."[19]

---

[14] Currently some flexible BIPV products exist based on metal foils that are sufficiently durable for outdoor applications requiring long life. NanoMarkets LC, *Flexible Substrate Markets: Special Report for the FlexTech Alliance,* April 2012, 5. PELG and ESP KTN, *Capability Guide: UK Plastic Electronics,* 2012, 4–5.

[15] "Flexible Electronics Is the Winner," *Printed Electronics World,* March 2, 2011.

[16] Chemical Sciences and Society Summit, *Organic Electronics for a Better Tomorrow*, 11.

[17] OE-A, *Organic and Printed Electronics,* 13.

[18] NanoMarkets, *Flexible Substrate Markets: Special Report for the FlexTech Alliance,* April 2012, 3.

[19] OE-A, *Organic and Printed Electronics,* 14.

- ***Sensors.*** Sensors embedded in plastic tags are already being used in radio frequency identification (RFID) applications, and many RFIDs use antennae that are printed on flexible substrates. Sensors embedded in uniforms and other forms of clothing can be used for similar purposes. The firm Mc10, a startup founded by a materials scientist from the University of Illinois at Urbana-Champaign, has entered into a partnership with the sportswear maker Reebok "to develop clothes that can monitor impacts on the body, strains on joints, heart rate, blood pressure, and sweat pH."[20]
- ***Medical devices.*** Flexible electronics have potential in many important applications, such as wearable health monitoring devices and medical implants. Medical patches and implants utilizing bendable and stretchable sensors can potentially monitor a wide range of biological functions.[21]
- ***Smart textiles.*** Flexible electronics technology may enable the creation of fabrics that can alter their characteristics in response to external stimuli, whether mechanical, thermal, electrical, chemical, or biological. "Currently much of this field is still in the development or prototype stage, with significant work going into areas such as stretchability and hybrid integration."[22]
- ***Defense applications.*** Flexible electronics devices printed onto uniforms, including electronic readers, displays featuring high-quality photographs, maps, and other information, and communications devices would drastically reduce the loads soldiers carry. Flexible sensors can be placed on aircraft and vehicles to perform multiple functions. Printed photovoltaic devices on tents, buildings, and vehicles could generate local power, reducing the logistical demand for fuel.[23] U.S. defense funds are being used to develop the "so-called Dick Tracy wristwatch, a flexible band to be strapped on soldiers' wrists to provide communication, satellite images and Google Earth-type maps."[24]
- ***Stretchable electronics.*** Electronic devices that can be stretched as well as bent without losing functionality have many potential applications. The company Mc10 is reportedly commercializing "stretchable skin" developed at the University of Illinois that contains sensors to monitor body functions and provide early warning of heart and brain problems

[20] "Stretchable Electronics: A Shapely Future for Circuits," *The Economist,* March 10, 2011.

[21] Brian Buntz, "TEDMED Update: Why the Future of Medical Electronics Is Flexible," *Electronic Components,* April 12, 2012.

[22] OE-A, *Organic and Printed Electronics,* 19.

[23] "Army Researchers Creating Electronic Devices with Flexible Screens," *The Huntsville Times,* March 10, 2013.

[24] The Dick Tracy device is designed to continue working after being penetrated by a bullet, with the only loss of functionality occurring where the hole is located. "E-fabric Spools Bring Bullet-Proof Watches, Paper-Thin Batteries," *The Christian Science Monitor,* April 15, 2011.

before they occur, enabling real-time intervention by a doctor.[25] In the automotive sector, rigid electronics components could be fabricated by stretching them to form during plastic molding, yielding lighter, smaller, and lower cost parts such as overhead lighting systems. Mercedes Benz is reportedly experimenting with stretchable electronic fabrics for use in vehicle interiors that could serve functions such as heating, cooling, and sensing presence.[26]

- ***Flexible batteries.*** Highly flexible batteries would enable the design of a broad range of products freed from the constraints of rigidity and weight that characterize conventional batteries. Flexible batteries could power wearable electronics, including wristbands and clothing, enable the design of slimmer electronics products, and provide a power source for digital smart labels, such as freshness detectors on food packaging.[27]

Current Commercial Applications

Although flexible electronics technology is still in its developmental stage, numerous applications are already present in the market. Korea's LG Electronics and Samsung introduced curved OLED 55-inch televisions in 2013, designed to replicate the viewing experience of an IMAX movie.[28] The two firms made announcements in 2013 of smartphones with curved displays.[29] However, currently available "flexible" displays feature a flexible OLED layer beneath a rigid glass or plastic cover—the "rollable" screen does not yet exist in commercial application.[30] Flexible e-paper displays are being used in e-books for e-readers such as Amazon's Kindle, although sales have declined along with the sales of e-readers.[31] Ohio's Kent Displays reports brisk sales of its electronic e-Writer, a paperless erasable writing tablet known as the "Boogie Board" utilizing flexible liquid crystal display technology.[32] Flexible organic photovoltaic solar panels have been commercially available since 2010 and are finding applications in

---

[25] "MC10 Develops Stretchable Skin to Monitor Health," *Flexible Substrate,* November 2012.

[26] Peter Harrup, "Stretchable Electronics and Electrics for Electric Vehicles," *Flexible Substrate,* January 2013.

[27] "Batteries Get Flexible," *Chemical and Engineering News,* May 6, 2013; "A New Battery That Could Revolutionize Wearables," *GIGAom,* January 8, 2013; "Flexible Battery-Like Films Could Someday Power Your Wearable Device," *GIGAom,* April 28, 2014.

[28] "LG Throws the OLED Competition a 55-inch Curve," *JoongAng Daily Online*, April 30, 2013. "Samsung Unveils Curved OLED TV," *The Korean Times,* June 27, 2013.

[29] Samsung indicated it would introduce the Galaxy Round, a smartphone with a curved 5.7-inch OLED screen. LG has indicated it will begin mass producing "unbreakable," bendable displays for smartphones. "Curved, Then Flex . . ." *Flexible Substrate*, October 2013. HSBC Global Research, *Flexible Display: Fantastic Plastic—A Shape Shifting Game Changer,* April 2013, 4–5.

[30] "Financiers Focus on Flexible Displays," *Flexible Substrate,* November 2013.

[31] "Printed, Flexible and Organic Technology Sees 15.2% CAGR Over the Next Decade," *Solid State Technology,* May 2013.

[32] "Production Line Planned," *Akron Beacon Journal,* March 2, 2012.

buildings.[33] E Ink, a startup based on research originating at MIT, later acquired by a Taiwanese firm, reportedly will go into mass production of Mobius, "a large format (13.3″) digital paper display based on flexible thin-film transistor (TFT) technology developed by Sony."[34]

Potential for "User Innovation"

The development of future markets for flexible electronics technologies is likely to be substantially influenced by what MIT's Eric von Hippel has termed "user innovation"—that is, new products developed by users through improvisations on existing products to meet their particular needs, some of which may have much broader potential application.[35] If past experience is a guide, users will innovate new products from existing flexible electronics products along a path that is not presently foreseeable. The process is slow in the early stages because "there aren't a lot of users." In that phase, small companies start up, founded by individuals with good user connections, and

> then eventually big companies come in because, they say, now we have enough information about this market and we'll get into it, maybe by acquiring one of those little companies—because new we know there's a market of sufficient scale for us.[36]

MARKET GROWTH

Forecasting the future size and growth rates of flexible electronics markets is necessarily an inexact exercise. For many anticipated applications of flexible electronics technologies, no commercial products have entered the market or have done so in a very limited manner, so that no meaningful historical sales data exist upon which forecasts can be predicated. Past forecasts in this field have been repeatedly confounded by events.[37] That said, a number of surveys have been undertaken that canvass large numbers of knowledgeable individuals, companies, and research organizations, and arguably represent an informed conventional wisdom. These forecasts envision the evolution of robust markets for flexible electronics products over the next 10-15 years.

---

[33] OE-A, *Organic and Printed Electronics,* 13–14.

[34] "The Glass Is Half Full. . . ," *Flexible Substrate,* May 2013.

[35] An example of user innovation cited by von Hippel is the skateboard, created by users who pulled apart roller skates and fastened them to boards, creating a new sports/recreational technology. "The User Innovation Revolution," *MIT Sloan Management Review,* Fall 2011.

[36] Ibid., quoting interview with von Hippel. See generally Eric von Hippel, *The Sources of Innovation* (New York and Oxford: Oxford University Press, 1988); Eric von Hippel, *Democratizing Innovation* (Cambridge, MA: The MIT Press, 2005).

[37] See essay by industry analyst Chris Williams, "Last Word: Who? Where? When? . . ." *Flexible Substrate,* December 2013. "Look at 3D TVs—seasoned professionals love the technology, and hype the benefits of it for consumer use, but the consumers themselves have had a look and said 'no thanks.' "

IDTechEx, an extensively cited consultancy that has covered the field of printed, flexible, and organic electronics since 1999, released a report in mid-2013 that estimated that revenues from flexible electronics products would total about $16 billion in 2013, with most of the income derived from OLED displays, e-paper, flexible photovoltaics, and conductive ink.[38] IDTechEx estimates that between 2013 and 2023, the total market for these products would grow to $76.8 billion, or a compound annual growth rate of 15.3 percent.[39] Other estimates envision a much higher growth rate.[40] For example, Smithers Pira, the global authority on packaging, paper, and print industry supply chains, forecasts that the market for plastic and printed electronics will be $190.0 billion in 2025.[41] Looking further ahead, IDTechEx predicts that the market for printed and potentially printed electronics could be "larger than the silicon semiconductor industry, which is not a surprise given that it is applicable to so many things."[42] The consultancy forecasts a market valued at $340.0 billion by 2030. (See Table 2-2.)

Industry observers caution that in addition to the dearth of historical sales data to buttress such forecasts, the market potential for flexible electronics products will be determined, in significant part, by the performance characteristics of the devices themselves in competition with conventional rigid electronics products.[43] Cost relative to that of competing conventional devices is currently a major hurdle for flexible electronics consumer applications and may act as a longstanding drag on market growth unless current equipment, process, and yield challenges are surmounted.[44]

A 2011 study by Germany's National Academy of Science and Engineering concluded that "organic and large area electronics are forecast to have a global

---

[38] IDTechEx engages in consultancy in the fields of organic, inorganic, and hybrid electronics, which are manufactured through printing processes or have such potential. It studies related topics such as smart packaging, silicon photovoltaics, and RFID. IDTechEx annually stages the world's largest conferences on printed electronics topics in North America, Asia, and Europe. It publishes periodic reports on topics and subtopics associated with printed electronics and an online newsletter, *Printed Electronics World*, *<http://www.idtechex.com>*.

[39] "Printed, Flexible and Organic Electronics Sees 15.3% CAGR," *Printed Electronics World*.

[40] In 2013, the consultancy IHS Electronics & Media forecast that the global flexible display market alone would reach a value of $67.7 billion by 2023, nearly equivalent to the $76.8 billion forecast by IDTechEx for that year for the entire flexible electronics industry. "Flexible Display Market to Reach $67.7 Billion by 202, Says HIS," *Flexible Substrate,* November 2013.

[41] Adam Page, Smithers Pira, "Market for Organic and Printed Electronics," in OE-A, *Organic and Printed Electronics,* 34.

[42] IDTechEx, *Printed, Organic & Flexible Electronics*, 5.

[43] "Flexible displays are going to face a tough challenge if design groups focus on creating devices that use them in applications that already successfully use rigid displays. Unless there is a very compelling reason, people rarely are willing to sacrifice to display performance they are accustomed to viewing. So long as making displays flexible results in diminished visual performance, it's going to be a tough sell to convince people that a rollability or foldability feature is justification. . . ." Mark Fihn, "Visual Excitement . . . ," *Flexible Substrate* (September 2012).

[44] "Slim Leap Forward," *South China Morning Post,* September 28, 2012; "Flexible AMOLEDs: At the Tipping Point?" *The Emitter: Emerging Display Technologies,* March 10, 2014.

TABLE 2-2 IDTechEX Market Forecasts for 2030

Segment	Projected Sales (Billions of Dollars)
Logic/memory	100
OLED display	100
Photovoltaic	70
OLED light	30
Electrophoretic	14
Sensors	6
Battery	5
Conductors (ink only)	4
Other display	2
Electroluminescent	.5
Electrochromic	.5
Other	8
TOTAL	340

SOURCE: IDTechEX, *Printed, Organic & Flexible Electronics Forecasts*, 7.

market volume of several hundred billion euros in the medium and long term, corresponding about to the economic importance of the current conventional silicon-based electronics."[45]

OE-A is a leading international industry association for printed and organic electronics. Although based in Europe, it is comprised of more than 200 companies that include firms based in North America, Asia, and Australia representing various segments of the value chain in the field of flexible electronics.[46] OE-A publishes periodic applications "roadmaps," prepared with the participation of its membership, which forecast what is seen as the likely evolution of the organic and printed electronics industry.[47] (See Table 2-3.) Given the breadth and depth of OE-A's membership, its most recent roadmap, released in June 2013, is likely to reflect current industry views about how the industry's applications will develop in the coming decade. As the roadmap indicates, although some significant new

[45] acatech, *Organic Electronics in Germany: Assessment and Recommendations for Further Development,* 2011, 12.

[46] OE-A members include 167 organizations in Europe, 11 in Asia, 28 in North America, and 1 in Australia. Andrew Hannah, *Building a Global Network for the Printed Electronics Industry,* 2013.

[47] The 2013 OE-A Roadmap, the fifth in the series, is based on the work of teams of experts in each of five applications clusters, who developed roadmaps for their sectors. These results were presented to and discussed with OE-A members during the association's meetings. "The resulting roadmap is a synthesis of those results representing common perspectives of these groups." OE-A, *Organic and Printed Electronics,* 10.

TABLE 2-3 OE-A Roadmap for Organic and Printed Electronics Applications (2013)

Theme	2013	2014-2016	2017-2020	2021 +
Organic photovoltaics	Portable chargers	Consumer electronics	Specialized building integration, off grid PV	Building integration, grid-connected PV
Flexible displays	Integrated into smart cards, price labels, bendable color displays	Bendable OLEDs, plastic LCD, large area signage, rollable color displays	Rollable OLEDs, transparent rollable displays, flexible consumer electronics	Rollable OLED TVs, telemedicine
OLED lighting	Design projects	Transparent and decorative lighting modules	Flexible lighting	General lighting technology
Electronics	Single-cell batteries, memory for interactive games, ITO-free transparent conductive films	Rechargeable single-cell batteries, transparent conductors for touch sensors, printed reflective displays	Printed multi-cell batteries, printed logic chips, integrated flexible multi-touch sensors	Directly printed batteries, active and passive devices to smart object
Integrated support systems	Garments with integrated sensors, anti-theft, brand protection, printed test strips, physical sensors	Integrated systems on garments, large area physical sensors arrays and mass market smart packaging	Textile sensors on fiber, dynamic price displays, NFC/RFID smart labels, disposable monitoring devices	OLEDs on textiles, fiber electronics, health monitoring systems, smart buildings

SOURCE: OE-A, *Organic and Printed Electronics*.

applications are likely to become available commercially in the next 1-3 years, such as "intelligent packaging," the most dramatic applications, such as rollable televisions and grid-connected organic photovoltaic systems, may be a decade or more away. OE-A comments that organic electronics technology

> is still in its early stage; while increasing numbers of products are available and some are in full production, many applications are still in lab-scale development, prototype activities, or early production.[48]

[48] OE-A, *Organic and Printed Electronics,* 10.

A basic challenge facing companies seeking to commercialize flexible electronics technologies is the fact that the conventional technologies they are challenging continue to evolve and improve. By the time the UK firm Plastic Logic was ready to announce its e-reader Que, featuring flexible e-paper in early 2010, "Steve Jobs unveiled the iPad and with it effectively blew Plastic Logic out of the water."[49] At a 2011 industry conference on organic photovoltaics (OPV), assumed to be far less costly than silicon and thin-film PV, it was observed that the costs of the latter were falling rapidly and that "by the time we have large scale production of OPV, crystalline silicon PV could be living without subsidies."[50]

CHALLENGES

Flexible electronics are widely acknowledged to have generated "a great deal of hype," particularly in the years preceding the onset of the global financial crisis in 2008.[51] Optimistic forecasts of explosive commercial growth did not materialize.[52] New product introductions have been announced but then delayed, usually for technological reasons.[53] Lawrence Gasman, founder of the consultancy NanoMarkets LC, commented in early 2013 that "the history of flexible displays over the past decade has already been one of the broken promises, mostly because of technological issues, and in 2012, it became apparent that the supposed 'killer app' for flexible displays probably didn't have as much potential as was once thought."[54] Although the financial crisis was a factor affecting the lower than normal rate of commercialization of flexible electronics products, technological hurdles have proven more daunting than was recognized a decade ago. Kenneth Warner, founder of the consultancy Nutmeg Consultants, specializing in displays, commented in November 2013:

[49] "Why Plastic Logic Failed—Despite the E-book Boom," *GigaOM,* May 17, 2012.

[50] "Organic PV: Closing the Reality Gap," *Optics.org,* September 6, 2011.

[51] Adam Page, Smithers Pira, "Market for Organic and Printed Electronics," in OE-A, *Organic and Printed Electronics*, 34. *The New York Times* predicted in 2001 that "thin flexible batteries may soon be plastered on cardboard or plastic surfaces, producing novelty packaging items like cereal boxes that twinkle with light-emitting diodes or containers that advertise their wares by playing brief jingles." "What's Next: Batteries Push Paper Into Electronics Age," *The New York Times,* May 24, 2001.

[52] "Earlier roadmaps for printed electronics have been almost entirely erroneous. It is not primarily about cost reduction, nor is there a trend toward organic versions taking over most applications." "Printed Electronics—Many New Directions," *Printed Electronics World,* February 21, 2011. The principal factor frustrating the realization of the roadmaps has been underestimation of the technological obstacles entailed.

[53] UK-based Plastic Logic announced an e-Reader with a flexible display in 2010, but subsequently announced that the commercial release would be delayed indefinitely. In 2011, Korea's Samsung told industry analysts that it would introduce handsets featuring flexible displays "sometime in 2012, hopefully the earlier part than later," but a year later reported that the technology was still "under development." "Samsung's Courtship of Apple Parts Suppliers May Drive Up Costs," *The Argus,* May 16, 2013; "Flex is Next," *San Jose Mercury News,* January 17, 2011.

[54] Lawrence Gasman, "Notable Developments in Flexible Glass," *Flexible Substrate,* January 2013.

> What is delaying the introduction of flex-many products? The answer is almost everything. Not only must the display itself (including the backplane and barrier film) be flexible and not deteriorate over many flexing cycles, but so must the other components of the device, including the touch screen, circuit boards and battery. Hard switches and buttons must either be eliminated or designed to work with a flexing substrate and, perhaps, case. None of this is easy, but solving the problems are where the product and investment opportunities lie.[55]

Encapsulation

Flexible electronic circuits are likely to be made of organic materials that are highly vulnerable to contamination from exposure to oxygen and water vapor. Although rigid electronic devices can be protected by encapsulating them in glass or other hard materials, encapsulation of flexible circuits poses greater challenges. Plastic tends to be permeable and may degrade in demanding environments. Flexible glass may be an alternative but, at present, is expensive, as are other specialized encapsulation alternatives.[56] Flexible electronics products such as OLED lighting systems and devices incorporating flexible memory, logic, and battery functions "still need better flexible barrier films to extend their useful lifetimes."[57] OE-A declared in its roadmap that "barrier properties of flexible, low-cost encapsulation need to be strongly improved to enhance the lifetime of the devices."[58]

Functional Materials

Conventional electronic devices are fabricated with conductive elements made of metal or metal oxides on rigid substrates, usually processed at high temperatures. The transition to flexible electronics requires the creation of circuits that do not lose functionality when bent, folded, or rolled. The use of flexible plastic substrates precludes fabrication processes involving temperatures so high that they deform the substrate. Identification of materials and processes capable of surmounting these physical limits has proven a universal challenge.

Indium tin oxide is one of the most widely used conductors in conventional electronic devices, with applications in displays, lighting, and photovoltaics.

---

[55] "Financiers Focus on Flexible Displays," *Flexible Substrate,* November 2013.

[56] NanoMarkets, *Flexible Substrate Markets: Special Report for the FlexTech Alliance,* April 2012, 9–10.

[57] Several promising efforts to address this challenge have been reported. These include flexible glass developed by Corning and novel coatings being developed by Beneq Oy and Vitriflex. Beneq Oy reports that thin atomic layer deposited (ALD) inorganic coatings "have shown to provide excellent moisture barrier properties on flexible substrates." "Sealing Up ALD Moisture Barrier Technology," *Flexible Substrate,* May 2013. "Printed Electronics Sector Takes a Hard Look at the Flexible Future," *Solid State Technology,* March 2013.

[58] OE-A, *Roadmap for Organic and Printed Electronics*, reproduced in OE-A, *Organic and Printed Electronics,* 28.

However, although ITO circuits can bend slightly without loss of functionality, they are probably too brittle for use in rollable or foldable applications, and ITO has other major disadvantages.[59] An alternative is the use of printed metallic conductive inks, but these, too have drawbacks—silver, which is highly conductive, is expensive, and various alloys of aluminum, copper, and nickel (including silver-coated variants) involve trade-offs in workability, conductivity, and reactivity.[60]

Ordinary forms of carbon are barely conductive compared to silver, although carbon pastes can be blended with silver to reduce cost, adjust conductivities in specialized applications, and inhibit electromigration. However, in terms of performance carbon-based conductive materials are not competitive with silver. Exotic carbon nanoparticles, including graphenes and carbon nanotubes, are highly conductive and have the potential to form flexible circuits, but "the production of carbon nanomaterials is still pretty complicated, at least of the conductive forms."[61] In many respects ranging from energy efficiency to materials resource requirements, flexible electronics promises to be more environmentally friendly than silicon-based electronics. However, "many polymers require carcinogenic solvents, including some not allowed in the EU printing industry because of their toxicity."[62] Although the extent of the problem of toxicity associated with flexible electronics is not clear, given the incipient state of the industry, a general challenge in flexible electronics is developing polymers that are soluble in nontoxic solvents and "rely on more benign methodologies in general."[63]

Manufacturing Hurdles

The manufacture of flexible electronics devices offers numerous potential advantages in terms of lower cost and high volume throughput. However, the fabrication of complex flexible electronic devices poses challenges for which solutions have not yet been found and which could delay the widespread commercial introduction of many products. Even the manufacture of flexible displays on conventional thin-film transistor production lines poses an array of technical

[59] China, one of the world's principal suppliers of indium, has maintained export restrictions on rare earth metals including indium for a decade, driving up its cost and raising concerns about its availability. "S. Korea to Speed Up Overseas Development of Indium," *Yonhap,* May 11, 2011; "Scrambling for Raw Materials: EU Plans Measures to Tackle Resource Crunch," *Spiegel Online,* November 20, 2010. "ITO, if flexed too much, tends to crack, so for truly flexible displays, ITO cannot be used." *NanoMarkets, Special Report for the FlexTech Alliance: Transparent Conductor Market,* December 2012.

[60] NanoMarkets, *Silver Inks and Pastes Markets: Special Report for the FlexTech Alliance,* February 2012, 810.

[61] Ibid., 8-9.

[62] Royal Society for Chemistry, "Organic Electronics for a Better Tomorrow" September 2012. Access at *http:www.rcs.org.*

[63] NanoMarkets, *Silver Inks and Pastes Markets: Special Report for the FlexTech Alliance,* op. cit., p. 8-9.

difficulties.[64] In early 2014, the chronically low yield rates in the production of flexible AMOLED (active matrix organic light-emitting diode) displays was cited as the principal factor underlying the huge cost differential between AMOLED and conventional rigid displays.[65]

Roll-to-roll processing techniques for flexible electronics are widely discussed, but "layer-to-layer registration with a conventional deposition and patterning process remains a formidable challenge when the film is handled in this way."[66] The U.S. Army–supported Flexible Display Center at Arizona State University encountered extreme difficulty in immobilizing plastic substrates sufficiently to enable deposition of amorphous silicon transistor arrays, and overcoming this hurdle was the object of a dozen separate research programs undertaken by the Center and its research partners. A National Institute of Standards and Technology–sponsored workshop in September 2011 convened a group of 30 U.S. experts in roll-to-roll manufacturing to identify the "key barriers to the adoption of nanofabrication processes within R2R manufacturing processes."[67] The workshop concluded that although many U.S. companies are engaged in R2R manufacturing, there was a general lack of standardized infrastructure, most universities did not have R2R fabrication facilities, and data were lacking on what could and could not be achieved using R2R processes. In addition, certain process tools and core capabilities presented "challenges," and "the supply chain has not been established."[68]

---

[64] Nick Colaneri from the Flexible Display Center at Arizona State University recently noted that the heat sensitivity of polymer films required the development of systems for handling them and process steps to avoid exceeding their temperature limits. The Flexible Display Center has developed potential solutions, and he indicates that the first flexible high-resolution displays to enter the commercial market will utilize "materials handling techniques that have been developed to allow use of this existing film transistor fabrication facility." However, he notes, "these techniques are still under intensive development, including the evaluation of the relative merits of different design trade-offs." Nick Colaneri, "Manufacturing Flexible Displays: The Challenges of Handling Plastic," *Solid State Technology,* May 1, 2013.

[65] "[M]anufacturing costs for flexible AMOLED displays are a multiple of equivalent size rigid panels." "Flexible AMOLEDs: At the Tipping Point?" *The Emmitter: Emerging Display Technologies,* March 10, 2014.

[66] Colaneri, "Manufacturing Flexible Displays."

[67] InterNano, "Accelerating Progress for Advanced Manufacturing," June 28, 2012.

[68] Equipment and competency challenges identified included large-area, cost-effective, e-beam patterning tools/capabilities; plasma etching tools for large-area, uniform R2R processing; ink jet applicators compatible with wide range ultraviolet monomers; development of high-quality nickel metal electroforming processes for high aspect ratio, large pattern volume structures; high durability, low-cost transparent imprinting of molds, or inexpensive/fast replacement transparent molds; fabrication of seamless cylindrical imprint molds; large area, real-time metrology and process characterization, etc. Jeffrey D. Morse, *Nanofabrication Technologies for Roll-to-Roll Processing: Report from the NIST-NNN Workshop,* September 27-28, 2011.

Substrates

The use of bendable substrates for electronic devices offers numerous advantages, including the potential for manufacturing on a roll-to-roll basis and applications requiring flexibility. However, although the use of traditional electronic conductive materials requires heating, plastic substrates cannot be processed in high-temperature environments. Moreover, the bendability that makes flexible substrates attractive, as well as surface characteristics, makes it more difficult to ensure the adhesion of inorganic and organic coatings. Paper, sometimes discussed as an alternative to plastic, is rough, porous, dimensionally unstable, and characterized by poor barrier characteristics.[69]

Device Performance

The consultancy IDTechEx released a report in 2013 indicating that although many types of electronic components are being produced through various printing processes (including batteries, PV devices, interconnects, memories, antennas, and energy harvesters), "in many cases the performance is not as high as their non-printed counterparts and therefore businesses are leveraging their other characteristics, including potential for low cost, large-area coverage and flexibility."[70] Current flexible displays "are almost always compromises on visual performance."[71]

The vast preponderance of government financial support for flexible electronics that is catalogued in this report is flowing toward research projects to address these technological challenges, usually through industry-university-governmental collaborations.

Market Uncertainty

Flexible electronics is an emerging field characterized by a multiplicity of potential applications, manufacturing processes, and base materials, and the direction in which the industry or industries will evolve is not at all clear. There are no commonly accepted base materials and methods that characterize most of the semiconductor industry (e.g., complementary metal oxide semiconductor devices fabricated through photolithography). These uncertainties have deterred investment in flexible electronics and will continue to do so in the future, a factor

---

[69] Department for Business Enterprise and Regulatory Reform, *Plastic Electronics in the UK: A Guide to UK Capability,* 2008/09, 9.

[70] "The performance of printed transistors is also below par today. Their mobility is limited to $<0.1\ cm^2/Vs$ in most cases. Their lifetime and stability are poor and they often need expensive encapsulation layers. Indeed their performance may not yet be adequate today for the obvious target markets of large-sized display backplanes and RFID tags." "IDTechEx Releases Printed and Thin Film Transistor and Memory Report," *Flexible Substrate,* February 2013.

[71] Mark Fihn, "Visual Excitement …" *Flexible Substrate,* September 2012.

underlying the conclusion by the President's Council of Advisors on Science and Technology (PCAST) that flexible electronics is a sector that may well not develop if left to market forces.[72]

[72] PCAST, *Report to the President on Ensuring American Leadership in Advanced Manufacturing* (June 2011), 19–20. Stephen Freilich, Director of Materials Science and Engineering for DuPont Central Research and Development, observes that "[f]rom a materials supplier's standpoint [in a new technology field] there can be a disincentive to do truly revolutionary work when you see this rapid change in markets and technologies. We can do it, but the investment is so great, and rate of return so dependent on the longevity of the technologies that you're not going to see the kind of innovation you need." He indicates that in such situations companies will limit themselves to incremental changes in existing materials and technologies. Steven C. Freilich, "DuPont Reflections on Photovoltaics," in National Research Council, *The Future of Photovoltaic Manufacturing in the United States* (Washington, DC: The National Academies Press, 2011), 68.

3

Government Promotional Efforts

Flexible electronics is a relatively new field involving many unproven technologies and processes, with a potentially long and uncertain path to market for commercial applications. Although private capital has supported some early commercial ventures in the field, most of the initial development of the industry worldwide has involved extensive government support, particularly with respect to the funding of applied research to translate scientific knowledge into commercial products and industrial processes. The scope of these promotional efforts varies substantially between regions.

In the United States, federal government support is generally regarded as appropriate for basic, generic research and basic infrastructure, whereas public support for the development of products and industrial processes can be controversial. At the state level, public funding of research closely associated with commercial activities has been common and widely accepted. In countries such as Germany, Taiwan, and others examined in this study, government support for applied research with explicitly commercial objectives is the norm.[1]

In the European Union, an extremely broad and deep effort to promote an indigenous capability in flexible electronics has been under way at multiple governmental levels for more than 10 years. This effort is far broader than the promotional efforts in North America and East Asia. The European Commission

[1] See generally David C. Mowery and Nathan Rosenberg, "The U.S. National Innovation System," in Richard R. Nelson, *National Innovation Systems: A Comparative Analysis* (New York: Oxford University Press, 1993), 35–36; National Research Council, *Best Practices in State and Regional, Innovation Initiatives* (Washington, DC: The National Academies Press, 2013); National Research Council, *21st Century Manufacturing: The Role of the Manufacturing Extension Partnership Program* (Washington: The National Academies Press, 2013).

is funding numerous research consortia with an emphasis on establishing manufacturing competency within Europe. National governments have committed hundreds of millions of dollars to supporting research efforts, particularly the United Kingdom (UK), Germany, the Netherlands, and Finland. Regional governments are providing financial and infrastructure support to "organic electronics" innovation clusters, and a 2011 European Union (EU)-sponsored survey identified no fewer than 17 organic and large area electronics ("OLAE") clusters in the EU. The survey team observed that

> [t]he conditions created in terms of public funded support, particularly EU support, in favor of OLAE have never been more favorable. This has resulted in a flood of demonstrators and prototypes of the new devices out of these publicly-funded research projects, maximizing the technology push and helped moving up Europe's technology readiness level.[2]

Europe is "certainly . . . looking at a far broader range of printed components than is pursued in East Asia." European programs are promoting an extraordinary range of flexible electronics–based applications, including "smart packaging," printed flexible rechargeable batteries, radio frequency identifications, lighting systems, ubiquitous sensor networks for health care, smart textiles, and touch screens.[3]

In the East Asian countries Japan, Korea, China, and Taiwan, government promotional efforts in flexible electronics are heavily concentrated on development of flexible displays for consumer applications such as smartphones, TVs, and e-readers. Asian government promotional programs are built on the legacy of earlier major government programs to develop industrial capabilities in semiconductors, high definition television, optoelectronics, liquid crystal displays, and photovoltaics. South Korea, China, Taiwan, and Japan are engaging the same government research organizations, the same types of promotional policies, and in most cases the same companies and industrial groups that achieved success in these other sectors. With sustained government support, Asia-based producers have achieved near-total global dominance in displays for consumer electronics applications, facilitating their moves into flexible electronics displays.

In North America, the Department of Defense recognized the multiple potential defense applications of flexible electronics technology in the 1990s and has been providing major funding for research and development (R&D) in the field for more than a decade. The foremost U.S. center for flexible electronics R&D, the Display Technology Center at Arizona State University (ASU), represents a $100 million investment by the U.S. Army. Other federal agencies, including the National Institute of Standards and Technology (NIST) and the National Science

---

[2] The FP7-ICT Coordination Action OPERA and the European Commission's DG INFSO Unit G5 "Photonics," *An Overview of OLAE Innovation Clusters and Competence Centres,* September 2011, 9.

[3] Peter Harrop, "Printed Electronics—Europe is Different," *Printed Electronics World,* March 9, 2011.

Foundation (NSF), have been funding basic and applied research and technology transfer, and federal laboratories, including facilities at NIST and the Department of Energy's National Laboratories, are performing and sponsoring research in flexible electronics, often in collaboration.[4] A number of U.S. states are supporting the establishment of flexible electronics innovation clusters to commercialize research results from local universities. In October 2013, the Canadian government announced a new promotional effort and an industrial consortium to foster the development of printed electronics technologies, involving a $40 million investment by the National Research Council of Canada over a 5-year period.[5]

GOVERNMENT FUNDING LEVELS

It is probably impossible to construct an accurate tabulation of government financial support for flexible electronics around the world from public sources. Funding data for flexible electronics is commonly not segregated and is widely dispersed under headings such as "nanotechnology," "new materials," "high technology equipment," and "green energy." A European Union project, PolyMap, was undertaken in 2008-2011 to determine public funding for OLAE within EU Member States. It reported in 2010 that "very few questionnaires have been returned so far . . . ," various organic electronics-related projects were "scattered through a large number of different national programs," few programs were identified with "primarily OLAE context," databases of funding agencies lacked suitable search tools, and "regional funding is virtually impossible to get information on."[6] Funding levels in countries such as South Korea are even more opaque.

Table 3-1 is an attempt to establish a rough perspective on comparative funding levels by presenting the handful of relatively large government expenditures in flexible electronics based on reasonably reliable sources. As can be seen, even with these figures the timeframes are not coextensive, and it is therefore difficult to make meaningful comparisons. The myriad smaller governmental outlays by various entities are not depicted, a fact that understates the figures for all countries shown, but particularly Germany, where the national innovation system is characterized by multiple small government grants and loans. German figures would be further increased by inclusion of state and regional financial support

[4] In 2012, for example, Georgia Tech's Center for Organic Photonics and Electronics disclosed discovery of a universal technique to reduce the work flow of a conductor using commercially available polymers that are inexpensive, environmentally friendly, and consistent with existing roll-to-roll mass production techniques. The discovery is expected to facilitate lower cost and more flexible electronic devices. The project was jointly funded by the National Science Foundation, the Office of Naval Research, and the U.S. Department of Energy. "Stable Electrodes for Improving Printed Electronics," April 19, 2012, <*http://www.cope.gatech.edu/news/release.php?nid=124901*>.

[5] "Next Generation Printing Innovations with Electronic Intelligence," *Printed Electronics World,* October 14, 2013.

[6] PolyMap Report, *WP1: Survey of National and Regional Funding with OLAE Context* (February 23, 2010), 3.

TABLE 3-1 Known Major Government Funding—Flexible Electronics

Government Entities	Time Frame	Reported Funding (Millions of Dollars)
United States		
Army[a]	2004-2014	100
Ohio 3rd Frontier[b]	2003-2010	56
NSF-Bioflex	2012	20
Europe		
EU Seventh Framework[c]	2007-2013	158
UK—EPSRC[d]	2009	104
Netherlands/Belgium (Holst)[e]	2013-2016	95
Netherlands/Noord Brabant (Solliance)[f]	2013	37
Germany BMBF[g]	Through 2011	264
East Asia		
Japan NEDO (OLED projects)[h]	2008-2012	274
Taiwan[i]	2006-2011	200

NOTE: Conversion at prevailing exchange rates as of July 25, 2013.
SOURCES: [a]"Army Invests $50 Million in Flexible Displays," *CNET News,* January 29, 2009; "Army Partners with Arizona State University to Develop Flexible Displays," *Military and Aerospace Electronics,* January 2009; [b]NorTech, "A State's Initiative: Advancing Flexible Electronics in Northeast Ohio" (2010); [c]The FP7-ICT Coordination Action OPERA, *An Overview of OLAE Innovation Clusters and Competence Centres,* September 2011; [d]House of Commons, Universities, Science and Skills Committee, *Engineering: Turning Ideas Into Reality* I (March 18, 2009), 33; [e]"Public and Industrial Agreements Enable Further Growth of Holst Centre," Holst Centre News Release, April 5, 2012; [f]"Building Activities Start on High Tech Campus Eindhoven," Solliance News Release, April 15, 2013; [g]Germany National Academy of Science and Engineering, *Organic Electronics in Germany: Assessment and Recommendations for Further Development,* 2011, 8; [h]House of Commons, Innovation, Universities, Science and Skills Committees, *Engineering: Turning Ideas Into Reality* 1 (March 18, 2009), 41, based on site visits in Japan. Figure represents NEDO funding of OLED R&D; [i]Presentation of Janglin Chen, ITRI, National Research Council, *Flexible Electronics for Security, Manufacturing and Growth in the United States: Report of a Symposium,* 2013.

for research and the development of innovation clusters and of state and federal support for flexible electronics research carried out by the Fraunhofer Institutes.[7]

Table 3-1 shows $158 million in expenditures between 2007 and 2013 for flexible electronics by the EU pursuant to its Seventh Framework Programme as reported by OPERA, an EU-backed research alliance promoting "competitiveness

[7] The German research ministry, BMBF, contracts with the Fraunhofer Institutes for research, and funds used to support flexible electronics research may be reflected in the $264 million figure shown for the BMBF. However, the federal and state (Land) governments also provide roughly one-third of the Fraunhofer Institutes' "core" funding, which is not linked to specific research projects but which conveys advantages (such as advanced equipment and well-staffed laboratories) to Fraunhofer research projects for private companies. In 2010, this government core funding totaled €553 million euros ($734 million). Fraunhofer Gesellschaft Annual Report 2010.

clusters" in organic and plastic electronics. This figure does not include expenditures by the European Regional Development Fund (ERDF) in flexible electronics that have been used to support creation and expansion of flexible electronics research facilities in the EU.[8] Finally, Table 3-1 depicts funding levels for the EU member states that have made the largest public investments in flexible electronics; to this should be added smaller, but not insubstantial, public investments by other Member States, such as France and Finland.[9]

The U.S. figures for funding by the Army and Ohio's Third Frontier fund do not include smaller grants and research contracts from federal agencies (e.g., NIST, NSF, Office of Naval Research, Air Force) or other state and regional U.S. governments. In 2010, Andrew Hannah, then CEO of Plextronics, cited an industry estimate that aggregated funding for printed electronics from U.S. various government programs and concluded that U.S. spending in the sector was less than $50 million in 2009.[10]

South Korea releases information about government funding of high-technology R&D under the rubric of broad thematic categories such as "World Premier Materials" and "Convergence Technologies," indicating very substantial aggregate funding levels but without a breakdown of funding allocation by sector or technology within these categories. Budget figures are sometimes given for public research institutes conducting R&D relevant to flexible electronics without indicating what percentage of the funding is derived from government sources. Andrew Hannah, the CEO of Plextronics, indicated in a 2010 presentation that no data were available with respect to the level of Korean government funding for flexible electronics, but that it was "assumed to be greater than Taiwan," which was $200 million between 2006 and 2013.[11]

Incomplete as the data in Table 3-1 are, with respect to government funding of commercially relevant flexible electronics technologies, it is difficult to escape the conclusion that the United States is being outspent by both Europe and Asia.[12]

---

[8] The Centre for Process Innovation, Ltd. (CPI) is a nonprofit organization established in north-east England in 2004 to conduct research in advanced manufacturing. Between 2004 and 2012 CPI received £17 million ($26 million) from the ERDF, nearly all of which was used for projects involving its Printed Electronics Technology Centre. Written Evidence Submitted by the Centre for Process Innovation, House of Commons, Communities and Local Government Committee, April 2012.

[9] The EU OPERA project, which mapped the EU's OLAE clusters, reported in 2011 that €20 million in French public funds had been invested in relevant polymer R&D projects and that Finland had invested €10 million in PrintoCent, a printed electronics research center, between 2009 and 2011. The FP7-ICT Coordination Action OPERA, *An Overview of OLAE Innovation Clusters and Competence Centres,* September 2011, 13.

[10] Andres Hannah, *The Global View of Printed Electronics and What it Could Mean to the U.S.,* September 24, 2010. In 2014, Solvay, a Belgian company, completed the acquisition of U.S.-based Plextronics, Inc.

[11] Andrew Hannah, *Global View of Printed Electronics.*

[12] The same conclusion has been reached by some representatives of the U.S. flexible electronics industry. Andrew Hannah, CEO of Plextronics, observed in 2010 that "the U.S. is being outspent" by foreign governments in printed electronics. Ibid.

Moreover, the U.S. Army's investment in the Display Technology Center at ASU will undoubtedly result in commercial as well as military applications, as is the intent, but the primary objective of that investment is the development of displays for incorporation in military equipment, systems, and uniforms. Moreover, the foreign participants in the internationalized ASU projects will reap some of the benefits of the research—also as is intended—whereas the international spillover effects of the European and Asian efforts are likely to be more limited.

COOPERATIVE RESEARCH CENTERS

In all regions, most government spending in flexible electronics is directed toward university-industry-government consortia conducting applied research at government or university research centers. The mission of virtually all of these centers is to translate basic research into commercially relevant products and processes, and to play a silo-breaking and integrational role between various scientific and engineering disciplines and between individual companies, university departments, and government organizations. Such cooperative research centers, known variously as centers of excellence, joint laboratories, engineering research centers, and university-industry research centers, have been characterized as "the organizational solution to the problems team science poses for disciplinary and bureaucratically structured institutions like universities."[13] In addition, by fostering collaboration, the centers mitigate the cost and risks associated with research by individual companies, which typically enjoy expertise in only a portion of the disciplines required to engage in the manufacture of flexible electronics products.

Cooperative research centers in flexible electronics are typically equipped with research, measurement, and simulation tools and are staffed with scientists, faculty, engineers, and students with expertise in various relevant competencies, including materials science, electrical engineering, circuit design, and process technologies. Many of the centers also operate pilot manufacturing lines in collaboration with equipment makers to test and prove manufacturing processes, validate prototypes, and develop new lines of processing equipment. Virtually without exception, the centers are either owned and operated by governments, or substantially dependent on government financial support. (See Table 3-2.)

Research collaborations with commercial objectives involving multiple firms and research organizations are inherently challenging, particularly in countries such as the United States, with a deeply embedded tradition of individualism, or South Korea, characterized by intense rivalry between industrial groups—which underscores the value of examining other national models. In continental Europe, longstanding industrial traditions of collaboration are currently buttressed by institutional arrangements for applied research, involving powerful financial

[13] Craig Boardman and Denis Gray, "The New Science and Engineering Management: Cooperative Research Centers as Government Policies, Industry Strategies and Organizations," *Journal of Technology Transfer,* February 2010, 447.

TABLE 3-2 Government-Supported Centers for Applied Research with Major Flexible Electronics Programs

Country	Research Center	Source(s) of Government Support	Themes
US	ASU Flexible Electronics and Display Center	Army, Arizona	Military and dual use applications
US	Akron Polymer Innovation Center	Ohio Third Frontier	Substrate, R2R, displays
US	CAMM Binghamton U	DoD, New York	R2R manufacturing
US	Western Michigan U CAPE		Printed electronics
US	Georgia Tech COPE	DoD, DoE, NSF	Organic photonic materials/ devices
Neth/Bel	Holst Centre	TNO, EU, Flanders	OLED displays, lighting
Neth/Bel	IMEC	TNO, EU, Flanders	RFID, PV, foils
Neth	Solliance	Brabant, TNO	PV
Fin	PrintoCent	VTT, Oulu	R2R manufacturing
UK	Centre for Process Innovation	TSB, One North East	OLED lighting, PV, displays
UK	Printed Electronics Technology Centre	ERDF, TSB, One North East	OLED lighting, PV
UK	Welsh Centre for Printing and Coating	EPSRC, TSB, FP7	Printing for flexible electronics
UK	Organic Materials Innovation Centre	EPSRC	Biomaterials, packaging
UK	Cambridge Integrated Knowledge Centre	EPSRC	PV, manufacturing
Ger	Fraunhofer COMEDD	BMBF, ERDF, Saxony	OLED lighting, R2R
Ger	Fraunhofer IAP	BMBF, Land	PV, OLED, signage, security applications
Ger	Fraunhofer ISC	BMBF, Land	Encapsulation technology
Ger	Fraunhofer FEP	BMBF, Land	R2R for flexible displays
Ger	Fraunhofer ENAS	BMBF, Land	RFIDs, flexible antennae, batteries
Ger	Fraunhofer EMFT	BMBF, Land	PV, sensors
Ger	Fraunhofer IIS	BMBF, Land	Textiles for medical, sports applications
Ger	Fraunhofer 1ZM	BMBF, Land	Electronic textiles
Ger	Fraunhofer ISIT	BMBF, Land	Bendable displays with memories

continued

TABLE 3-2 Continued

Country	Research Center	Source(s) of Government Support	Themes
Ger	Fraunhofer IWS	BMBF, Land	Flexible thermoelectric generators
Ger	Fraunhofer IPA	BMBF, Land	Electronic foils
Taiw	ITRI Display Technology Center	MOEA	Flexible displays
Jpn	Flexible Electronics Research Center	AIST	Displays, tags, sensors
SKor	Korean Printed Electronics Center	KETI	Printed electronic lighting, signage, PV, automotive sensors
SKor	Korea Institute of Machinery & Materials		PV, equipment, manufacturing
SKor	Korea Institute for Chemical Technology	MOTIE	PV, materials
SKor	Korea ElectroTechnology Research Institute		Electrodes for displays, PVs, touch screens, sensors
China	Industrial Institute of Printed Electronics	Municipal government of Changzhou	RFID, RZR manufacturing
China	Nano and Advanced Materials Institute	Government of Hong Kong	Transparent conductive films, silver nanowire, scalable printing techniques

SOURCE: Chapters 5-7 of this study.

incentives to cooperate and intellectual property and cost-sharing practices that have fostered pervasive, sophisticated research collaborations of the kind cited with approval by the visiting WTECH panel of U.S. experts in flexible electronics in 2010.[14] In Taiwan the government's leverage relative to industry is sufficiently great that it can shepherd companies into de facto research and rationalization cartels that mature in the form of complete industry chains.[15]

Government-supported cooperative research centers are frequently proactive in forming and shaping consortia. Taiwan's ITRI has not only organized industry alliances in flexible electronics but also assigned specific roles to individual participating companies.[16] The European Union's Flex-o-Fab project, a 3-year

[14] WTECH, *European Research and Development in Hybrid Flexible Electronics* (2010), 7. See generally National Research Council, *21st Century Manufacturing*, "Appendix A2, Fraunhofer Gesellschaft: The German Model of Applied Research," 224–284.

[15] See National Research Council, *21st Century Manufacturing*, "Appendix A3, Taiwan's Industrial Technology Research Institute: A Cradle of Future Industries," 285–336.

[16] Interview with Dr. Janglin Chen, Director ITRI Display Technology Center, Hsinchu, Taiwan, February 14, 2012.

effort to demonstrate processes for organic light-emitting diode (OLED) lighting foils, is being directed by the Holst Centre, a Dutch/Belgian flexible electronics research center supported by national and regional government entities in the Netherlands and Belgium.[17] Germany's Fraunhofer institutes, which derive much of their revenue from public sources, commonly encourage their industrial partners to form consortia when confronting major research and development challenges.[18]

PROMOTION OF INNOVATION CLUSTERS

Innovation policies in the regions surveyed in this study emphasize the formation of innovation clusters, geographically localized groups of companies in related sectors that do business with each other and share needs for skilled workers, research infrastructure (i.e., universities and public laboratories, supportive industry associations, community colleges with relevant training programs), and new technology. As Michael Porter famously stated in his influential 1990 book *The Competitive Advantage of Nations*, regional clusters, rather than individual firms or industries, are the primary determinant of competitiveness. Governments seek to promote clusters through economic incentives to firms to locate in a particular geography, the establishment of supporting research infrastructure (including cooperative research centers), provision of networking services and events, provision of incubation and other business services, and assistance in securing financing for startups.[19]

In the European Union, the EU-sponsored OPERA project released a survey in 2011 of OLAE innovation clusters and competence centers in Europe. The study observed that there were more than 17 OLAE clusters in Europe with more emerging and that at least 400 firms and institutions active in the field were located in these clusters.[20] In a 2010 presentation on Korean initiatives in flexible and printed electronics, Professor Changhee Lee of Seoul National University emphasized the nine innovation clusters of companies, research institutes, and universities that have emerged in South Korea in this sector.[21] The

[17] "European Project Develops Flexible OLED Lighting Production Process," *Plastic Electronics,* February 8, 2013.

[18] Interview with Fraunhofer Institute for Process Engineering and Packaging IVV, Friesing, Germany, June 13, 2012.

[19] See generally Stefano Breschi and Franco Malerba, eds., *Clusters, Networks and Innovation* (Oxford: Oxford University Press, 2005); National Research Council, *Clustering for 21*st *Century Prosperity,* Charles W. Wessner, rapporteur (Washington, DC: The National Academies Press, 2012).

[20] The FP7-ICE Coordination Action OPERA, *An Overview of OLAE Innovation Clusters and Competence Centres,* September 2011.

[21] Professor Lee identified these as Seoul, Daejeon City, Pohang City, Sunchon City, Jeonbuk Province/Jeonju City, Kumi City, Paju City, Suwon City, Kihueng, and Cheonan City. Changhee Lee, "Flexible Electronics—A Korean Initiative."

state of Ohio's promotional effort in flexible electronics is based on a cluster strategy developed through observation and study of foreign cluster policies.[22]

Some flexible electronics clusters have already achieved substantial scale and sophistication. The largest organic electronics cluster in Europe is in Dresden, where about 40 companies, 17 research institutes, and more than 950 employees (as of 2012) are pursuing various research and innovation themes in organic electronics.[23] (See Table 3-3.)

---

[22] Presentation by John West, "The Genesis of a New Cluster," National Research Council, "Building the Ohio Innovation Economy: Summary of a Symposium," April 25-26, 2011.

[23] "Organic Electronics for Saxony OES," *<http://www.colae.eu/companies/organic-electronics-saxony-oes-2/>*.

TABLE 3-3 The Dresden Flexible Electronics Cluster

R&D	**Production**	**Supply Chain**				
Organic Displays/PV/lighting		***Materials***	***Substrates***	***Tools/Plant Engineering***	***Process***	***Analytics***
TUD IAPP	Novaled	TU Dresden IAPP	TU Dresden IAPP,	FHR Anlagenbau	Novaled AG	Fraunhofer IZFP,COMEDD
Fraunhofer COMEDD	Plastic Logic	Fraunhofer, COMEDD, IPF, IFW	Fraunhofer, COMEDD, FEP	VON ARDENNE	Heliatek GmbH	
Fraunhofer FEP, IWS	Heliatek	Liebnitz		DTF Technology	LEDON	TU Dresden
Leibniz IPF	LEDON Lighting		Leitbniz IPF, IFW	3DMicromac	Plastic Logic	SEMPA SYSTEMS
Organic electronics/RIFD		Novaled AG		Sunic System	WOLFRAM	
Fraunhofer IPMS	SAW Components	Plastic Logic		KSG Leiterplatten	Design/ Engineering	SURAGUS GmbH
Fraunhofer COMEDD	Smartrac	Heliatek		Pm TUC	Fraunhofer COMEDD	Fraunhofer IWS
	PE	Sim4tec		Fraunhofer FEP, IWS, COMEDD		Leibniz IPF, IFW
	Ortner, AIS	IHM, Sensient		IWS, COMEDD	AVT	
Plant Engineering				TU Chemnitz	T printechnologies	
Fraunhofer FEP, IWS	VON ARDENNE					
	DTF Technology					

continued

TABLE 3-3 Continued

R&D	Production	Supply Chain				
Organic Displays/PV/lighting		***Materials***	***Substrates***	***Tools/Plant Engineering***	***Process***	***Analytics***
	FHR Antagenbau					
	Creaphys					
	3DMicromac					
	Sunic System					
Transport electrodes						
TUD IAPP						
Fraunhofer COMEDD, IWS, FEP						
Encapsulation						
TUD IAPP						
Fraunhofer COMEDD, FEP						
Printed Batteries						
TU Chemnitz						
TUD IAPP						

SOURCE: City of Dresden, Dresden—Europe's Largest Cluster for Flexible Electronics (April 2012).

4

The Emerging Competitive Landscape

Because flexible electronics products have only begun to enter the market, most assessments of emerging national and regional advantage examine the apparent strengths and weaknesses of each country/region based on its existing human and institutional assets and, in some cases, recent performance in related sectors. At present, the three zones in which major developmental efforts in flexible electronics are under way are Europe, East Asia, and North America. Although countries located in each zone enjoy significant competitive strengths, no region is clearly poised to dominate the new industry.

In 2011, Germany's National Academy of Science and Engineering published a study of Germany's global competitive position in organic electronics. It surveyed more than three dozen German experts in the field asking, among other things, for an assessment of the regions of the world regarded as holding leadership in various thematic areas associated with organic electronics. (See Table 4-1.)

The acatech survey has not been updated, as of this writing, and while it is a useful benchmark of relative regional competitiveness, it has been overtaken by a number of events:

- The market for flexible photovoltaic electronics products, an area of U.S. and European strength, has been devastated by a global industry shakeout.[1] A once promising U.S.-based organic photovoltaics (PV) maker, Konarka, went bankrupt in 2012, an event which has undermined prior U.S. leadership in organic PV.[2]

[1] "Investments, Upheaval in the PV World," *Printed Electronics Now,* September 2011.

[2] "Konarka Technologies Files for Chapter 7 Bankruptcy Protection," *Printed Electronics Now,* May 31, 2012.

TABLE 4-1 acatech Survey of Competitive Positions in Organic Electronics

Thematic Area	Leading Region
Materials	Europe
Plant technology	Europe/Asia
Device technology	Europe/US/Asia
Products	US/Asia
OLED displays	Asia
OLED lighting	Europe
Organic FET	US/Europe
Organic PV	US/Europe

SOURCE: acatech—National Academy of Science and Engineering, eds., *Organic Electronics in Germany: Assessment and Recommendations* (Munich: National Academy of Science and Engineering, 2011).

- Technological difficulties associated with the new flexible electronics materials and processes have delayed product introductions in all regions, including East Asia.[3]
- Asian producers of flat panel displays, including Samsung, LG, and AU Optronics, have moved to commercialize handsets, TVs, and e-readers utilizing flexible displays. They have developed techniques for converting their existing conventional liquid crystal display (LCD) manufacturing lines to produce flexible displays, consolidating Asian leadership in displays.
- Taiwanese and Japanese firms are joining hands to challenge emerging Korean leadership in displays.
- Small U.S. and European firms are entering niche applications markets, but no major device aggregators have emerged in either region comparable to the Asian makers of consumer displays, and Asian firms arguably lead in the "products" category.
- Numerous small U.S. and European firms that have developed significant proprietary technologies have been acquired by firms based outside of their regions, usually in Asia, strengthening Asia's competitive standing in factory technology, materials, displays, and products.

[3] In 2013, Samsung indicated that the introduction of smartphones with flexible displays by the end of the year might be delayed because of technical issues associated with Vitex System encapsulation technology, which reportedly slowed down the production process. "Samsung Delays its Flexible Displays," *PhoneArena.com,* April 17, 2013; "Samsung Delays Flexible OLED Displays," *Plastic Electronics,* October 2, 2012. Problems associated with manufacturing yields reportedly induced LG and Samsung to delay introduction of 55-inch organic light-emitting diode (OLED) televisions in 2012. "LG and Samsung Face Delay with OLED Televion," *Plastic Electronics,* October 18, 2012. In 2012 Taiwan's AUO announced a delay in its mass production of OLED displays for smartphones citing "issues with the fabrication." "AUO Latest to Delay OLED Mass Production," *Plastic Electronics,* October 31, 2012.

THE EAST ASIAN EDGE IN DISPLAYS

The shadow that looms over any competitive assessment of global competition in flexible electronics is the fact that East Asian firms manufacture virtually all of the world's rigid consumer displays and are moving quickly to translate this leading position into a lasting competitive edge in flexible consumer displays. The Asian display makers possess deep, relevant manufacturing competencies, established supply chains that can be deployed to support new investments in flexible electronics, a proven ability to bring attractive products to market quickly, and the financial strength needed to make large, risky investments. (See Table 4-2.) Asian firms have acquired most of the critical intellectual property associated with OLED displays from Western sources.[4] The prospect that large Asian industrial groups will enter the market for flexible displays is a factor underlying the apparent reluctance of most U.S. and European firms to enter commercial production of consumer displays, with the exception of proprietary niche technologies.

The 2011 study by the German National Academy of Science and Engineering drew a picture of a global market for organic electronics in which Asian firms were acquiring a virtually insurmountable leadership position in OLED displays. The study warned that

> the initial situation as described is characterised by acute danger to Germany's position in all facets of organic electronics. Material and factory technology from the Asian area are gaining an increasingly strong role on the global market, driven by technology synergies of the ambitious OLED displays. Even now, most of the materials for the displays are developed and produced in Asia as well.[5]

Asian prospects for success in flexible displays are enhanced by the fact that most of the enterprises that are preparing to commercialize flexible electronics technologies belong to large diversified industrial groups with extensive manufacturing capabilities. The groups are in a position to provide sustained financial support for investments in promising but risky new products and also frequently possess relevant technological and manufacturing expertise, experience, and resources in areas such as semiconductors, optoelectronics, rigid displays, chemistry, photovoltaic energy, batteries, specialty materials, inks, and printing. Several of these companies enjoy global market dominance with respect to current generation technologies with direct application to flexible electronics.[6] The

---

[4] IDTechEx, *Printed, Organic & Flexible Electronics Forecasts: Forecasts, Players & Opportunities 2011-2021* (2011), 44–45.

[5] Ibid., 24.

[6] In the first quarter of 2011 Korea's Samsung Mobile Displays (now merged into Samsung Display) held nearly 70 percent of the world market for OLEDs and nearly 99 percent of the global supply of active matrix organic light-emitting diodes (AMOLED) panels in the first quarter of 2011. "Samsung Mobile Display Starts Operation of New AMOLED Plant," *Yonhap,* May 31, 2011; "Samsung, LG Display to Invest in High-end Display Market," *Yonhap,* May 3, 2010.

TABLE 4-2 Asian Industrial Groups Commercializing Flexible Electronics Products

Group	Country	Group Product Strengths	Flexible Electronics Group Member	Technology
Samsung Group	Korea	Semiconductors, LCDs, batteries, solar cells	Samsung Mobile Display	Flexible displays, resins
Samsung Group	Korea	Semiconductors, LCDs, batteries, solar cells	Samsung Electro-Mechanics	Inkjet print heads, copper ink
Toppan Printing	Japan	Photomasks, LCD filters	Toppan Forms	Nano silver printed electrodes
LG Group	Korea	OLED, displays, solar cells	LG Display	Flexible displays, e-paper
Fuji Electric Group	Japan	Power semiconductors, solar cells	Fuji Electronic Systems	Flexible solar cells
Chie-Mei Group	Taiwan	Materials, lighting technology, displays	Chin Lin Technology	e-paper, smart labels
Yuen Fuong Yu Group	Taiwan	Displays, paper	Prime View International (E Ink Holdings)	e-paper
Hanwha Group	Korea	Solar cells, batteries	Hanwha Chemical	Carbon nanotubes
Asahi Kasei Chemicals	Japan	Polymers, specialty chemicals	Asahi Kasei Finechem	Dopant for conductive polymers
BOE Technology Group	China	LCDs, photovoltaics	Ordos Yuansheng Optoelectronics Co.	AMOLED displays

industrial groups enjoy distribution channels and customer relationships that will facilitate the development of markets for new flexible electronics products.

Particularly noteworthy are South Korea's industrial groups, the chaebol, large conglomerate groups held together by cross-shareholdings, family ties, and inter-firm agreements.[7] In 2012, Korea's top 10 chaebol accounted for 52 percent of the output of all listed Korean companies, a figure that has grown from 44.9 percent in 2008, indicating increased economic concentration in the large groups.[8] The larger chaebol have enormous financial resources—Samsung's revenues, for example, exceed those of world-class U.S. information technology firms. (See Table 4-3.)

The chaebol have drawn global attention during the past decade with bold, risky investments and decisive execution that have repeatedly paid off in dramatic fashion. Samsung, perhaps the most widely studied of the chaebol groups, "spots markets that are about to take off and places huge bets on them," a strategy that succeeded in dynamic random access memory (DRAM) devices, flash memory, LCDs, and mobile phones.[9] An investment analyst commented in 2012 that

> Samsung has set itself to be a dominant global force in every industry it enters, from smartphones to consumer electronics. The strength of its balance sheet will allow it to outspend most rivals in research, development and marketing while its ability to own the whole supply chain makes it unique. It is at the forefront of technology and I expect to see a phone with a flexible screen in the next year or so.[10]

Samsung is making massive investments in technologies applicable to flexible electronics products. In late 2012, it indicated it would invest 2 trillion won (about $1.8 billion) to increase its output of OLED screens, of which 300 billion won ($268 million) would be allocated to flexible displays. The *Korea Times* reported in November 2012 that

---

[7] For recent academic perspectives on the chaebol, see Stephan Haggard, Wonhyuk Lim, and Euysung Kim, *Economic Crisis and Corporate Restructuring in Korea: Reforming the Chaebol* (Cambridge: Cambridge University Press, 2010); Seung-Rok Park and Ky-hyang Yuhn, "Has the Korean Chaebol Model Succeeded?" *Journal of Economic Studies* 39, no. 2 (2012); Charlotte Marquerite Powers, "The Changing Role of Chaebol: Multi-Conglomerates in South Korea's National Economy," *Stanford Journal of East Asian Affairs,* 2010, 105–116. Powers observes that notwithstanding well-publicized instances of corruption and arguments that the chaebol are obsolete, "by virtue of their size and capital reserves, the chaebol will be Korea's key asset as globalization brings international firms into constant contact with one another. . . . In the globalized economy size matters, and the chaebol conglomerates enjoy the distinction of being among the largest in the world. . . . [T]he chaebol have ready access to the capital and manpower required to constantly develop new and more efficient ways of manufacturing goods to export [and] are also large enough to absorb potential market-entry failure; smaller firms might just collapse" (ibid., 112–113).

[8] "Chaebol's Economic Concentration at Record High," *The Korea Herald Online,* February 6, 2012; "Chaebol Owners Tighten Their Grip," *JoongAng Daily Online,* May 31, 2013.

[9] "Asia's New Model Company," *The Economist,* October 1, 2011.

[10] "Asia Major," *Money Marketing,* March 1, 2012. "Samsung Electronics Striving to Win Absolute Superiority in Global Market Share This Year," MK English News Online, January 6, 2010.

TABLE 4-3 2012 Revenues of Leading Electronics Firms

Company	2012 Revenues (Billions of Dollars)
Samsung Electronics	178.6
Apple	156.5
Hewlett-Packard	120.4
IBM	104.5
Dell	56.9

SOURCE: Fortune Global 500 (2012).

> [Samsung] is expected to roll out finished flexible products not prototypes from the latter half of next year [2013] while its biggest rival LG Display plans to start mass-producing plastic-based flexible OLED screens from the end of next year.[11]

OLED technology, which is self-illuminating and does not require an external light source for viewing, "is considered to be the solution for the next generation of flexible display."[12] Korean firms, which have already begun to commercialize OLED technology, are in the best position to dominate the emerging market for flexible displays.

LG Electronics, another Korean chaebol firm, is mounting a vigorous challenge to Samsung, entering production in 2013 of a 55-inch OLED TV with a curved screen, is expected to compete with Samsung in the flexible display smartphone market as well. In April 2013, the consultancy HSBC Global Research predicted that "we expect [LG Electronics] to be the biggest beneficiary of the commercialization of flexible display."[13] In January 2014, LG unveiled the world's first flexible OLED TV, with a screen with degrees of curvature that can be modified by viewers using the TV remote.[14]

The other potential challenge facing Samsung in flexible displays comes from China, where an extraordinary national effort to establish a presence in conventional rigid displays has enabled domestic producers to capture about 20 percent of the global market.[15] Chinese LCD makers, powerfully backed by local governments, are currently working to invest in the production of small AMOLED displays, aiming at the smartphone market where indigenous makers represent a

[11] "Samsung to Invest $1.8 Billion on OLED," *Korea Times,* November 14, 2012.

[12] HSBC Global Research, *Flexible Display: Fantastic Plastic—A Shape-Shifting Game Changer,* April 2013, 22.

[13] Ibid., 40. Competition between LG and Samsung has been sufficiently ferocious that the Korean government has reportedly taken a mediating stance between the two groups. "Government Expected to Step in to End Display Feud Between Samsung and LG Display," *OSA Direct,* January 21, 2013.

[14] "LG Unveils World's First Flexible OLED TV," *Flexible Substrate*, January 2014.

[15] "China to Account for More Than 70% of Flat Panel Display Equipment Spending," *CTimes,* February 7, 2014; "Rise of Chinese LCD Makers Threatens Local Makers," *Taipei Times,* November 12, 2012; Tain-Jy Chen and Ying-Hua Ku, "Indigenous Innovation vs. Teng-Long Huan-Niao: Policy Conflicts in the Development of China's Flat Panel Industry," *Industrial and Corporate Change,* 2014.

potentially huge source of demand.[16] China lacks a sophisticated, relevant basic research base to support flexible electronics, and most of the supply chain resides outside of China, but its move into AMOLED displays is seen as a competitive threat to Samsung.[17]

Korea's recent successes in emerging electronic technologies tend to obscure systemic weaknesses that could limit its long-term prospects in flexible electronics. Professor Changhee Lee's 2010 description of Korean flexible and printed electronic initiatives noted that Korea's weaknesses were "lack of fundamental research and core IPs" and "materials."[18] He might have added that small businesses, which in many countries are the drivers of innovation, are seen as underperforming in Korea.[19] Moreover, for the foreseeable future Korea's emerging flexible electronics industry will remain dependent on technology, intellectual property, materials, and equipment that originate outside of Korea.

The ascendancy of South Korean chaebol firms in displays has been paralleled by the veritable implosion of the Japanese electronics giants that once dominated global markets for semiconductor memories, televisions, and displays. The Japanese electronics majors are seen as having failed to adjust to the digital revolution.[20] In the fiscal year ended March 2012, Japan's eight largest electronics firms suffered a combined net loss of more than $20 billion, roughly the equivalent of the gross domestic product of Paraguay.[21] These firms, engaged in massive facilities downsizing and workforce reductions, were characterized by *The Wall Street Journal* in 2012 as "dinosaurs stumbling around after the asteroid hit."[22] The Japanese government has intervened with a $2 billion bailout, consolidating the displays operations of Sony, Toshiba, and Hitachi in a new entity, Japan Display Corporation, which is majority-owned by the government. (See Figure 4-1.) The new entity is expected to introduce OLED-based displays and e-paper products in an effort to counter the Korean competitive challenge.[23] In late 2013, a joint venture between Sony and Panasonic to make OLED TVs

---

[16] "iSuppli Sees China Targeting AMOLED Sector," *Optics.org* (November 17, 2011); "China Aims to be Major Player in AMOLED Industry," *Yonhap* (November 18, 2011).

[17] "Potential Threat to Samsung: Chinese AMOLED Panels to Pour into Market," *Business Korea*, May 2, 2014; "Can China Break Samsung's AMOLED Grasp?" *China Focus*, April 8, 2014.

[18] Changhee Lee, "Flexible and Printed Electronics—A Korean Initiative," September 24, 2010.

[19] "Non-Chaebol Firms Losing Ground to Chaebol in S. Korea," *Yonhap*, July 3, 2013. "Do the Chaebol Choke Off Innovation?" *Bloomberg Business Week*, December 3, 2009.

[20] "What Happened to Japan's Electronics Giants?" *BBC News Asia*, April 1, 2013.

[21] "Reshaping Japan's Tech Sector a Struggle," *Financial Times*, June 4, 2012; "Japanese Electronics Sector Teetering on the Brink of Collapse," *Dong-A Ilbo Online*, May 30, 2012.

[22] "Japan's Electronic Giants Struggle to Reboot," *The Wall Street Journal*, November 1, 2012.

[23] "It's Japan vs. Korea with Launch of Samsung Display, Japan Display," *PC Mag*, April 2, 2012; "Japan Takes a Gamble on Displays," *The Wall Street Journal*, April 10, 2012; "Japan Display Begins Production, OLED from 2013," *FlatpanelsHD*, April 3, 2012; "New Prototype e-Paper Shown Off by Japan Display," *Good Reader*, November 6, 2012.

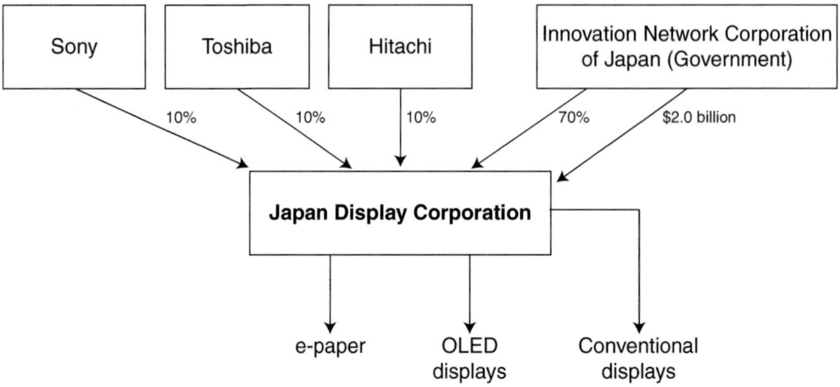

FIGURE 4-1 The creation of Japan Display Corporation (2012).
SOURCE: "Panasonic to Sell TV Panel Factory to New Japan Display Venture," *CIO* (November 15, 2011).

collapsed when the two firms failed to develop cost-effective production processes or durable panels.[24]

Notwithstanding the massive government resources deployed on behalf of Japan Display, the new company's daunting market position underscores the extent of the challenge that it faces. (See Table 4-4.)

Taiwanese industry and government leaders, like those in Japan, have become increasingly concerned about the competitive challenge from Korea across a spectrum of electronics technologies.[25] Japanese and Taiwanese organizations

---

[24] The two companies did not provide details of the manufacturing cost issues that had proven insurmountable, but early in 2013 Sony issued a press release citing technical "challenges" confronting the joint venture. Smaller OLED TVs used low-temperature polysilicon thin-film transistors (TFTs) to force light through the OLED layer. Sony indicated, however, that "there were some challenges inherent in the manufacture of large OLED displays." Accordingly, the company was attempting to fabricate larger displays using oxide semiconductor TFTs and incorporating Sony's "Super Top Emission" technology, which "has a high aperture ratio and enables light to be extracted efficiently from the structure's OLED layer." While the combination of these two technologies enabled the development of a promising prototype, the two firms were apparently unable to develop cost-efficient production processes. The low sales achieved by LG and Samsung with respect to the rollout of 55-inch OLED TVs in 2013 were also cited as a possible factor underlying the Sony/Panasonic decision. "Sony Develops the World's First and Largest '56-inch 4K OLED TV' Prototype to be Exhibited at 2012 International CES," Press release, January 7, 2013; "Technical Difficulties Foil Sony-Panasonic OLED Effort," *Flexible Substrate,* January 2014.

[25] "Samsung Cannot 'Kill Taiwan'—CEOs," *Taipei Times,* March 23, 2013. In 2013, Taiwan's HTC Corp. said it had learned a lesson from its supply relationship with Samsung. HTC launched a mobile phone "HTC Desire" using an AMOLED display supplied by Samsung. "But once the HTC Desire was welcomed by global consumers and telecom operators at the time [2010] Samsung 'strategically declined to supply its AMOLED displays to the smartphone maker.'" Jack Tang, President of HTC

TABLE 4-4 The Global Market Share for Displays (2012)

Firm	2012 Market Share (Percent)
Samsung Display (K)	24.5
LG Display (K)	23.4
Innolux (T)	12.8
AU Optronics (T)	11.8
Sharp (J)	8.3
Japan Display Inc. (J)	3.5
Other	15.7

SOURCE: NPD Display Search, In "S. Korean Display Makers Increase Global Market Share," *Yonhap,* February 7, 2013.

are responding to what is seen as a powerful challenge by Korean producers in electronics by forming Taiwan/Japan technology alliances, featuring Japanese technology and Taiwanese manufacturing competency and ability to bring products to market quickly. (See Table 4-5.) Terry Gou, Chairman of Taiwan's Hon Hai Precision Industry (Foxconn), explained his company's 2012 decision to acquire a 9.9 percent equity stake in the Japanese display maker Sharp, indicating that in his view Sharp's technology was superior to that of Samsung:

> The Sharp deal will help us defeat Samsung. . . . I respect the Japanese and especially like their execution and communication styles. Unlike the Koreans, they will not hit you from behind.[26]

The rapid emergence of a display industry in China is a final wild card in the emerging Asian competition in flexible displays. In the past decade numerous Chinese electronics firms, usually strongly backed by local governments, entered the LCD industry, placing strong downward price pressure on established Asian producers.[27] A number of these firms, most notably BOE Technology, are undertaking multi-billion-dollar investments in the production of flexible AMOLED displays, challenging Korean dominance in this technology.[28]

North Asia, said that "we found that key component supply can be used as a competitive weapon." "HTC Learns Lesson from Samsung Display Row," *Central News Agency,* May 28, 2013.

[26] "Hon Hai Vows to Beat Samsung Display in 3-5 Years," *Chosun Ilbo Online,* June 20, 2012; Chosun Ilbo, a major Korean daily newspaper, attributed Gou's apparent "personal animus" to "The Koreans" to an antitrust proceeding in which Samsung had reportedly presented testimony against Taiwanese firms. Ibid. On competitive tensions between Korea and Taiwan see generally "Anti-Korean Spleen in Taiwan," *Asia Sentinel,* November 25, 2010.

[27] "Analysis—China's LCD Industry Fears Overcapacity," *Asian Pulse.*

[28] "AMOLED Production: Entering a New Era?" *Information Display,* March/April, 2013.

TABLE 4-5 Flexible Electronics—Taiwan/Japan Tie-ups

Year	Entity		Scope of Collaboration
	Taiwan	Japan	
2013	ITRI	Kaneka Corporation	Flexible substrates
2013	E Ink	Sony	Flexible e-paper for e-readers
2012	AU Optronics	Idemetsu Kokan	OLED materials for displays
2012	AU Optronics	Sony	OLED televisions
2011	Industrial Technology Investment Corporation (ITIC)	Mitsui Sumitomo Insurance Venture Capital	Investments in Taiwanese and Japanese firms

SOURCES: ITRI and Kaneka Unveil New Flexible Display Technology," *ITRI Today,* April 3, 2013; "The Glass is Half Full," *Flexible Substrate,* May 2013; "AUO Signs Accord with Idemetsu to Manufacture OLEDs," *Taipei Times Online,* February 3, 2012; "AUO, Soney Co-develop World's 1st 4K OLED TV Panel," *Central News Agency,* January 9, 2013; "Top Industrial Research Body Praised for Helping Japanese Firms," *Central News Agency,* February 5, 2013.

INITIAL U.S. AND EUROPEAN ADVANTAGE IN MATERIALS

The manufacture of flexible electronics products requires a broad range of exotic materials, which are often proprietary and difficult to produce, as well as the competency to develop new materials as the demands of an evolving industry require. Although Asian firms are likely to dominate the manufacture of flexible displays, at least initially, they are dependent on U.S. and European firms for key materials. The U.S.-based Universal Display Corporation, for example, is the world's leading supplier of phosphorescent emitter materials, the key elements in an OLED device.[29] Corning produces Gorilla Glass, which is used for the front of current smartphones, and is introducing Willow Glass, a flexible glass product with many potential applications in flexible display products.[30] The 2011 competitive assessment by Germany's National Academy of Science and Engineering observed that "the organic materials, their synthesis and optimization, are of particularly high importance for the technology field of organic electronics" and were difficult to develop and produce. Germany's leadership in chemistry and new materials was seen as its greatest asset in the emerging industry.

[29] Universal Display Corporation Form 10-K, filed February 27, 2013, 5.

[30] "Corning Willow Glass Used to Make Flexible Solar Power Roofing Shingles, Could Lower the Cost of Solar Power Significantly," *ExtremeTech,* July 3, 2013. Corning received a FLEXI Award for Willow Glass from the FlexTech Alliance in 2013, an award given to recognize leading developments in flexible and printed electronics. Using a research grant from the FlexTech Alliance, Corning partnered with other firms to demonstrate the compatibility of a flexible glass web in R2R processing and printing of organic PV devices. "Top Flexible Electronics Developments Win 2013 FLEXI Awards," Flexible Substrate, February 2013.

TABLE 4-6 Key Firms Supplying Materials for the Flexible Electronics Products

U.S. Company	Relevant Material	Application
Corning	Mother glass Fnt seal	TFT OLED encapsulation
Kodak	Fluorescent/phosphorescent materials	OLED emission materials
UDC	Fluorescent/phosphorescent materials HIL, HTL, ETL, EIL	OLED common layer
DuPont Teijin	HIL, HTL, ETL, EIL PET, PEN film	OLED common layer Flexible substrate

SOURCE: HSBC Global Research, *Flexible Display: Fantastic Plastic—A Shape Shifting Game Changer,* April 2013; Universal Display Corporation Form 10-K filed February 27, 2013, 5.

The U.S. and European strength in materials means that both regions benefit economically, to a degree, from Asian sales of displays, because U.S. and European firms are integral parts of the supply chains of Asian firms. The existence, competencies, and intellectual property portfolios of these materials suppliers represent an important asset with respect to any future U.S. or European initiative to contest the consumer flexible display market.

MILITARY APPLICATIONS—U.S. EDGE

The U.S. defense establishment has a long history of funding research by universities and industry for defense applications, and many military research programs have given rise to thriving commercial industries.[31] U.S. defense organizations have been sponsoring research efforts in flexible electronics since the late 1990s, and, at present, the most substantial center for flexible electronics R&D anywhere in North America is the Flexible Electronics and Display Center of Arizona State University, established with $100 million in support funding from the U.S. Army. No known program outside the United States has devoted a comparable level of sustained public support for flexible electronics technologies for military use, and it is highly likely that the United States and U.S.-based companies will dominate this field, notwithstanding the fact that some flexible electronics products for the military will probably be manufactured offshore.

The U.S. Army and other service branches that are supporting flexible electronics research represent an element that is often missing in this emerging field—that is, customers who know and can specify exactly what they want and are

[31] See generally Stuart W. Leslie, "The Biggest Angel of All: The Military and the Making of Silicon Valley," in Martin Kenney, ed., *Understanding Silicon Valley: The Anatomy of an Entrepreneurial Region* (Stanford: Stanford University Press, 2000); National Bureau of Standards, *The Influence of Defense Procurement and Sponsorship of Research and Development on the Development of the Civilian Electronics Industry*, June 30, 1977.

willing to pay for it according to an agreed fixed schedule. The U.S. armed services currently utilize conventional displays in aircraft, vehicles, and infantry units, but glass displays are relatively heavy, breakable, and require many extra pounds of protection around the glass, as well as batteries. Flexible OLED-based displays do not need a backlight, potentially are thinner, lighter, more durable, and do not need to receive power except when a soldier needs to change an image.[32] The military service branches are funding a broad array of research projects intended to develop specific items of equipment based on flexible electronics technology.

In the past, U.S. military R&D programs have resulted in developments of devices for military use that are manufactured in small batches of extremely high cost relative to commercial products based on similar technologies. In flexible electronics, however, U.S. defense organizations are promoting the establishment of commercial industries that can supply devices for military use at the same price levels as those prevailing in the commercial marketplace.[33] As a result, U.S. defense-related investments in flexible electronics may translate into commercial opportunities for U.S.-based industries.

U.S./EUROPEAN LEADERSHIP IN ORGANIC PHOTOVOLTAICS

An initial U.S./European edge in organic photovoltaics has been largely nullified by the collapse of world markets for solar panels after 2008. The 2011 assessment of global competition by the German National Academy of Science and Engineering observed that "while Asia has covered the field of organic displays, the U.S. seems to be striving for a leadership position in organic photovoltaics." The assessment acknowledged that the United States was "an organic photovoltaic pioneer," that a U.S. firm, Konarka, was "the current technology leader" and that the presence in the United States of other innovative companies, such as Plextronics, Solamer Energy, and Global Photonic Energy, "clearly show the intensity with which the U.S. drives commercialization of organic photovoltaics." The study observed that

> [t]he U.S. efforts of conquering the market for organic photovoltaics are mainly countered by the activities of the company Heliatek and its outstanding know-how in Germany. Even though Germany has a good general initial situation for organic photovoltaics, the intense activities in the U.S. are a threat to it.[34]

However, beginning in 2011, global demand for photovoltaic modules—whether conventional silicon-based or organic—declined substantially, reflecting decisions by some European countries to reduce subsidies for solar power

[32] "Army-Backed Flexible Display Effort: A Symbol of Public-Private Partnership," *IEEE Computer Society,* July-September 2006.

[33] Dr. Eric W. Forsythe, "Flexible Communications," *Army AL&T Magazine,* July-September 2012.

[34] acatech—National Academy of Science and Engineering, eds., *Organic Electronics in Germany: Assessment and Recommendations* (Munich: National Academy of Science and Engineering, 2011), 24.

production.[35] At the same time, producers in China, which had undertaken major investments in PV capacity, were undertaking a major export push, driving down prices of PV modules and placing severe economic pressure on U.S. and European manufacturers.[36] The price of a silicon panel fell from $3.40 per watt in 2008 to $1.28 per watt by the end of 2011, and as of mid-2013 it was reportedly "heading toward 50 cents."[37] In 2012-2013, the United States and EU imposed antidumping duties on imports of Chinese solar panels.[38]

The collapse of global solar PV markets adversely affected organic PV makers as well as producers of silicon-based PV modules. Most dramatically, the U.S. technology leader, Konarka, went bankrupt in 2012.[39] Plextronics, a U.S. startup that planned to develop solar ink cells with photovoltaic applications, filed for Chapter 11 protection in early 2014.[40] In the UK, the Cardiff-based developer of flexible PV films, G240 Innovations, entered bankruptcy administration in December 2012.[41] At present, industry analysts foresee modest growth in organic PV markets going forward, with demand reaching "only a few hundred million dollars in market size in the next decade."[42]

THE FLOW OF TECHNOLOGY TO ASIA

The United States and Europe are centers for some of the world's most advanced basic research into themes applicable to flexible electronics. That fact does not appear to place Japan, South Korea, and Taiwan at much of a disadvantage in the field, not only because these countries are developing their own research programs, but also because their companies can readily access the new materials, device, and process technologies being developed outside of Asia. In

---

[35] "Twilight of an Industry: Bankruptcies Have German Solar on the Ropes," *Spiegel Online,* April 3, 2012. According to a report by Forbes in mid-2012, supply of solar cells was nearly double total demand. "As a result, some companies are looking to sell below cost to gain market share, and ride at the losses as other companies go out of business." "Solar Market Continues to Evolve," *Printed Electronics Now,* July 2012.

[36] "Sun Burn 2: Global Changes Slow Solar Growth," *Toledo Free Press,* July 26, 2012; "China's Photovoltaic Industry: Exporting on the Cheap," *Energy Tribune,* September 3, 2009; "Cloudy Skies Remain for Taiwan's PV Industry," *Taiwan Economic News,* November 30, 2012; "PV Armageddon: The Rapid Market Swings Concealed Major Efficiency Gains," *Printed Circuit Design & Fab,* April 2012.

[37] "China Eclipsing Colorado Makers—Cheap Loans, Subsidies Giving Chinese Companies a Star Role," *The Denver Post,* August 18, 2013.

[38] "China Sun Panels Face EU Levies," *Financial Times,* May 6, 2013; "U.S. Sets Antidumping Duties on China Solar Imports," *Bloomberg,* October 10, 2012; "REC Regrets Escalation in Solar Industry Trade War, Denies Any Wrongdoing," *Printed Electronics Now,* July 20, 2012.

[39] "Solar Shakeout: Konarka Technologies Files for Bankruptcy," *PVTech,* June 2, 2012.

[40] "Local Tech Darling Files for Chapter 11," *Pittsburgh Post-Gazette,* January 26, 2014.

[41] "Restructured G24: Emerges Under New Ownership," *Flexible Substrate,* May 2013. G24: has reportedly secured funding and emerged from administration under the name G24: Power Limited. Ibid.

[42] "Flexible PV: Three-Fold Growth in the Next 5 Years," *Flexible Substrate,* February 2013.

some cases, Asian companies and research organizations are direct participants in U.S. and European flexible electronics research programs.[43] In addition, U.S. and European startups that have developed flexible electronics technologies and intellectual property are seeking Asian partners to manufacture and sell their products, or are being acquired outright by Asian manufacturers.

U.S. companies that have developed proprietary flexible electronics technologies commonly turn to arrangements with established Asian manufacturing firms to produce and commercialize their technologies from production bases in Asia, risking long-term loss of control of the technology as well as most of the value added.[44] (See Table 4-7.) In April 2011, for example, U.S.-based Nova Centrix entered into an agreement with Japan's Showa Denko (SDK) pursuant to which SDK will manufacture and sell nanoparticle inks developed by Nova Centrix:

> Nova Centrix is one of several nanomaterials suppliers working with Japanese and other Asian partners to support production and commercialization of their technology. Experience of industrialized production methods can be leveraged as these technology developers try to commercialize their technologies, and much of the world's display and electronics manufacturing occurs in Asia.[45]

In 2013, Ascent Solar Technologies, Inc., a Colorado-based company founded in 2005 producing thin-film flexible photovoltaic modules, entered into an agreement with the municipal government of Suqian, in China's Jiangsu Province, to establish a factory in Suqian to manufacture Ascent's proprietary technologies. Suqian will contribute capital, a factory site, and various incentives; Ascent will

---

[43] In 2010, Arizona State University's Flexible Display Center (FDC), established in conjunction with the U.S. Army Research Laboratory to accelerate the development of flexible displays in the United States, entered into a strategic research partnership with AU Optronics Corporation, the largest manufacturer of thin-film transistor LCDs in Taiwan. Another FDC strategic research partner, U.S.-based E Ink, was acquired by Prime View International, a Taiwanese company that is the largest manufacturer of electronic paper in the world. "The Flexible Display Center and AOU Enter Strategic Partnership to Accelerate Flexible AMOLED Development," *Nanowerk,* November 16, 2010. MIT has conducted joint R&D with the Korea Institute of Science and Technology (KIST) to develop flexible transistors that could be incorporated in "wearable computers." "Transistor Could Lead to Wearable Computers," *JoongAng Ilbo,* October 9, 2007.

[44] In 2010, Japan's Konica Minolta Holdings entered into a partnership with U.S.-based Konarka Technologies Inc., a world-leading producer of organic thin-film photovoltaics with a broad portfolio of patents and technology licenses and a skilled technical, scientific, and manufacturing team. Pursuant to the agreement Konica Minolta invested $20 million in Konarka and into a R&D collaboration with Konarka to improve organic thin-film photovoltaic performance. The two companies agreed to establish a joint venture in Japan that would manufacture organic thin-film photovoltaic panels. "Konica Minolta and Konarka Join Forces to Develop Organic Thin Film Photovoltaics," *Nanowerk,* March 4, 2010.

[45] "Nanomaterials Firms Turn to Asia for Commercial Opportunities," *Plastic Electronics,* April 15, 2011.

TABLE 4-7 U.S. Flexible Electronics Companies with Asian Production Arrangements

U.S. Company	Technology	Asian Partner	Country
Nova Centrix	Nanoparticle inks	Showa Denko	Japan
Applied Nanotech	Copper ink for thin-film substrates	Ishihara Chemical	Japan
Plextronics	Photovoltaic panels	Korea Parts & Fasteners	Korea
Konarka	Thin-film photovoltaic panels	Konica Minolta	Japan
Ascent Solar	CIGs modules on flexible film	Municipal Government of Suqian	China

SOURCES: "Nanomaterials Firms Turn to Asia for Commercial Opportunities," *Plastic Electronics,* April 15, 2011; "Konica Minolta and Konarka Join Forces to Develop Organic Thin Film Photovoltaics," *Nanowerk,* March 4, 2010.

contribute $1.6 million, its technology, and "certain equipment from its Colorado facility."[46]

Companies based in Korea, Taiwan, and Japan have obtained important flexible electronics technologies through acquisition of, or significant equity investment in, U.S. and European companies with proprietary technologies.[47] (See Table 4-8.) Most recently, in July 2013, Samsung indicated its intention to acquire Germany's Novaled AG, a major developer of OLED technology, through an affiliate, Cheil Industries, for a reported price of more than $200 million.[48] The

[46] "China Eclipsing Colorado Makers—Cheap Loans, Subsidies Giving Chinese Companies a Star Role," *Denver Post,* August 18, 2013. Ascent reported an accumulated deficit of $247.8 million as of December 31, 2013. Its survival is substantially attributable to large infusions of capital from TFG Radiant, a Shenzhen-based joint venture between China's Radiant Group and Singapore-based Tertius Financial Group. Ascent Solar Form 10-K for 2013.

[47] In 2009, the Taiwan Cement Group indicated that it planned to manufacture carbon nanotubes, which will be used as conductors in flexible electronics devices. According to Taiwan Cement Chairman Leslie Koo, "carbon nanotubes cost up to US $2,000 per gram, while cement costs about NY $2,000 per metric ton. Therefore the group plans to acquire a U.S. nano technology firm and work with local universities to develop and produce integral materials that are commonly used by the 3C and optoelectronic industries." "Academic-Enterprise Cooperation Pact Inked to Promote Green Energy," *Central News Agency,* September 24, 2009.

[48] "Samsung to Buy Germany's Novaled, Raising Bet on Next-Generation Success," *The Wall Street Journal*, July 30, 2013. Samsung has been working with Novaled since 2005 on advanced OLED technology. In 2012 the two companies signed a strategic purchase agreement pursuant to which Samsung committed to purchase dopant materials used in the transport layers of its AMOLED display modules from Novaled, while the latter provided its PIN OLED technology for use in the manufacture of AMOLED display modules by Samsung Mobile Display. Jong-Woo Park, CEO of Cheil Industries, commented with respect to the acquisition that "[l]eadership in future display market will be determined by technological capacity. This acquisition is expected to generate significant synergy in new-generation OLED materials R&D and will play a critical role in enhancing Cheil Industries' market position as a global leader in electronic materials." "Samsung and the Growing Market for OLEDs," *Printed Electronics Now,* August 2013.

Asian acquisitions appear to be based on the assumption that the Asian firms will commercialize and manufacture products based on technologies developed in the United States and Europe.[49] Dr. Janglin Chen, Director of Taiwan's ITRI Display Technology Center, the principal focus of flexible electronics technology development in Taiwan, observed in 2010 that

> [r]ecent financial difficulty drove a wave of western start-up firms to seek fund infusion, or manufacturing partners in Asia. This trend has helped to bring to Taiwan a few important technologies in the flexible electronics area.[50]

Prominent examples of this phenomenon are the absorption of the U.S. firm E Ink, arguably the world leader in e-paper materials and intellectual property, into the Taiwanese Yuen Fuong Yu (PVI) Group and the acquisition of U.S.-based SiPix by Taiwan's AU Optronics Corp. "E Ink had amassed a portfolio of hundreds of patent applications, including 150 in the United States, and the IEEE Spectrum ranked E Ink's patent portfolio as number three worldwide for computer peripherals and storage."[51] E Ink had also developed technology alliances with numerous partners globally in the development of e-paper displays. Thus, with the acquisition, PVI "gained substantial intellectual property and employee talent, while securing a supply of critical components during the rapid growth phase of the market, and adding alliances and relationships across the e-paper and flexible display industry."[52] Thus between 2009 and 2012 much of the world's e-reader technology became concentrated in Taiwan. Dr. Chen observes that "one firm's demise happens to be the other firm's fortune."[53]

Recent European actions in the conventional photovoltaics sector may portend similar European initiatives in flexible and printed electronics, including flexible photovoltaics. Germany's publicly supported Fraunhofer Institute for Solar Energy Systems (ISE) had collaborated with Schott Solar AG for more than 20 years in the development of crystalline silicon photovoltaics. In early 2013, as Schott Solar experienced increasing economic distress, Fraunhofer ISE purchased Schott's portfolio of 111 patent families "covering the entire value chain of silicon photovoltaics, from crystallization to system installation" for an undisclosed

[49] The CEO of NanoGram, a U.S.-based developer of nanomaterials design technology, commented on the 2010 acquisition of his company by Japan's Teijin Limited, a chemicals manufacturer, that "our nanoparticle manufacturing technology, and the materials we have developed using the technology, have proven to have superior functionality in a variety of applications for solar, flexible displays and printed semiconductor applications. What NanoGram needs to take those materials to market is application expertise and the ability to scale quickly. Teijin recognized the potential in NanoGram and its technology. It will be the company that takes us to the next level." "Teijin Acquires NanoGram Corporation—Will Accelerate Nanoparticle R&D for Printable Electronics," *Nanowerk,* August 9, 2010.

[50] Dr. Janglin (John) Chen, "Flexible Electronics Development in Taiwan," September 24, 2010.

[51] "Prime View Reaches Agreement to Acquire E Ink," *Flexible Substrate,* June 2009.

[52] Ibid.

[53] Presentation of Dr. Janglin Chen, National Research Council, "Flexible Electronics for Security, Manufacturing and Growth in the United States," 2013.

sum. Fraunhofer ISE indicated that "with this transfer, both partners ensure that comprehensive know-how in the photovoltaic sector remains in Europe."[54] (See Table 4-8.) Schott reportedly entered into the deal "rather than sell to the Asians . . . the purpose seems to be specifically to protect European expertise in the solar sector." Fraunhofer ISE's director, Eicke Weber,

> has been busy over the past year trying to coordinate the European PV sector and EU funders to produce gigawatt-scale PV plants in Europe as a way of competing with the gigawatt-scale plants in Asia. . . . If Europe does not somehow protect this expertise during the current shakedown of the PV sector, Europe will quickly lose the entire industry.[55]

Japan is to some extent undergoing an experience similar to the United States and Europe relative to Korea and other Asian countries. In a growing number of cases, Japanese universities have conducted groundbreaking research but have found that the only companies willing and able to commercialize their ideas are in Korea. A Japanese professor, Sujio Ijima, discovered carbon nanotubes in 1991, flexible conducting carbon-based structures that can be used to create flexible electronics devices.[56] But by 2010 Ijima, who is often cited as a possible Nobel Prize winner, was developing a graphene-based flexible touch screen panel with a team of researchers at Korea's Sungkyunkwan University. Ijima commented that Japanese electronics companies are "lagging behind their South Korean rivals." He has approached Japanese state-run research funding bodies but found their response was slow:

> Even if research funds are provided in the next fiscal year's budget, it would take a year and a half for the project to get underway. That is about the time frame in which Samsung could come out with a new product.[57]

In August 2013, Ijima delivered the opening address at the Nanocarbon Application Forum in Osaka, ruefully showing his audience the world's first smartphone with a touchscreen made of carbon nanotubes, made not in Japan but in China, exemplifying the "loss of Japan's leading edge regarding nanotube research."[58]

Asian firms are now developing proprietary technologies in the field of flexible electronics that are potentially of interest to companies in North America and Europe. An ongoing subject of interest will be whether U.S. and European firms

---

[54] Fraunhofer ISE, "Fraunhofer ISE Acquires Patent Portfolio from Schott Solar—Expertise on Crystalline Silicon Photovoltaics Remains in Europe," Press Release, January 10, 2013.

[55] "Fraunhofer Fights to Keep Solar Industry in Europe," *Renewables International,* January 11, 2013.

[56] Sujio Ijima, "Helical Microtubules of Graphitic Carbon," *Nature,* November 7, 1991.

[57] "South Korean Electronics Companies Are Beating Japanese Competitors to the Punch in Getting Cutting Edge Technologies Developed by Japanese Researchers into Their Product Ranges," *Asahi Shimbun,* July 9, 2010.

[58] "Japan Losing Out in Latest Applications for Carbon Nanotubes," *Nikkei Asian Review,* November 20, 2013.

TABLE 4-8 Asian Acquisitions of U.S./European Flexible Electronics Enterprises

Year	Investor	Country	Target	Country	Technology	Investment	Reported Sale Price (Millions of Dollars)
2013	Samsung	Korea	Novaled AG	Germany	OLED	Acquisition	354
2009	PVI	Taiwan	E Ink Corp.	US	e-paper displays	Acquisition	215
2012	E Ink	Taiwan	SiPix	US	e-paper	31.58% equity	50
2010	Teijin	Japan	Nano Gram Corp	US	Inks, silicon nanoparticles, process technology	Acquisition	NA
2011	Hanwha Chemical	Korea	XG Sciences	US	Graphenes	19% equity	3
2011	POSCO	Korea	XG Sciences	US	Graphenes	20% equity	NA
2009	LG	Korea	Kodak OLED business	US	OLED	Acquisition	
2007	Sumitomo Chemical	Japan	Cambridge Display Technology	UK	Polymer OLED	Acquisition	285

SOURCES: "Samsung to Buy Germany's Novaled," *The Wall Street Journal,* August 9, 2013; "Prime View International Acquires E-Ink for $215 Million," *ZD-Net,* June 1, 2009; "E-Ink to Acquire Shares of SiPix, SiPix Imaging," *Telecompaper Asia and Africa,* August 9, 2012; "Nanostart-Held Nano Gram to be Fully Acquired by Major Japanese Corporation," *Nanotechnology Now,* August 9, 2010; "S. Korea's Hanwha Chemical Buys Into U.S. Graphene Maker," *Asia Pulse,* January 28, 2011; "POSCO Buys Into U.S. Graphene Maker," *Asia Pulse,* June 8, 2011; "Sumitomo Chemical to Acquire Cambridge Display Technology Inc.," *JCN Network,* July 31, 2007.

can obtain such technologies on a reciprocal basis. In 2013, Janglin Chen, the Director of Taiwan's ITRI Flexible Display Center, said in an interview in 2013 that "ITRI being a government-funded institute, priority is given first to domestic companies when it comes to transfer of research results or technology."[59] In 2012, following the reported theft of Korean technology for AMOLED and white OLED displays from Samsung and LG Electronics by Israeli operatives, *Yonhap*, Korea's semiofficial news agency, commented that "South Korea rigorously prohibits leakage of the technologies as it tags them as the nation's core industrially strategic tech."[60]

PERSPECTIVES FROM RECENT SWOT ANALYSES

The SWOT (strengths, weaknesses, opportunities, threats) analysis or matrix is a widely utilized analytic tool developed by Albert Humphrey, a business consultant, while working at the Stanford Research Institute. SWOT analysis identifies the business objectives of a company, industry, or country and the internal and external factors that are favorable or unfavorable to achieving that objective. Reflecting the fact that flexible electronics is a new and highly promising industry sector, SWOT analyses have been undertaken in Europe, North America, and Taiwan with an eye to assessing the prospects for national and regional business success in the field. SWOT analyses are necessarily subjective and do not quantify or assign relative value to individual strengths/weaknesses. However, the development of SWOT analyses in flexible electronics has usually engaged academic and industry leaders and other experts in the field in intensive strategic assessments of the emerging global competitive landscape. These analyses thus represent the informed "conventional wisdom" of individuals with deep knowledge of the subject. (See Table 4-9.)

TABLE 4-9 SWOT Analysis Matrix

	Helpful	**Harmful**
Internal Origin	Strengths	Weaknesses
External Origin	Opportunities	Threats

[59] "Q&A: Dr. Janglin Chen of ITRI (Part 1)," *The Emitter: Emerging Display Technologies,* July 3, 2013.

[60] A Korean official commented that the pilfered technologies were likely transferred to foreign rivals, and that "this may expectedly deal a massive economic blow to the entire nation and cause a sea change in the landscape of the global display market." "Samsung, LG's Key Display Technologies Leaked by Israeli Firm," *Yonhap,* June 27, 2012.

European SWOT Analysis

In 2009, a SWOT analysis of the European competitive position in organic and large area electronics (OLAE) prepared by the European Commission was published under the auspices of the Commission's Seventh Framework OPERA program, an initiative to "define a coherent strategy for the future of OLAE in Europe."[61] (See Table 4-10.)

The Commission credited Europe with a large market and established infrastructure and strong competencies in relevant materials and equipment. The OPERA task force that developed a strategic vision for Europe observed that the "level of R&D on organic materials is very high in Europe and companies like Merck, Novaled, CDT, AGFA, H.C. Storck and BASF are leading suppliers in the various OLAE markets." Europe was also viewed as the global leader in organic device design: "Both in organic LED and solar cells, current record values are held by European groups. In transistors and other circuit development, Europe is leading as well."[62] The 2011 study by Germany's National Academy of Science and Engineering echoed the OPERA findings from a German perspective:

> One of Germany's strengths is seen in the globally established and committed chemicals industry, which supports both development and production of materials for organic electronics. . . . The leadership position of Germany in printing technology and the printing machines industry should benefit successful economic implementation or organic electronics.[63]

The Commission's SWOT analysis saw as a significant European weakness the fact that neither "giant" European companies nor startups were entering the market, raising the risk that "external companies will benefit from the research and investment done in Europe." Similarly, the OPERA analysis observed that while Europe was strong in materials, equipment, and device design, "when it comes to manufacturing, the picture is less rosy. Although European toolmakers have made significant innovations like inline coating, novel printing techniques, and OVPD, the actual organic device manufacturing (being [at that time] only displays) is 100% Asian."[64] OPERA observed in 2011 that "only a few large-[European] companies [were] active in OLAE" and that the "organic display market [was] dominated by Asia."[65] The 2011 German study concurred:

> A development that has already happened in other technologies before is becoming apparent: While Germany has great know-how in rather application-remote stages of technology development implementation of the excellent research

[61] OPERA, *Towards Green Electronic in Europe,* December 28, 2009.

[62] OPERA, *Towards Green Electronics in Europe,* 8–9.

[63] acatech, *Organic Electronics in Germany*, 21.

[64] OPERA, *Towards Green Electronics in Europe,* 9.

[65] The FF7-ICT Coordination Action OPERA and the European Commission's DG INFSO Unite G5 "Photonics," *An Overview of OLAE Innovation Clusters and Competence Centres,* September 2011, 20.

TABLE 4-10 European Commission OLAE SWOT for Europe

Strengths	Weaknesses
• Already in possession of materials and production machinery	• Lack of startups and entrepreneurs with clear view from research to manufacturing
• Huge European market	• Committed giants are needed
Opportunities	**Threats**
• Formidable research on OLEDs, printed RFIDs and transistors	• European market will be taken over by foreign manufacturers
• Field of applications steadily growing	• External companies will benefit from the research and investment done in Europe

SOURCE: OPERA, *Towards Green Electronics in Europe,* December 28, 2009.

> results in products successful on the market is not driven with the required perseverance.[66]

The American World Technology Evaluation Center (WTEC) team that visited European flexible electronics research centers in 2010 was particularly impressed with the numerous European innovation centers working to develop flexible electronics systems and manufacturing processes. The centers leveraged the technical expertise and financial resources of multiple companies, academic laboratories, and national and EU government organizations. At the technical level

> the centers foster a highly synergistic and interdisciplinary environment in which the complementary expertise of industrial, government and academic scientists is combined to achieve new systems design goals (e.g., ultra-low-power systems in foil) enhanced device performance, broader materials devices, and practical, low cost manufacturing approaches.[67]

The WTEC team was favorably impressed with the fact that multiple flexible electronics research projects were spread across Europe involving many university groups working in a complementary fashion. The EU had already prioritized flexible electronics for about a decade, the Europeans took a long-term perspective on the industry, and they had created strong research groups that worked together for many years. European basic research was particularly strong, and close collaboration existed between universities and basic research organizations, on the one hand, and applied research organizations and European industry, on the other. Facilities for prototyping and pilot-scale manufacturing existed at numerous research centers.[68]

[66] acatech, *Organic Electronics in Germany,* 23.

[67] WTEC Panel Report, *European Research and Development in Hybrid Flexible Electronics,* July 2010, 7.

[68] WTEC, *European Research and Development,* xv–xvi.

Notwithstanding these strengths, European assessments acknowledge that in the face of the Asian competitive challenge, their efforts may prove futile. The 2011 study by the German National Academy of Science and Engineering stated that

> [i]n the competition for the leadership position in organic electronics, Asia holds a decisive advantage: the local OEM like Samsung, LG or Sony rules the consumer electronics industry. Asia also holds a global market share of nearly 60 percent on the level of contracted work [contract manufacturing] for electronics, giving it a better initial situation than Germany has. These basic structural benefits are expressed in mainly compatible value-added chains and structures, e.g. in the production of LCD- and OLED-TVs. Quick commercialization of organic displays in Asia proves that the local industry there is able to play its advantages. Many experts assume that in particular mass production of OLED displays may move entirely to Asia in the future. . . . Even German pilot plants, e.g., in the COMEDD at the Fraunhofer IPMS in Dresden, are realized with Korean coating plants due to the advantages of Asian plant technology.[69]

Taiwan SWOT Analysis

In 2006, Taiwan's research organization responsible for developing its flexible electronics industry released a SWOT analysis for Taiwan in flexible electronics. Although this analysis by the Industrial Technology Research Institute (ITRI) is now 7 years old, most of the factors cited remain pertinent today. The analysis also applies, at least to a degree, to South Korea. (See Table 4-11.)

American SWOT Analyses

The members of the American WTEC team that visited Europe in 2010 to study European research efforts in flexible electronics noted that their European colleagues were "nearly unanimous" in their praise of leading U.S. research universities and their Ph.D. programs, and the close connections between university-based research groups and the venture capital community, enabling "a well-developed process for moving innovation out from the academic laboratory." European scientists saw as U.S. strengths the practical knowledge in creating startup companies, a highly developed venture capital infrastructure, the ability to attract talent from everywhere, and strong support from federal agencies such as the National Science Foundation, the Office of Naval Research, and the Departments of Defense and Energy.[70] However, the WTEC team observed that

[69] acatech, *Organic Electronics in Germany,* 23–24.

[70] The 2011 organic electronics study by Germany's National Academy of Science and Engineering faulted the German system of government research funding relative to that of the United States: "Foreign funding like the U.S. funding by the Department of Energy (DoE) is considered more dynamic and flexible in general by the experts." acatech, *Organic Electronics in Germany,* 41.

TABLE 4-11 ITRI SWOT Analysis for Flexible Electronics Technology Development in Taiwan

Strengths	Weaknesses
• Design of critical electronic components • High-quality professional personnel in the semiconductor and display industries • Mature manufacturing technology and cost-down capability of end-products • Rapid and mature vertical integration • Complete semiconductor and display industry supply chain • Rapid and flexible business models • Tax breaks for new companies	• Innovations in product, application, and business models • Intellectual property (IP) strategy and legal system • R&D in critical raw materials • Ability to integrate leading-edge technologies and products
Opportunities	**Threats**
• Excellent potential in flexible electronics • Potential for two decades of market growth • Patent deployment for critical technologies already in the early stages • High added value for integrating multi-discipline industries • Elevation to traditional industries to increase added value	• Progressive movement in related technologies and product deployments in Europe, the United States, and Japan • Leading positions in technology development and patent deployment in developed countries • Risks in circuit-printing technology • Investors remain hesitant about making long-term investments in emerging technologies

SOURCE: Electronics and Optelectronics Research Laboratory, ITRI.

it was troubling that few of the groups the panel visited actually considered the United States as a threat in any sense, reserving that for Asian countries, particularly Korea and Japan, and organizations and companies based in these countries.[71]

In 2010, NorTec, a highly regarded public economic and innovation development organization based in Ohio, conducted a SWOT analysis in connection with the creation of a strategic roadmap for a flexible electronics innovation cluster in northeast Ohio, FlexMatters. Although the SWOT was concerned with the particular cluster, not the United States as a whole, the findings are arguably applicable to a considerable degree to an assessment of the American competitive position.[72]

[71] WTEC, *European Research and Development,* xvi.

[72] The SWOT was conducted with the assistance of Daniel Gamota, Chair of the International Electronics Manufacturing Initiative (iNEMI) Large Area Flexible Electronics Roadmap. Gamota, a leading expert in the field, was part of the WTEC team that surveyed European flexible electronics research efforts in 2009-2010. The SWOT analysis was based on online surveys followed up by one-

The NorTech SWOT analysis analyzed the three principal groups that comprised the FlexMatters cluster, small companies, large companies, and academia. The matrix in Table 4-12 is an abbreviated summary of the main findings.

The NorTech assessment noted the existence of a strong local university research base and innovative large and small companies capable of capitalizing on the opportunities offered by flexible electronics, but struggling with workforce issues and worried about competition from Asia and uncertainties with respect to regulation and availability of funding. Consistent with this assessment, the WTEC team that surveyed European flexible electronics laboratories in 2010 summed up the challenge facing the United States in this field as follows:

> What the panel discerned from this study is that the relatively low prevalence of actual manufacturing and advanced systems research and development in the United States has led to an incomplete hybrid flexible electronics R&D scenario for this country: it is strong in basic research and in innovation but weak in advanced development for manufacturing, mirroring trends in some other sectors as well. Although the United States may be doing what it does best, manufacturing is moving to regions of the world that provide greater investment and commitment to product development. It then becomes questionable as to whether this approach is a healthy one and can be sustained in the long term.[73]

---

on-one interviews in which interviewees discussed "critical issues that affect their business growth and sustainability." NorTech, *FlexMatters Strategic Roadmap,* November 2010.

[73] WTEC, *European Research and Development,* xvi.

TABLE 4-12 NorTech SWOT Analysis—Northeast Ohio Flexible Electronics Cluster

Strengths	Weaknesses
Academia	Academia
• Fundamental discoveries	• New business methodologies
• Relevant technologies—materials, devices, design	• Promotion of technical leadership
• Ability to secure funding	Big Companies
• Patents and know-how	• Workforce issues
• Talent	• Talent pipeline from universities
Big Companies	• Market intelligence
• Innovation methodology	Small Companies
• Product portfolios	• Funding
• IP	• Patent infringement
• System development/design layout	• Technology intelligence
• Global footprint	• Partnering and teaming opportunities
• Large customers, robust sales channels	• Market intelligence
Small Companies	
• Innovative products	
• Manufacturing competency	
• IP strategy	
Opportunities	**Threats**
Academia	Academia
• New technologies	• Loss of first mover advantage
• Industry alliances, hub of cluster	• Sustainability of funding long term
Big Companies	Big Companies
• New products/systems	• Global competition
• New manufacturing strategy	• Aggressive Asian patent filing
• Product line extensions	• Funding uncertainty
Small Companies	• Disruptive technology introduction
• New product components/devices	• New legislation/standards
• New processes (R2R)	Small Companies
• Novel manufacturing platforms	• Asian dominance of low-cost manufacturing
	• Lack of funding
	• Technology shift/loss of customers
	• Extinction

SOURCE: NorTech, *FlexMatters Strategic Roadmap,* November 2010.

5

European Initiatives

The European Union (EU) is engaged in the world's most comprehensive and multifaceted developmental effort in flexible electronics, engaging multiple levels of government and dozens of public research institutes. The European effort has been under way since the late 1990s and has given rise to a number of fledgling organic electronics innovation clusters across the European continent. The European effort is characterized by an extremely broad range of research projects sponsored at the EU level, reinforced by very substantial national and regional efforts in the United Kingdom (UK), Germany, the Netherlands, Belgium, and Finland.

EUROPEAN UNION PROGRAMS

The EU is implementing a number of initiatives, including significant levels of EU funding, to support the development of organic and large area electronics (OLAE) in the EU, with an emphasis on integration of national and regional programs at the European level. Funding levels for the period 2007-2013 exceed $150 million.

Research policy in the EU is characterized by a tension between EU authorities and national and regional governments. At the EU level, policy makers take the perspective that national and regional research promotion efforts are duplicative and insufficiently integrated, and EU policies are designed to coordinate national and regional research efforts and to promote transborder research cooperation between the member states. "Member States, on the other hand, demonstrate considerable resistance towards EU level policy intrusion."[1] The EU launched the PolyMap project in 2008, a survey of public funding for OLAE research at the national and subnational level, which led off with the presumption that although OLAE research was being funded throughout the EU, "there seems to be a lot of duplication and 'catch-up' research" with respect to national budget spending levels estimated at €300-500 million.[2]

FET Flagship Projects

The European Union's Future and Emerging Technology (FET) Flagship projects are large-scale science-driven research efforts aimed at achieving visionary goals. One of the first FET Flagship projects announced was the Graphene Flagship, a 10-year, €1 billion effort to develop applications and manufacturing processes for graphenes. Envisioned applications include "fast, flexible and strong consumer electronics such as electronic paper and bendable personal communications devices."[3] From a start point in 2013 and with an initial 30-month budget of €54 million, a consortium is being formed of 126 academic and industrial research groups in 17 European countries. The Graphene Flagship will be coordinated by Chalmers University of Technology in Gothenburg, Sweden. During the 30-month ramp-up phase the Flagship will focus on energy technology and sensors with applications in the communications and transport sectors.[4] The ramp-up phase will be followed by a steady-state phase under Horizon 2020, the successor program to the Seventh Framework, with anticipated EU expenditures of €50 million per year beginning in 2016. Paralleling the Graphene Flagship effort, the member states and "associated nations" will coordinate national funding efforts in grapheme through an ERA-NET, a scheme established under the Sixth Framework in 2002 to minimize duplicative effort and facilitate transnational research coordination.[5]

EU Framework Programmes

The European Union Framework Programmes for Research and Technological Development provide EU funding to support research within the EU. The

---

[1] Merli Tamtik, "Rethinking the Open Method of Coordination: Mutual Learning Initiatives Shaping the European Research Enterprise," *Review of European and Russian Affairs* 7, no. 2 (2012), 2. See also EU Commission, Capacities Part 6: Support for the Coherent Development of Research Programmes, Work Programme 2013, C (2012) 4526 as of July 9, 2012.

[2] Herman Schoo, "Introduction of PolyMap," June 14, 2010.

[3] European Commission, "FET Flagships: Frequently Asked Questions," Press Release, January 18, 2013, Brussels.

[4] "Graphene Appointed EU Future Emerging Technology Flagship," Press Release, January 28, 2013, *<http://www.graphene-flagship.eu/GFfiles/130124_PresseText_A4.pdf>*.

[5] "Graphene Flagship Sets Sail at Chalmers University of Technology," *Printed Electronics Now,* October 11, 2013; EU Commission, *Networking the European Research Area: Coordination of National Programmes, <http://cordis.europa.eu/coordination/era-net.htm>*.

Seventh Framework Programme, which began in 2007, emphasizes collaborative transnational research in industry-academia consortia, and included a "European Investor Gate" (EIG) program to address existing shortfalls in early-stage investment and to enable startups to bridge the "valley of death."[6]

During the first 4 years of the Seventh Framework (2007-2013) the EU reportedly financed research and development (R&D) projects in the field of OLAE at a level of more than €120 million.[7] Europe's leading firms and public research organizations in the sector typically participate in multiple Seventh Framework consortia focusing on flexible electronics themes.

A common criticism of the EU R&D programs is that they involve too much bureaucracy and are difficult for small and medium enterprises (SMEs) to access. A 2011 analysis of the prospects for European entrepreneurship in OLAE by the Finnish research organization VTT observed that the EU-funded research programs

> are designed from the point of view of large research institutes and companies. Planning and administration takes time and energy, participants' interests are diverse, projects are long, and the distance between project objectives and market needs may grow wide. R&D grant, gained through a tedious application process to fund a particular technology development project, is not a very flexible instrument in the SME context. Moreover, they cannot be used to finance investments or market activities.[8]

In early 2014, the EU announced the €80 billion "Horizon 2020" program as the successor umbrella R&D effort succeeding the Seventh Framework Programme. Although Horizon 2020 will support basic research "because we can never be sure where it may lead us or what the applications could be," "more money will be available for testing, prototyping, demonstration and pilot-type activities, for business-driven R&D, for promoting entrepreneurship and risk-taking, and for shaping demand for innovative products and services."[9]

OLAE Projects

The Seventh Framework flexible electronics projects address a broad variety of themes in the field, including organic lighting systems, manufacturing technology, applications, organic photovoltaics, and systems integration. (See Table 5-1.) Most project budgets are in the range of €4-15 million, with the EU normally

---

[6] "European Investor Gate," <*http://www.startupeuropehub.eu/index.phpl/about/projects/32-eig-european-investor-gate-project*>.

[7] The FP7-ICT Coordination Action OPERA, *An Overview of OLAE Innovation Clusters and Competence Centres,* September 2011, 8.

[8] VTT, *Promoting Entrepreneurship* (2011) op. cit., 32.

[9] Marie Geoghegan-Quinn, EU Commissioner for Research, Innovation and Science, *Launch of Horizon 2020 in Greece,* January 10, 2014; "EU Cash on the Horizon for Innovators," *The Independent,* January 24, 2014.

TABLE 5-1 EU Seventh Framework OLAE Projects

Acronym	Theme	Timeframe	Budget (Millions of Euros)	EU Funding (Millions of Euros)
FLEXIBILITY	Flexible multifunctional integrated ultra-thin systems	2011-2015	6.9	4.9
ROTROT	Roll-to-roll production of organic tandem cells	2010-2012	4.6	3.1
ORICLA	RFID tags based on organic thin-film technology	2010-2012	4.7	3.0
OLAE+	OLAE	2011-2016	18.4	6.0
SCOOP	OLED microdisplay with enhanced brightness/color performance	2011-2014	5.0	3.5
INNOPRIO 21	Photonics innovation/ implementation strategy	2011-2014	2.3	2.0
COLAE	OLAE commercialization clusters	2011-2014	5.1	3.8
POLARIC	Printable, organic large area integrated circuits	2010-2014	13.8	9.9

SOURCE: CORDIS.

supplying well over half the amount. A number of Seventh Framework–funded projects are designed to promote European development of "organic large area electronics," a term that is largely synonymous with "flexible electronics." The Community Research and Development Information Service (CORDIS), which is the European Commission's information service, states the goal of the EU's OLAE projects is as follows:

> Our mission is to contribute to Europe's leading position in this disruptive technology through the support of R&D activities, crucial for the development and progress in organic and large area electronics. Today, the issues of particular interest for the R&D community in this area are reliability, stability, device performance and device architecture, together with heterogeneous integration. The current stage of development of organic and large area electronics emphasizes the importance of the creation of a European critical mass in the area and overcoming national fragmentation. The ultimate aim is to establish a pan European fertile research substrate and succeed in converting European R&D leadership into innovation and socio-economic growth.[10]

The OLAE project is a call for proposals for transnational R&D pursuant to the Commission's ERA-NET Plus scheme, an initiative to improve the

[10] CORDIS, Organic and Large Area Electronics website, <*http://cordis.europa.eu/fp7/ict/organic-elec-visual-display/*>.

TABLE 5-2 EU Seventh Framework OLED Projects

Acronym	Theme	Timeframe	Budget (Millions of Euros)	EU Funding (Millions of Euros)
FLEX-O-FAB	Pilot R2R, sheet-to-sheet manufacturing for flexible OLEDs	2012-2015	11.2	7.1
TREASORES	Transparent electrodes for large area large scale production of optoelectronic devices	2012-2015	14.0	9.1
OLED100.EU	OLED lighting in European dimensions	2008-2011	19.7	12.5
IMOLA	Light management for OLEDs on foil applications	2011-2014	5.1	3.4
FLAME	Flexible organic active matrix OLED displays	2008-2011	4.1	3.0
FAST2LIGHT	High throughput large area cost effective OLED production technologies	2008-2011	15.4	10.0

SOURCE: CORDIS.

coordination of national research projects with transnational funding by "topping up" with EU funds as an incentive for collaboration.[11]

Lighting

The Seventh Framework is also sponsoring projects to develop European capabilities in organic light-emitting diode (OLED) technology, which has extensive potential application in the field of flexible electronics in applications such as displays and lighting systems. One of the projects, Flex-o-Fab, is intended to enable commercialization of OLED lighting systems utilizing roll-to-roll (R2R) manufacturing processes by 2018. (See Table 5-2.)

In early 2013, the Holst Centre, a Netherlands-based, government-supported research institute, was designated the project director of an EU Seventh Framework project, "Flex-o-Fab," a 3-year effort to demonstrate a reliable manufacturing process for OLED lighting foils. The €11.2 million project is intended to enable market introduction of the manufacturing process within 3 years of the

[11] "The EU Launches a €18 Million Organic Large Area Electronics Funding Competition," *OLED-Info.com,* September 25, 2011; "ERA-NET Plus to Hold European Competition for Collaborative R&D Funding for Plastic Electronics," *Flexible Substrate,* September 2011.

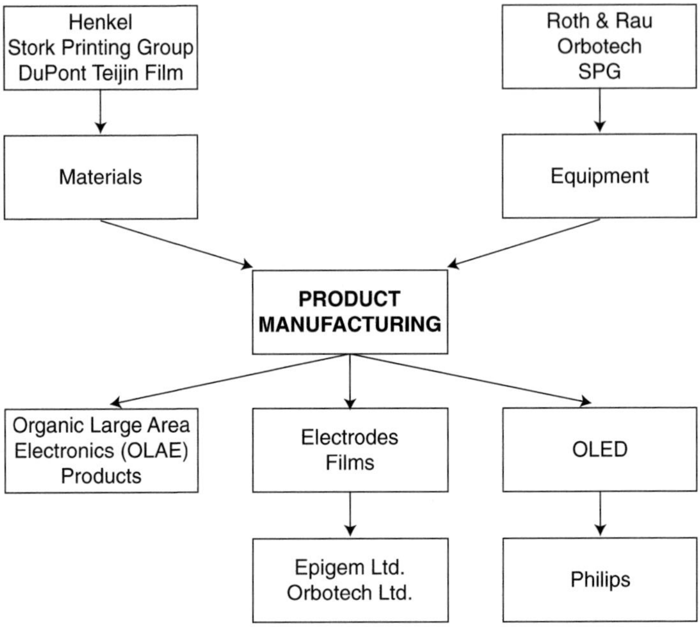

FIGURE 5-1 EU Flex-o-Fab project.
SOURCE: EU Seventh Framework, *Flex-o-Fab,* November 25, 2012.

end of the project in September 2015.[12] Flex-o-Fab will draw on technologies and skills currently used to produce glass-based OLEDs and flexible displays and will seek to migrate existing sheet-to-sheet production processes to roll-to-roll production, reducing costs, and permitting high-value production. Flex-o-Fab involves a collaboration with public and private entities, who are leaders in their respective fields of relevant expertise.[13] (See Figure 5-1.)

In strategic terms Flex-o-Fab builds on the strong competitive position of the European lighting industry and seeks to introduce disruptive new technologies that will create long-term European manufacturing jobs because of their high level of technical novelty and specialization. "[T]he intellectual property generated will protect these advances from Asian and U.S. competition."[14]

[12] "European Project Develops Flexible OLED Lighting Production Process," *Plastic Electronics,* February 8, 2013. EU funding will cover €7.1 million of the project cost. Project Overview (for CORDIS Fact Sheet), *Flex-o-Fab: Pilot-Scale Hybrid Roll-to-Roll/Sheet-to-Sheet Manufacturing Chain for Flexible OLEDs*, EU Seventh Framework, November 25, 2012.

[13] "EU Project to Take OLEDs from Lab to Fab," *Hearst Electronic Products,* January 20, 2013.

[14] Project Overview (for CORDIS Fact Sheet), *Flex-o-Fab*, 8.

TABLE 5-3 EU Seventh Framework OPV Projects

Acronym	Theme	Timeframe	Budget (Millions of Euros)	EU Funding (Millions of Euros)
HIFLEX	Development of indium-free OPV module for R2R processing	2010-2013	5.0	3.7
X10D	Development of tandem organic solar cells with conversion efficiency of 12%	2011-2014	11.9	8.6
DEFOTEX	Development of textile solar cells for flexible photovoltaic fabrics	2008-2011	4.2	3.1
THIME	Thin-film measurements on OPV layers for plastic, paper, and textile substrates	2012-2014	1.5	1.1
ESTABLIS	Ensuring stability in organic solar cells for strong, flexible, low-cost application	2012-2015	3.9	3.9
R2R-CIGS	CIGS (copper gallium selenide) deposition on flexible polymer films for manufacturing of flexible CIGS PV modules	2012-2015	9.6	7.0
LARGE CELLS	Development of large area thin-film solar cells based on polymers	2010-2014	2.2	1.6

SOURCE: CORDIS.

Photovoltaics

The Seventh Framework is supporting a number of joint R&D projects in the area of organic photovoltaics (OPV), generally providing the bulk of the funding for each project. (See Table 5-3.)

Materials

Finally, the Seventh Framework is funding a variety of research projects involving materials science, component, industrial processes, and applications directly relevant to the field of flexible electronics. (See Table 5-4.)

OLAE Innovation Clusters

European innovation policy emphasizes the promotion of innovation clusters at both the national and the EU level.[15] National and regional governments in

[15] The European Commission defines "cluster" as "a group of firms, related economic actors, and institutions that are located near each other and have reached a sufficient scale to develop specialized

TABLE 5-4 EU Seventh Framework Flexible Materials/Processes/Applications Projects

Acronym	Theme	Timeframe	Budget (Millions of Euros)	EU Funding (Millions of Euros)
CONTEST	Training in electronic skin technology	2012-2016	3.8	3.8
OLATRONICS	Integration of processes for producing OLAE	2008-2011	6.2	3.8
MOMA	Embedded organic memory arrays	2010-2012	4.7	3.0
LOTUS	Low-cost highly conductive structures for flexible large area electronics	2010-2012	5.5	3.7
APOSTILLE	Reinforcement of research potential in nano and organic/flexible/printed electronics	2010-2013	1.2	1.1
MOWSES	Nanoelectronics based on 2-dimensional dichalconentides for flexible electronics	2013-2017	3.7	3.7
PASTA	Integration of electronics and textiles	2010-2014	8.9	6.5
CLEAN4YIELD	Nanoscale detection and inspection techniques	2012-2015	7.1	10.6
INTERFLEX	Interconnection systems for flexible foil systems with electronic components	2010-2013	5.3	3.5
PLASTRONICSPEC	Automated inspection system for printed electronics	2011-2013	1.4	1.1
ORICLA	RFID tags made at low temperatures on thin films	2010-2012		3.0

SOURCE: CORDIS, <*http://www.clean4yield.edu*>.

Germany, France, the UK, and Finland, among others, have implemented policies to promote innovation clusters in various thematic areas. The EU sees its role as to "facilitate and add to such efforts, notably by improving the framework conditions, promoting research and education excellence and entrepreneurship,

expertise, services, resources, suppliers, and skills." EU Commission, *Towards World-Class Clusters in the European Union: Implementing the Broad-Based Innovation Strategy* (Communication from the Commission to the Council, COM, 2008) 652 final/2/, 2.

fostering better linkages between industry (especially SMEs) and research, and encouraging mutual policy learning and cluster cooperation across Europe."[16] In May 2013, the European Commission announced that it would launch a campaign to mobilize €10 billion in private, regional, national, and EU funds for investment in micro- and nanoelectronics, which would reinforce Europe's three world-class electronics clusters—Dresden, Eindhoven, and Grenoble.[17]

Coordinating European Efforts in Flexible Electronics

The EU has undertaken several initiatives to promote cooperation and coordination within Europe with respect to various technologies relevant to the field of flexible electronics.

- ***PolyMap.*** PolyMap was a small (€600,000) project undertaken in 2008-2011 to map public funding for organic electronics within the EU and set up an open database with respect to materials, devices, standardization, and applications.
- ***OPERA.*** The Organic/Plastic Electronics Research Alliance (OPERA) was established by the EU to create the conditions for competitiveness clusters in organic and plastic electronics within the EU.
- ***PRODI.*** The Coordination Action "Manufacturing and Production Equipment and Systems for Polymer and Printed Electronics" (PRODI) was a 2008-2010 EU project to integrate European manufacturing systems for R2R polymer and printed electronics.
- ***Photonics21.*** The European Technology Platform Photonics21 convenes relevant stakeholders in photonics, including topical interest areas of OLED and OLAE.

UNITED KINGDOM

The British commitment to developing a national capability in flexible electronics has been sustained and extensive, involving numerous public and private sector actors and more than 20 universities. Many current activities in the field of flexible electronics trace their origins to groundbreaking research in the 1980s at the Cavendish Laboratory at the University of Cambridge. Researchers there demonstrated that conjugated polymer diodes could emit light when electrically stressed, a discovery that suggested that organic semiconductors could be made into flexible large area displays and led to the spinoff company Cambridge

---

[16] Ibid.

[17] "Commission Proposes New European Industrial Strategy for Electronics—Better Targeted Support to Mobilise Euros 10 Billion in New Private Investments," *Targeted News Service,* May 25, 2013.

Display Technology in 1992 to pursue this concept.[18] A famous 1990 paper in which the researchers publicized their findings today remains the most widely cited in the field of organic microelectronics.[19] The British government provided research funding support through the 1990s and 2000s, enabling the creation of plastic electronics centers of excellence.[20] A pioneering British firm, Plastic Logic, established the world's first plant for mass producing plastic semiconductor devices in 2008.[21] British universities have launched an array of promising startups in the field.[22] At present, the competencies of British universities in plastic electronics (the term commonly used for flexible electronics in the UK) are broad and deep. (See Table 5-5.)

In 2007, Britain's Council for Science and Technology identified plastic electronics as a "high risk/high reward" priority technology for the UK.[23] In 2008, the government's Economic and Social Research Council sponsored a study of the potential for plastic electronics in the UK that concluded that "the UK is well positioned to become a global leader in the innovative development of high functionality products that incorporate plastic electronics."[24] A "Capability Guide" sponsored by PELG in 2008 demonstrated the existence of an extensive ecosystem of university programs, research organizations, and companies within the UK capable of supporting the commercialization of plastic electronics technologies.[25]

If anything, British policy makers' interest in plastic electronics intensified after 2007. The field of plastic electronics was used as a case study in 2008-2009 by the House of Commons Innovation, Universities, Science and Skills Committee review of engineering in the UK.[26] In 2009 the Department for Business Innovation & Skills released a "UK Strategy for Success" in plastics electronics, developed by an industry-led group that recommended actions to facilitate and coordinate investment, development of demonstrators and other mechanisms to showcase the commercial potential of plastic electronics, and workforce and

[18] J.H. Burroughes, D.D.C. Bradley, et al., "Light-Emitting Diodes Based on Conjugated Polymers," *Nature,* 1990.

[19] PELG and ESP KTN, *UK Plastic Electronics Capability Guide,* 2012, 8.

[20] Logystyx UK, Ltd., *Capability Guide: UK Plastic Electronics,* 2008, 8–9.

[21] "First Plastic Chip Plant to Open in Germany," *Chosun Ilbo,* January 4, 2007.

[22] In addition to Plastic Logic and Cambridge Display Technology, both spun out of the University of Cambridge, other startups included MicroEmissive Displays (1999, University of Edinburgh), OLED-T, (1999, South Bank University), Molecular Vision (2001, Imperial College, London), Nano e-Print (2006, University of Manchester), and Lumicure (St. Andrews University). House of Commons, Innovation, Universities, Science and Skills Committee, *Engineering: Turning Ideas into Reality,* vol. 1, March 18, 2009, 43.

[23] Council for Science and Technology, *Strategic Decision Making for Technology Policy: A Report by the Council for Science and Technology,* 2007.

[24] Zella King, AIM Research, *Plastic Electronics: Putting the UK at the Forefront of a New Technological Revolution,* 2007, 33.

[25] Logystyx UK, Ltd., *Capability Guide: UK Plastic Electronics,* 2008.

[26] *<http://www.publications.parliament.uk/pa/cm20090/cmselect/cmdius/50/5002.htm>*.

TABLE 5-5 Plastic Electronics Competencies—British Universities

	Organic materials	Inorganic materials	Devices	Processes	Characterization & test	Equipment	Commercialization
University of Bangor, Plastic Electronics Research Centre	x		x	x	x		
University of Birmingham, School of Metallurgy and Materials				x	x		
Brunel University, Cleaner Electronics Research Group		x	x	x			
Brunel University, Centre for Phosphors and Display Materials		x		x	x		
Brunel University, Organic Electronic Group	x		x				
University of Cambridge, Cambridge Integrated Knowledge Centre			x	x		x	
University of Cambridge, Cavendish Laboratory	x	x	x	x			
University of Cambridge, Electronic Devices and Materials Group		x	x	x	x		
University of Cambridge, Inkjet Research Centre				x			
Cranfield University, Precision Engineering						x	
Durham University, Organic Electroactive Materials	x		x		x		
University of Hull, Organophotonics	x	x	x				
Imperial College, Centre for Plastic Electronics	x	x	x	x			x
Loughborough, Innovative Electronics Manufacturing Research Centre	x		x	x		x	
University of Manchester, Organic Materials Innovation Centre	x	x	x				
METRC, Centre for Nanotechnology	x		x				
Nottingham Trent University		x	x	x	x		
University of Oxford, Department of Materials	x	x		x			x
Queen Mary London, Centre for Condensed Matter & Materials Research		x	x		x		
University of Sheffield, Biomaterials Science & Tissue Engineering Group			x	x			
University of Sheffield, Electronic and Photonic Molecular Materials Group	x		x				
University of St. Andrews, Organic Semiconductor Centre	x	x			x		
University of Strathclyde, Department of Pure and Applied Chemistry	x		x				
University of Surrey, Advanced Technology Institute	x	x	x	x	x		

continued

TABLE 5-5 Continued

	Organic materials	Inorganic materials	Devices	Processes	Characterization & test	Equipment	Commercialization
University of Swansea, Centre for Innovative Functional Industrial Coatings			x	x			
University of Swansea, Welsh Centre for Printing & Coating			x	x		x	x
University College, London Centre for Nanotechnology	x		x	x			
University of the West of Scotland, Thin Film Centre	x				x		
University of York, Liquid Crystal Group	x				x		

SOURCE: UK Plastic Electronics Leadership Group, <*http://www.ukplasticelectronics.com/downloads/*>.

training initiatives.[27] The Technology Strategy Board identified "plastic and printed electronics" as one of the "five pillars" of the UK strategy for electronics in 2008, and it has remained a priority through 2013.[28] In 2011, the EU's OPERA project, established to promote innovation clusters in OLAE observed that the only country in Europe in which national platforms for OLAE existed was Britain, "where national initiatives are coordinated by the Technology Strategy Board and complemented by regional cluster initiatives."[29]

Despite Britain's prowess in science, supportive government policies and propensity to spawn innovative startups, the outlook for its aspirations in plastic electronics manufacturing must be regarded as uncertain at best. Plastics Logic, one of the most promising British startups, established its first manufacturing operations in Germany rather than Britain, citing the locational advantages of Dresden over potential sites in the UK. Other university spinoff companies have gone bankrupt or have been acquired by foreign companies. British university research centers in the field are a major national asset but face challenges from abroad. A Parliamentary committee reported in 2009 that in a visit to Imperial College London, faculty members told the committee that

---

[27] Department for Business Innovation & Skills (BIS), *Plastic Electronics: A UK Strategy for Success—Realizing the UK Potential,* 2009.

[28] Innovate UK website, <*http://www.innovateuk.org/electronics-sensors-and -photonics*>.

[29] The FF7-ICT Coordination Action OPERA and the European Commission's DG INFSO Unit G5, "Photonics," *An Overview of OLAE Innovation Clusters and Competence Centres,* September 2011, 11.

> capital equipment used for plastic electronics research in UK university laboratories was not globally competitive. In particular, Swiss, US and German research groups were considered to be better provided for, and several researchers maintained collaborations with research groups in other EU countries such that their students could access state-of-the-art equipment.[30]

Thus despite an impressive record of innovation by British companies, it is not clear that flexible electronics technologies will transform domestic manufacturing or contribute a significant number of jobs in the UK. In August 2013, Chris Williams, the co-founding Director of the UK Displays and Lighting Knowledge Transfer Network, lamented the British failure to take domestically developed flexible electronics technologies and "put it all together to make an integrated system that is ready for commercial exploitation." He complained with respect to missed opportunities in OLED lighting systems:

> Once again, the UK Government and its partner agencies show it can mentor the scientific goose, and encourage it to research and develop its golden egg up to the point where it is almost ready to be laid, but then government funding is taken away and focused on newer areas that seem to be sexier and more soundbite worthy. It happened in the past with composite materials, and it is happening now with plastic electronics.[31]

Government and Public Support

Until very recently British Conservative and New Labor governments have demonstrated ambivalence toward "industrial policy," an attitude that has limited the government's financial support for British industry. Since the onset of the global financial crisis in 2008, the government has demonstrated a renewed interest in promoting innovation in manufacturing, although at present, government austerity measures inhibit financial support.[32] Against this background, and by the standards of recent British governments, the effort to develop a domestic flexible electronics manufacturing sector has enjoyed comparatively strong government backing for a sustained number of years. Government financial assistance has flowed through a number of institutional channels.

Department of Trade and Industry

The Department of Trade and Industry (DTI) was a government department responsible for regional economic development, innovation, science, and business growth. In 2007, DTI was reorganized into two departments, Innovation, Universities and Skills (DIUS) and Business, Enterprise and Regulatory Reform

---

[30] House of Commons, *Engineering*, 41.

[31] "Last Word: Missed Opportunities," *Flexible Substrate,* August 2013.

[32] HM Treasury, *Spending Review 2010* Cm 7941, October 2010, presented to Parliament by the Chancellor of the Exchequer by Command of Her Majesty.

(BERR). DIUS and BERR were recombined in 2009 to form the Department for Business, Innovation and Skills (BIS). Through 2009, DTI and its various successor entities provided £52 million in funding for collaborative R&D in themes relevant to plastic electronics.[33]

Technology Strategy Board

The Technology Strategy Board (TSB) is a nondepartmental public body that operates at arm's length from the government but is funded by the government through the BIS. Staffed largely by individuals with business experience, it directs government funds toward applied research in technologies in which the UK possesses world-class science and in which there has been "clear market failure" in commercial development of that science. The TSB invests about £350 million per year in research projects.[34] As of June 2008, the TSB had provided about £25 million in funding for various plastic electronics projects. Richard Price, founder of Nano e-Print, a maker of printed logic circuits that spun out of the University of Manchester in 2006, told Parliament in 2009 that the TSB's support had been "incredibly important":

> Firstly, it brings together consortia that would not necessarily have come together unless there was government support to share the risk. Secondly, it helps us in terms of our cash flow and enables us to further develop before we have to go back to the market for more investment. It also helps us build relationships with some of the knowledge transfer networks and to grow organically some of our networks within industry.[35]

Engineering and Physical Sciences Research Council

The UK's seven research councils support basic research and postgraduate training with funds from the government's science budget directed by the BIS. The Engineering and Physical Sciences Research Council (EPSRC) has historically been the principal source of public funds for plastic electronics research in the UK. In 2009, a Parliamentary committee reported that the EPSRC was spending about £68 million per year on research, training, and knowledge transfer activities of "direct relevance to the area of plastic electronics."[36] The EPSRC operates and supports a network of "Centres for Innovative Manufacturing" run by universities. EPSRC funding supports research projects and several of the UK's Plastic Electronics Centres of Excellence (PECOEs). (See Table 5-6.)

---

[33] BIS, *Plastic Electronics,* 10.

[34] Interview with David Way, Director of Knowledge Exchange and Special Projects, Technology Strategy Board, London, June 12, 2012. The TSB originated in 2004 as an advisory body within DTI. It was spun off as an independent public organization when DTI was broken up in 2007.

[35] House of Commons, *Engineering*, 34.

[36] Ibid., 33.

TABLE 5-6 EPSRC Funding of Flexible Electronics Research Themes

EPSRC Reference	Project	Timeframe	ERSRC Funding (Millions of Pound Sterling)
EP/K017144	Graphene Flexible Electronics and Optoelectronics: Bridging Gap Between Academia and Industry	2013-2018	6.9
EP/G060738/1	Heterointerface Control of Organic Semiconductor Devices	2009-2014	6.7
EP/K03099X/1	EPSRC Centre for Innovative Manufacturing of Large Area Electronics	2013-2018	5.6
EP/K017160/1	Innovation in Industrial Inkjet Technology	2010-2015	5.1
EP/K017160/1	New Manufacturable Approaches to Deposition and Patterning of Graphenes	2013-2016	1.1
EP/G037515/1	Doctoral Training Centre in Science and Application of Plastic Electronic Materials	2009-2018	7.3
EP/K01711X/1	Graphene Flexible Electronics and Optoelectronics	2013-2018	3.0

SOURCE: Engineering and Physical Sciences Research Council, <*http://www.epsrc.ac.uk*>.

Regional Development Agencies

The Regional Development Agencies (RDAs) were government-funded, business-based organizations aimed at fostering regional economic development including science and innovation. Until their abolition in 2012, a number of RDAs provided funding and other support to the development of printed and plastic electronics. One North East, Yorkshire Forward, and the Welsh Assembly government were particularly supportive of the sector.[37] The three northern RDAs jointly funded the Northern Way Innovation Programme, which provided £6.2 million in funding to plastic electronics in 2008-2011.[38]

Knowledge Transfer Network

Knowledge Transfer Networks (KTNs) are networks funded and run by the Technology Strategy Board to raise awareness of specific technology areas, facilitate research and information sharing, and link businesses in need of technology with technology sources. The Photonics and Plastic Electronics KTN (PPE-KTN) had worked to facilitate the TSB's programs relevant to plastic electronics, including sponsoring symposia and trade shows. The UK Displays and Lighting

[37] BIS, *Plastic Electronics,* 10.

[38] SQW Ltd., *The Evaluation of the Northern Way 2008-11: Final Report—Full Report,* April 2011, 74.

TABLE 5-7 UK Plastic Electronics Centres of Excellence

Centre	Location	Main University Affiliation	Sources of Public Funding
Center for Process Innovation (CPI)	Wilton	Durham Newcastle	One North East TSB
Printable Electronics Technology Centre (PETEC)	Sedgefield	Durham	One North East ERDF TSB
Welsh Centre for Printing and Coating (WCPC)	Swansea	Swansea	EPSRC TSB FP7
Organic Materials Innovation Centre (OMIC)	Manchester	Manchester	EPSRC
Cambridge Integrated Knowledge Centre	Cambridge	Cambridge	EPSRC

KTN (UKDL KTN), based at Bletchley Park, addresses plastic electronics themes relevant to applications in displays and lighting systems.[39]

Plastic Electronics Leadership Group

The Plastic Electronics Leadership Group (PELG) is a volunteer group representing the UK plastic electronics community. Its members include representatives of the TSB, BIS, the EPSRC, the PECOEs, the Electronics, Sensors and Photonics KTN, universities, and companies such as Plastic Logic, 3M, Thorn Lighting, Cambridge Display Technology, and Hewlett-Packard (chair).

Centres of Excellence

The UK has five Plastic Electronics Centres of Excellence (PECOEs), each of which is affiliated with one or more major research universities and supported by public funding, usually from multiple sources. (See Table 5-7.) These centres, which are the foci of emerging innovation clusters in plastic electronics, represent the centerpieces of British developmental efforts in this field. In 2008, the PECOEs entered into a Memorandum of Understanding to "provide a focused cluster for technology development and prototyping, aiming to translate UK strengths into industries of the future."[40] The PECOEs have agreed to create national open access

[39] Chris Williams, "Introduction to the UK Displays & Lighting KTN," <*http://www.lboro.ac.uk/research/iemrc/documents/Community/Meeting/UK_Displays_And_Lighting.pdf*>.

[40] University of Cambridge, Cambridge Innovation and Knowledge Center, accessed February 11, 2014, <*http://www-cikc.eng.cam.ac.uk/category/networks/*>.

research facilities, develop a shared vision and competency map, and develop common methods of working and common terms and conditions.

Centre for Process Innovation Ltd.

The Centre for Process Innovation (CPI) is an independent nonprofit applied research organization located in northeast England conducting research in the fields of advanced manufacturing, printed electronics, industrial biotechnology, and high temperatures. It has an asset base of £55 million, largely derived from public sources, and more than 120 employees. CPI was established in 2004 patterned on Germany's Fraunhofer Gesellschaft, which was a member of CSI's pre-incorporation board and assisted in setting up the organization. CPI's initial funding was provided by One North East, the regional development organization for northeast England, to address the region's "innovation gap" in industrial processing. CPI has been incorporated in the UK's High Value Manufacturing Catapult Centre (HVM), an indication that the TSB regards it as one of the preeminent applied research organizations in the United Kingdom.[41]

CPI's original mandate was translational, to move university research into the market through university-industry collaborations. With respect to regional universities, Newcastle was extremely strong in chemical engineering but less so in chemistry, the region's main strength in that field residing in Durham University. One North East provided funds to upgrade the regional university research infrastructure, including facilities in Newcastle's School of Electrical, Electronic and Computer Engineering, which have supported CPI's subsequent work in plastic electronics. CPI's academic partnerships have expanded to include Cambridge, Imperial College London, York, Manchester, and Liverpool.[42]

CPI has partnered with industry to develop new materials with applications in flexible electronics. In May 2013, CPI disclosed that a joint project with DuPont Teijin Films had developed a new form of polymer barrier film for lightweight OLED lighting and photovoltaic cells that could be applied to engineered polymer substrates replacing glass.[43] The same month CPI disclosed a project involving SMEs to develop high-performing organic thin-film transistor materials for application in flexible OLED displays, an effort that had already led to product development utilizing the new materials.[44] In June 2013,

---

[41] Written Evidence Submitted by the Centre for Process Innovation Ltd. (CPI) (TIC 28), House of Commons Science and Technology Committee, *Technology and Innovation Centres: Second Report of Session 2010-11* (February 9, 2011), 103.

[42] John Goddard, Douglas Robertson, and Paul Vallance, "Universities, Technology and Innovation Centres and Regional Development: The Case of the North-East of England," *Cambridge Journal of Economics,* 2012, 9.

[43] "CPI & DuPont Teijin Films Work Together to Develop a Barrier Film as Plastic Substrate for OLED Lighting and Photovoltaic Devices," May 28, 2013, <*http://www.uk-cpi.com*>.

[44] "CPI Works with SMEs to Develop High Performing Organic Thin Film Transistor," May 18, 2013, <*http://www.uk-cpi.com*>.

CPI reported that its researchers had "produced defect-free OLED lighting demonstrators with emissive areas larger than 250 square centimeters."[45] With respect to process R&D, with funding from the TSB, CPI is working with the British firm Peratech to determine whether commercial printing machines can produce printable electronics, including the "QTC" sensors originally developed by Peratech.[46] In a TSB-funded collaboration with Plastic Logic, "project ROBOLED," CPI is addressing challenges associated with OLED displays utilizing organic thin-film transistors (OTFTs).[47] In 2013, CPI reported that pursuant to ROBOLED it had developed a new backplane fabrication process enabling the bending of OTFT arrays to radii as small as 1 mm without a significant reduction in device performance.[48] CPI is also investigating a number of possible methods for device encapsulation in collaboration with adhesive, getter, and barrier film suppliers.[49]

Printable Electronics Technology Centre

The Printable Electronics Technology Centre (PETEC) was established by CPI in 2008 with a joint investment of £6.3 million by One North East and the County of Durham Economic Partnership. The European Regional Development Fund (ERDF) contributed an additional £3.8 million for capital investment and the TSB provided £2.1 million for equipment.[50] PETEC was intended to function as an incubator in which experimental printed electronics processes were transformed into manufacturable products, enabling the UK eventually to capture 5 percent of the world market for printed electronics products. Targeted technology areas included "advanced material deposition processes, printable electronic materials, printable circuits for high resolution display and smart packaging applications, solid state lighting and organic photovoltaics."[51] PETEC connects research with commercial activity through use of proof-of-concept devices and

---

[45] "CPI Produces Large Area Small Molecule and Polymer OLED Lighting Demonstrators," *OSA Direct,* June 3, 2013.

[46] Quantum tunneling composite (QTC) materials change from insulating to conducting materials when placed under physical pressure. Among other things, QTCs may be incorporated into textiles to detect the presence of volatile organic compounds (VOCs), helping to monitor the wearer for signs of illness, fatigue or exposure to dangerous substances. "Peratech Works on New Ink Formulations with CPI to Enable QTC Sensors to be Printed," *Flexible Substrate,* May 2013.

[47] The TSB is funding ROBOLED ("Robust OLED"), a 30-month effort running from 2012 to 2015, with £493,792. <*http://tsb-projects.labs.theodi.org/projects?page=562*>.

[48] "CPI Presents Ultra-Flexible OTFT Device Array Suitable for Foldable AMOLED Displays," *Flexible Substrate,* January 2014.

[49] "CPI Manufacturers Flexible OLED Lighting Demonstrators," *Flexible Substrate,* March 2014.

[50] CPI submitted a report to Parliament in 2012 indicating that of the total £17 million CPI had received from the ERDF since its inception, nearly all of it (£14 million) was used on projects linked to PETEC. Written Evidence Submitted by the Centre for Process Innovation, House of Commons Communities and Local Government Committee, April 2012.

[51] CPI Written Evidence, February 9, 2011, para 3.11.

pilot-scale manufacturing, working with business clients to identify materials, processes, and investments necessary to bring products to market.[52] PETEC's prototyping and printed electronics facilities are located in North East Technology Park (NETPark) in Sedgefield, County Durham, at the time of PETEC's establishment one of the fastest growing science parks in the UK.[53]

The PETEC facility originally featured a 600 square meter class 1000 clean room, which was augmented in 2011 with a new building with a class 100 clean room. PETEC has a fully equipped formulations laboratory for developing organic polymers and an electrical testing laboratory. To support development of OLED and OPV technology in the UK, PETEC installed a £4.5 million fully automated batch production system in a new class 100 clean room, a line known as the LACE line ("large area coating equipment)." LACE offers industrial users the ability to run substrates ranging in size for 4 to 8 inches in a cassette-to-cassette operation featuring full automation and data logging. These systems minimize manual intervention and maximize product yields.[54]

PETEC acquired LACE pursuant to a 2010 initiative by the then Universities and Science Minister, David Willetts, to invest £8.4 million in the field of plastic electronics. This involved two projects in which PETEC participated:

- ***Thin Organic Prototypes, Design, Research, Applications with Enduser Recognition (TOPDRAWER).*** This project, led by Thorn Lighting, involved Durham University, Pilkington, Conductive Inkjet Technologies, Tridanic, and Cambridge Display Technology in an effort to demonstrate the ability to manufacture a printed lighting panel usable in aesthetic designs. The LACE line was installed at PETEC to prove and test the manufacturing process developed by the project participants. The project was intended to help build a comprehensive

[52] "PETEC: Pushing PE into the Supply Chain," *Printed Electronics Now,* August 2008. "A company or university will call on PETEC after it has demonstrated technology on a small scale in the laboratory. The facilities allow them to develop a scalable economic process to manufacture prototype products and develop a cost-effective manufacturing route. To construct a custom line for a business, PETEC offers a large selection of equipment, which, in the most part, is off the shelf, but some that are unique and that do not exist at that particular scale elsewhere in the UK. These can be put together 'Lego style' in different orders to suite the product being made, allowing the client to demonstrate it on a meaningful scale, to gather the associated manufacturing cost and efficiency data needed to design a manufacturing process and to make the case for further investment." "Plastic Surgery," *The Engineer,* April 20, 2009.

[53] "NETPark Invests in Future Success," *Newcastle The Journal,* November 2008. The NETPark Research Institute provides office and laboratory space for new companies and university spinoffs. The Incubator Business Support Centre provides mentoring and support in commercialization of R&D, IPR, international business development, business planning, manufacturing and market research, "Science Park Looks Forward to a Bright Future," *Newcastle The Journal,* January 29, 2009.

[54] Mike Claussen, Bela Green, Martin Walkinshaw, and Simon Ogier, "Introduction to the PETEC Printed Electronics Centre and Technology Challenges," CS Mantech Conference, May 16-19, 2011, Palm Springs, California.

UK polymer light-emitting diode supply chain and a UK capability in novel manufacturing.[55]

- ***Manufacture of Really Reflective Information Surfaces (Morris).*** The "Morris" project was a 3-year effort involving a collaboration by PETEC, Hewlett-Packard, and Timsons to develop large reflective information surfaces with applications in posters, signage, electronic wallpaper, electronic whiteboards, and command/control rooms. The project aim was to develop specifications for a pilot line and material set, estimated costs and yields, demonstration devices, components, processes and equipment, and to use these elements to secure investments in pilot and eventually full manufacturing.[56]

PETEC's business model was criticized in 2009 by Plastic Logic, one of the leading flexible electronics firms in the UK. Plastic Logic commented that PETEC's model was revenue-driven with a major focus on contract research for "a small number of giant Asian electronics companies" and that PETEC had "struggled to define and articulate a compelling vision of how it will benefit the UK plastic electronics community as a whole." PETEC responded that its funding arrangements required it to become financially self-sustaining within 5 years and that economic activity in the UK plastic electronics sector was not sufficient to support the Centre. In addition, in order to qualify for public grants in research competitions, PETEC was required to line up matching private funding, which was difficult or impossible without overseas participation[57]

The Cambridge Cluster

The area around Cambridge University, known as "silicon fen" for its concentration of high-technology companies, has been the locus of substantial research and entrepreneurial activity in the field of flexible electronics.[58]

The University of Cambridge

The University of Cambridge's Cavendish Laboratory has been responsible for some of the world's pioneering research in the field of organic electronics. In the late 1980s Cavendish's Optoelectronics Group discovered that semiconducting

[55] Technology Strategy Board, "Minister Announces £8.4m Investment in Plastic Electronics Technologies," Press Release, July 1, 2010. "Bright Future for UK's First Printed Lighting Panel," *Electronic Product Design & Test,* August 16, 2010.

[56] Technology Strategy Board, "Minister Announces £8.4m Investment."

[57] House of Commons, *Engineering*, 39.

[58] Roughly 1,400 technology-oriented companies operate in and around Cambridge, employing about 43,000 people. "Autonomy Fails to Dent Cambridge's Tech Status," *Financial Times*, November 26, 2012.

TABLE 5-8 Cambridge Spinoffs

Company	Technology Focus
Plastic Logic	Organic transistors for flexible displays
Cambridge Display Technology	Polymer LEDs for emissive full-color displays
Eight 19	Roll-to-roll manufacturing of polymer-based solar cells

conjugated polymers behave in a manner similar to that of inorganic semiconductors and can be used in semiconducting devices such as field-effect transistors, solar cells, and light-emitting diodes. The group has launched three spinoffs to commercialize these discoveries.[59] (See Table 5-8.)

The University's Centre for Advanced Photonics and Electronics (CAPE) was formed in 2004 by the University's electrical engineering department and four industrial partners.[60] At present, its research themes include OLEDs, organic active matrix technologies for plastic displays, and bright reflective color systems for electronic posters, signage, and print displays.[61] CAPE is entirely funded by its industrial partners.[62] In early 2013, the university launched the Cambridge Graphene Centre to develop the potentially revolutionary material with applications in flexible and transparent electronics.[63] In May 2013, a number of researchers at the Graphene Center launched a startup company, Cambridge Graphene Platform (CGP) to commercialize a scalable method of ink production from graphene and other layered nanomaterials with applications in printed and large area electronics.[64]

Cambridge Integrated Knowledge Centre

The Cambridge Integrated Knowledge Centre (CIKC) was established in 2008 at the University of Cambridge by the Cavendish Laboratory and CAPE to facilitate the commercialization of academic knowledge in the area of low-temperature thin-film processing using molecular materials (e.g., polymers, liquid crystals, nanostructures) for applications in photonics and electronics. Its expertise includes flexible photovoltaics, flexible electronics, printed electronics and

[59] University of Cambridge, "Optoelectronics," *<http://www.phy.com/ac.uk/research/oe/>*.

[60] CAPE's current industrial partners are Dow Corning, Jaguar Land Rover, and Disney.

[61] CAPE, "Devices and User Interfaces," *<http://www.cape.eng.com.ac.uk/technology/tfgs/devices-user-interfaces/>*.

[62] "Enhancing CAPE-abilities in Photonics and Electronics," *Research Horizons*, Autumn 2008.

[63] "Cambridge University Opening Graphene Centre to Take Material to the Next Level," *Engadget,* January 25, 2013. Cambridge has received £12 million from the EPSRC for graphene research. "The Graphene Revolution," *The Daily Telegraph,* June 8, 2013.

[64] "University of Cambridge Graphene Spin-out Funded," *Plastic Electronics,* May 1, 2013.

components, and transparent electronics and sensors.[65] The CIKC received financial support from the Engineering and Physical Sciences Research Council of £5.1 million during the period 2007-2012.[66] The EPSRC explained the rationale for its support of the launch of the CIKC as follows:

> This will bring together the main research activities in the field at Cambridge, namely in the Electrical Engineering Division (in particular within the Centre for Advanced Electronics and Photonics, CAPE) and the Cavendish. Together this research spans the MMM field and is recognized as having a world-leading position. A key to this proposed IKC however is that it will also allow much greater interaction and collaboration with those in business than has previously been possible for EPSRC funded research activities. Hence the IKC, if awarded, would allow the creation of tightly focused commercialization activities jointly with the Judge Business School, the Institute of Manufacturing (including the EPSRC Innovative Manufacturing Research Centre) and the Centre for Business Research. These will allow the creation of a range of innovative knowledge transfer activities spanning business research, training and specific product exploitation. Finally, the Centre will also allow the secondment of researchers from industry and other universities to the IKC, specifically for knowledge transfer (as opposed to research), and in its later stages make use of the provision of pilot manufacturing lines for prototyping.[67]

A WTEC team that visited the CIKC in 2010 reported that the center leveraged "substantial faculty expertise at Cambridge" in fields such as liquid crystals, zinc oxide, semiconducting, polymers, amorphous silicon, and carbon nanotubes. The team observed that the multimaterials approach to flexible electronic circuits and systems was a particular strength of the CIKC and boded well for future commercialization efforts. The team noted that a central feature of the CIKC was the access it offered to a 7,000 square foot Electrical Engineering Division clean room and equipment for processing devices and systems as flexible substrates.[68]

A 2012 assessment of the CIKC's work observed that it had worked directly with 45 industrial partners, producing "significant intellectual property" and technology transfer to industry, and had contributed to the formation of two startups with others under consideration. Technology developed by the CIKC in the 2007-2012 timeframe included an R2R printing process for OPV devices that led to the formation of a spinoff company; novel sputtering processes to deposit high-quality metal oxide materials at temperatures compatible with plastic substrates for large area electronic devices, and a scalable manufacturing process for organic

[65] <*http://www-cikc.eng.cam.ac.uk*>.

[66] EPSRC, "IKC in Advanced Manufacturing Technologies for Photonics and Electronics—Exploiting Molecular and Macromolecular Materials (MMM)." <*http://www.gow.epsrc.ac.uk/NG80ViewGrans,aspx?GrantRef=EP/E023614/1*>.

[67] Ibid.

[68] WTEC, *European Research and Development in Hybrid Flexible Electronics* (2010) op. cit., 12.

transistors using self-aligned inkjet printing with high yield and uniformity over an array of 70 film transistors.[69]

Welsh Centre for Printing and Coating

The Welsh Centre for Printing and Coating (WCPC) is a part of the University of Swansea's School of Engineering and is one of the world's foremost centers for R&D in printing and coating processes, including multi-roll processes. The WCPC's printing and competencies include flexographic, lithographic, rotogravure, digital, and pad printing. The WCPC operates laboratory facilities for sample analysis and characterization of material properties associated with printing, as well as a four station flexographic press, a two station sheet fed lithographic press and pad, and inkjet and screen printing equipment. This equipment is used for fundamental research, prototyping, and the development of materials. The WCPC was designated a Centre of Excellence for Plastic Electronics by BIS in 2009.[70]

The WCPC has developed plastic electronics technologies with medical applications in conjunction with the Swansea University Centre for Nano Health. In 2011, the WCPC's director, Tim Claypole, reported on a project that had developed a method for putting antibodies into an ink that could be printed onto sheets of plastic, creating disposable sensors allowing medical staff to carry out tests at bedside or in surgery using hand-held devices rather than sending samples to a laboratory.[71]

Organic Materials Innovation Centre

The Organic Materials Innovation Centre (OMIC) is located at the University of Manchester, which is located in a region with a high-technology specialty polymer industry that has traditionally had one of the biggest concentrations of polymer researchers in the UK.[72] The OMIC was originally one of five University Innovation Centres that the British government established in 2002.[73] Based at

---

[69] EPSCR, "IKC in Advanced Manufacturing Technologies."

[70] "Swansea Unveiled as UK Centre of Excellence for Plastic Electronics," *Tendersinfo,* December 10, 2009.

[71] Claypole characterized the printed antibody sensors as instruments, which could "be used in something like a dipstick for urine or blood samples or as enzyme sensors which plug into a machine reader." Research partners included Micropharm, Innovia Films of Cumbria, and Abertawe Bro Morgannwg University NHS Trust. The project received financial support from the Welsh Government's Academic Expertise for Business (A4B) an EU-funded initiative to stimulate academic-industry collaboration. "Antibodies in Ink Dots Could Revolutionize Health Tests," *Western Mail,* August 22, 2011.

[72] "GBP 3m Boost to Region's Chemics," *Manchester Evening News,* July 13, 2004.

[73] OMIC received initial funding of £4.3 million from DTI. OMIC, "About OMIC," <*http://www.omic.org.uk/about.php*>.

the Chemistry Department at the University of Manchester, the OMIC works with companies in the development of specialized organic materials for applications in electronics, biomaterials, packaging, and home and personal care.

The University of Manchester is currently building the £61 million National Graphene Institute, a research center that will pursue practical applications of graphenes. The center, which will become operational in 2015, is receiving funding from the EPSRC and the European Regional Development Fund.[74]

Centre for Plastic Electronics

The Centre for Plastic Electronics (CPE), based at Imperial College London, was established to pursue research themes in plastic electronics, involving faculty from the College's departments of chemistry, chemical engineering, physics, and materials. The CPE is closely associated with the Imperial Doctoral Training Centre in Plastic Electronics, in which future scientists and engineers in the field of plastic electronics are trained.[75]

EPSRC Center for Innovative Manufacturing in Large Area Electronics

In February 2013, the EPSRC announced it would award £45 million in grants to fund the startup of four flagship research centers in innovative manufacturing. One of these is a newly formed Centre for Innovative Manufacturing, a university consortium led by Chris Rider at Cambridge. The new center convenes four academic centers of excellence: the CIKC, the CPE at Imperial College London, the WCPC, and the OMIC at the University of Manchester. The new organization, scheduled to launch in October 2013, will receive £5.6 million from the EPSRC. The main theme of the new center will be systems integration, bringing together component functions, printed logic circuits, printed sensors, reflective displays, and printed interconnects to address applications in areas such as smart packaging, intelligent sensors, anticounterfeiting, and smart objects.[76]

Building a Work Force

In 2009, the EPSRC announced that it would award £3 million in grant funding over the period 2009-2018 to establish a "Doctoral Training Centre in Science and Application of Plastic Electronic Materials." The motive for the grant was the recognition that plastic electronics enjoyed "enormous potential" in the UK but that "growth is severely limited by the shortage of scientists and engineers capable

[74] "Manchester Leads the Way in Graphene Membrane Research," *Printed Electronics World,* April 8, 2013.

[75] BIS, *Plastic Electronics,* 11.

[76] EPSRC, "EPSRC Centre for Innovative Manufacturing in Large Area Electronics," EP/Ko3099x/1; "Cambridge at Heart of £45M Manufacturing Push," *Business Weekly*, February 27, 2013.

of carrying ideas forward to application." The new training center brought together "two leading academic teams in the PE area" from Imperial College London (with expertise in the physics, chemistry, and application of molecular electronic materials) and Queen Mary University of London (polymers). The curriculum featured a 4-year track to doctorate degrees involving practical training and professional skills training and interdisciplinary projects with industrial output.[77]

Companies

In flexible electronics the UK has given rise to some of the earliest and most prolific innovating companies in the world, reflecting in significant part spinouts from universities.[78]

Plastic Logic

Plastic Logic was founded in Cambridge in 2000 when researchers at the University of Cambridge spun the company off to commercialize research results developed at the University's renowned Cavendish Laboratory.[79] It was the first company anywhere to fully industrialize mass production of plastic electronics products on a commercial scale. The company succeeded in raising \$50 million in venture capital between 2000 and 2005 and another \$100 million in 2007. In 2008 it established the world's first plant for mass producing plastic semiconductors in Dresden, Germany, a decision that was a blow to British policy makers who had hoped that the company's innovations would lead to manufacturing jobs in the UK. Plastic Logic explained its decision in terms of the "high skills base of the work force (in Dresden), the presence of Fraunhofer facilities and the difficulties associated with planning and construction timescales in the UK."[80] The company also cited the availability of subsidies in Germany and the supportive stance of local authorities:

> When we arrived in Dresden we were met by the Burgermeister the Mayor, and all his team. He said: "We really want you here. We want plastic electronics. It is a key strategic imperative for us to have this here—what do you want?[81]

---

[77] EPSRC, "Doctoral Training Centre in Science and Application of Plastic Electronic Materials," EP/G037515/1.

[78] "Universities: Spin-Outs Put New Life into the Sector," *Financial Times,* September 18, 2007.

[79] The company's founders were Stuart Evans, Sir Richard Friend, and Henning Sirring-Haus. Hermann Hauser, a major figure in British high technology and a co-founder and director of Amadeus Capital Partners, managed the initial rounds of investment in the company and held a 25 percent equity stake in 2004. "Partners Join to Pioneer Plastic Electronics," *Financial Times,* April 6, 2004.

[80] House of Commons, Science and Technology Select Committee, *Bridging the Valley of Death: Improving Commercialization of Research,* 2012. Written evidence submitted by Engineering the Future (V79).

[81] House of Commons, *Engineering,* 45. Plastic Logic commissioned KPMG to select a site for its manufacturing operation. After considering 200 potential locations, KPMG narrowed the list to three:

Plastic Logic's principal initial product was e-paper to be used in a reader marketed by the company under the name Que. However, although launched in 2000 the company did not have a viable product on the market when Amazon introduced the Kindle in 2007. The Que was announced in January 2010, but "just three weeks later, at an event in San Francisco, Steve Jobs unveiled the iPad—and with it effectively blew Plastic Logic out of the water." The Que "never even made it into stores, killed eight months after it was unveiled, without shipping a single unit."[82] Plastic Logic sought to recover from this setback through an alliance with Rusnano, a Russian state-owned development corporation specializing in nanotechnology themes.[83] According to a statement by Rusnano CEO Oleg Kiselyov in 2013, Rusnano invested $240 million in Plastic Logic.[84] Plastic Logic planned to build a plant in Russia to produce e-readers, hoping to sell a less expensive version of Que to Russian schools. "Except Russian schools, it seemed, were not that interested either."[85] In May 2012, Plastic Logic announced that it was abandoning plans to manufacture e-readers, closing its product development operations in the United States and focusing on other applications for its proprietary technologies.[86]

Industry observers have not attributed Plastic Logic's setbacks to its technology, which is regarded as excellent, nor to the lack of backing by governments and private investors. The company's management has been faulted for the business decision to concentrate on the e-reader market, despite the fact that "its founders within the Cavendish Laboratory have always argued that [its technology] has much broader implications for revolutionizing industrial enhancements."[87] The company was criticized for its "inability to execute fast enough" with the result that competitors beat it to the market with superior products.[88] The company's

Dresden, Singapore, and New York State. The decisive consideration was insistence by the Laender government of Saxony that it had sufficient expertise to prepare a high-technology building capable of supporting Plastic Logic's manufacturing operation in a relatively short timeframe. "Hunt for Ideal Site Leads from Cambridge to Dresden," *Financial Times*, January 3, 2002.

[82] Bobie Johnson, "Why Plastic Logic Failed—Despite the E-Book Boom," *GigaOM*, May 17, 2012; "Plastic Logic Abandons Pioneering E-Reader: British Firm Leapfrogs to Second Generation Device, Competition from Kindle and iPad killed off Product," *The Guardian,* August 12, 2010.

[83] "Rusnano to Become Private Equity Fund, Remain State-Owned," *Interfax,* June 14, 2013.

[84] "Rusnano Provisions 22 bln Rubles for Failed Projects," *Interfax,* April 2, 2013; "Rusnano Raises Stake in Plastic Logic," *Plastics News,* October 18, 2011.

[85] Johnson, "Why Plastic Logic Failed."

[86] "Plastic Logic Changing Strategy, Puts Off Building Plant in Russia," *Interfax,* May 17, 2012; "Jobs Threatened as Plastic Logic Puts Down E-Reader," *Plastics News,* May 17, 2012; "Plastic Logic's New Strategy to Expand Market Reach," *Printed Electronics World.*

[87] "Plastic Logic Scraps E-Reader Development," *Business Weekly,* May 16, 2012. Although the company focused on e-readers, it has been obvious to investors for some time that operating within such strict tramlines in the face of intense global competition may not be the most profitable strategy" (ibid.).

[88] "[A]ll the money in the world can't save you if you can't execute." Bobby Johnson, "Why Plastic Logic Failed Despite the E-Book Boom," *GigaOM* (May 17, 2012).

erratic strategic focus was seen as a weakness, at least in retrospect—"pivot too many times and you fall over."[89]

However, it is facile to dismiss the early struggles of Plastic Logic as the consequence of one-off management decisions that in retrospect are easy to second guess. Plastic Logic sought to commercialize a consumer hardware product employing tricky new technology in the face of competition from industry giants with established portfolios of related and interactive products that enhanced the capabilities and consumer appeal of each new platform. Apple and Amazon

> already had functioning ecosystems that their new devices could plug into—something that the [Plastic Logic] Que had to put to one side in its attempt to produce hardware. That meant that while the Kindle and iPad offered a wealth of content at a click, Plastic Logic was left pitching its reader at business people.[90]

Finally, Plastic Logic remains a going concern and technological pioneer, and reports of its "failure" may prove to have been premature. In 2013, Plastic Logic received a FLEXI R&D award from the FlexTech Alliance, an acknowledgement of significant contributions to innovation and R&D in flexible and printed electronics. The award was given for Plastic Logics' development of a scalable color filter alignment process enabling the integration of color displays into flexible applications such as car dashboards and medical bracelets.[91] In early 2014, Plastic Logic was reportedly developing flexible OLEDs using OTFT backplanes, which the company indicates are manufacturable on flexible substrates.[92]

Cambridge Display Technology

Cambridge Display Technology (CDT) was spun out of Cambridge University in 1992 and was the originator of polymer OLED technology.[93] CDT was acquired by Japan's Sumitomo Chemical in 2007.

SmartKem. SmartKem is a startup based in North Wales, which produces proprietary semiconductor materials, particularly inks, for applications in the field of flexible and printed electronics, including smartphones, e-readers, tablets, and TVs.[94] The company was founded in 2009 and received venture capital funding from Finance Wales, a public regional fund investing in SMEs, which enabled it

---

[89] Ibid.

[90] Johnson, "Why Plastic Logic Failed."

[91] "Top Flexible Electronics Developments Win 2013 FLEXI Awards," *Flexible Substrate,* February 2013.

[92] "Organic Electronics and Flexible Displays," *Flexible Substrate,* January 2014.

[93] CDT was formed through a collaboration between the Cavendish Laboratory and the University of Cambridge Chemistry Department, creating what one of its founders termed "one of the most interdisciplinary companies ever [requiring] close cooperation between chemists, physicists and materials scientists, along with semiconductor, process and production engineers." Royal Society of Chemistry, *Profile: Cambridge Display Technology,* <*http://www.rsc.org/pdf/mcg/cdt.pdf*>.

[94] SmartKem website, access at <*http://www.smartkem.com*>.

to establish operations in the OpTIC Technium incubator in Wales.[95] SmartKem received subsequent funding support from the Welsh Assembly Government.[96] In late 2013 SmartKem disclosed the signing of a joint development agreement with an unnamed "major Asian display original equipment manufacturer for the manufacture of flexible displays."[97] Steve Kelly, SmartKem's CEO, said in a January 2014 interview that the company's customers were "predominantly OEM and chemical companies within Asia."[98] The company indicates that a source of its competitive advantage has been independent validation of its materials by Bangor University pursuant to a Knowledge Transfer Partnership (KTP).[99]

NETHERLANDS/BELGIUM

National, regional, and local governments in the Netherlands and Belgium are providing substantial support to research efforts in flexible electronics. These countries are host to two world-class research centers active in the field, IMEC and its affiliated Holst Centre, as well as Philips, one of the foremost electronics and information technology companies in Europe.

Netherlands Organization for Applied Scientific Research (TNO)

TNO is a statutorily created public research organization for applied research. It is funded through a combination of core government funding and contract research for government organizations and private companies. (See Table 5-9.)

Netherlands Ministry of Economic Affairs, Agriculture and Innovation (EL&I)

EL&I makes fixed annual contributions to a number of Dutch public research organizations, including TNO and the Holst Centre, both of which are engaged

---

[95] "Company Profile: Semiconductor Start-up," *Chemistry World,* August 2011. The Technium Centers are incubators for innovative businesses developed jointly by the Welsh Development Agency and Swansea University. They have provided facilities, business advice, telecommunications, and venture funding. Launched in 2001, most Technium Centers were subsequently closed, and the program has been widely criticized as a costly failure. "Six Technium Innovation Centers to Be Cut," *Western Mail,* November 19, 2010; "£100m and 10 Years On—Lessons from the Techium Experience," *Western Mail,* June 1, 2013; "Technium Centers a Waste, Says Price," *Western Mail,* March 26, 2012.

[96] WAG Supports a Smart Move for Electronics," *Liverpool Daily Post,* January 19, 2011.

[97] "SmartKem Joint Development Agreement with Asian Display Manufacturer," *Printed Electronics World,* December 10, 2013.

[98] "Printed Electronics Is the Next Big Thing," *Electronics Weekly,* January 30, 2014.

[99] The University, which enjoys expertise in transistor physics, worked with SmartKem to process, test, and validate its materials in thin-film transistor form and to define protocols for forming the materials into transistor arrays. CEO Kelly said, "This has been of immense value to the company in understanding the performance of our materials." "University Plays a Role in SmartKem's Organics Success Story," *Electronics Weekly,* March 19, 2014.

TABLE 5-9 TNO Income by Source (2011)

Source of Funding	Amount (Millions of Dollars)	Percent
Knowledge as an asset (core funding)	95.2	13
Policy and applied research (Ministries)	155.3	20
Public contracts (Netherlands)	124.7	16
Private contracts (Netherlands)	195.7	26
Private contracts (foreign)	143.8	19
International contracts	50.0	7

SOURCE: Consolidated TNO Annual Accounts.

in support for research in flexible electronics. EL&I also coordinates the funding of TNO within the government.[100]

Flanders Institute for the Promotion of Innovation through Science and Technology (IWT)

IWT is a government-funded public organization in the Belgian region of Flanders, which directly funds R&D activities and promotes technology innovation and transfer. In the field of flexible electronics it provides substantial support to IMEC, a public research organization located in Leuven. IWT's funding is in the form of grants.[101]

Inter-University Microelectronics Center (IMEC)

IMEC, located in Leuven, in Belgium's Flemish region, is the largest microelectronic R&D center in Europe, with 2,000 employees in 2012 and state-of-the-art research infrastructure.[102] Founded as a nonprofit organization in 1984, its original capitalization included a contribution of roughly $60 million from the Belgian government.[103] IMEC still receives annual direct contributions from

[100] Government of the Netherlands, *The Science System in the Netherlands: An Organisational Overview,* 2012. In 2007, EL&I funding accounted for 50 percent of the annual budget of the Holst Centre. *PolyMap Report, WP01: Survey of National and Regional Funding with OLAE Context,* February 23, 2010.

[101] See generally Eric Sleeckx, "Implementing and Monitoring the Flemish Innovation System," in National Research Council, *Innovative Flanders: Innovation Policies for the 21st Century: Report of a Symposium* (Washington, DC: The National Academies Press, 2008), 38–41.

[102] IMEC has two clean rooms, including one which is capable of supporting microelectronics R&D at the 450 mm wafer scale level. IMEC, "IMEC at a Glance," *<http://www2.imec.bc/content/user/File/Borchure%20Imec%20at%20a%20glance.pdf>.*

[103] "A European Silicon Valley in Flanders—Europe's Largest Independent R&D Center Comes into Being Near Brussels," *Frankfurter Allgemeinc/Blick Durch die Wirtschaft,* June 12, 1986. JPRS-EST-86-022, September 11, 1986.

TABLE 5-10 IMEC Operating Revenue by Source

	2011 (Millions of Euros)	2012 (Millions of Euros)
Revenue from contract research	300.1	320.6
Miscellaneous income (including contributions in kind)	12.7	15.2
Subsidies from the Flemish region	45.7	48.2
Subsidies from the Dutch government	7.2	8.2

SOURCE: IMEC Annual Report 2012.

the Flemish region and the Dutch government that account for about 14 percent of its total revenue. Revenue from contract research, which accounts for the preponderance of IMEC's income, includes revenue derived from contracts with public bodies and government contributions (EU, national, and regional) supporting contract research by industry. (See Table 5-10.) Within several years of its establishment, IMEC had become regarded as one of the best microelectronics research organizations in the world.[104]

IMEC's core research activities emphasize "early phases where potential commercial value starts to emerge out of basic science."[105] Early-stage research is conducted in collaboration with industrial partners. Under IMEC's Industrial Affiliation Programmes, personnel from industrial partners (as well as equipment as needed) are embedded at IMEC. Industrial prototypes are examined, tested, and further refined on site. In 2006, an executive from Texas Instruments, which collaborated with IMEC, cited a number of advantages associated with doing so, including the superb quality of IMEC's equipment, development collaborations with equipment suppliers, IMEC's focus on fundamental science, and public funding, which enabled IMEC to keep its research infrastructure state of the art.[106]

IMEC's research strategy has two basic elements. "More Moore" seeks to maintain the direction of established technological trajectories while attaining incremental improvements, usually through scaling. The other element, "More Than Moore," seeks to pursue radical innovation leading to the emergence of new micro- and nanoelectronic technologies and markets.[107] Under the auspices of the latter element, IMEC has pursued research themes in areas such as organic

[104] "Flanders Technology International: Flemish Chips on a World Level," *De Standaard,* May 8, 1987. JPS-ELS-87-004, August 12, 1987, 23.

[105] Andrea Mina, David Cornell, and Alan Hughes, "Models of Technology Development in Intermediate Research Organizations" (Centre for Business Research, Cambridge University, Working Paper No. 396, December 2009), 17–18.

[106] Bowling, "IMEC and Sematech," in National Research Council, *Innovative Flanders,* 78.

[107] Mina et al., 2009, op. cit.

electronics, biomedical electronics, and organic photovoltaics.[108] These efforts commonly involve collaboration with its affiliate, the Holst Centre.[109]

A WTEC delegation that visited IMEC in 2010 reported that it had an organic and hybrid electronics research group focusing on organic circuitry and organic photovoltaics. The organic circuitry R&D group had successfully developed a 64-bit RFID tag operating at 800 bps that included a 13.56 MHz transponder, utilizing organic-based diodes capable of rectifying current at that frequency. The OPV effort was seeking to develop high-performance organic-based cells that could deliver power at less than €0.50/wp (peak power in watts), with 10 percent record efficiency and 5-year lifetimes, based on IMEC's view that silicon-based PV systems cannot match such low cost-per-watt ratios.[110] Shortly after the WTEC team visit, it was announced that IMEC would head an EU Seventh Framework project, X10D, for development of organic photovoltaic cells that enjoy a longer lifetime, lower production cost, and superior conversion efficiency than do silicon-based PV cells.[111]

In 2009, research collaboration between IMEC, Hasselt University, and the Belgian printing company Artist Screen resulted in the creation of a spinoff in the field of plastic electronics. Lumoza NV developed and commercialized large area screen printed electronics.[112] Lumoza's main product was large outdoor banners, dozens on which certain surfaces lit up in an animated manner.[113] Lumoza subsequently ceased operations.[114]

Between 2010 and 2012, IMEC led a consortium funded by the EU Seventh Framework that developed the "world's first radio frequency identification circuit (RFID) made with low-temperature thin film technology that allows reader-talks-first communication."[115] The ORICLA project raised the prospect for intelligent RFID tags that are inexpensive to produce on a mass basis using printing technology.

IMEC reported in 2012 that it was collaborating with the Holst Centre's system-in-foil research effort focusing on technologies applicable to flexible plastic foil substrates such as polyethylene naphthalate (PEN) and polyether

---

[108] *<http://www.emm-nano.org/?page_id=76>.*

[109] Andrea Mina, Cornell, D., and Hughes, A. "Models of Technology Development in Intermediate Research Organizations, op. cit.

[110] WTEC, *European Research and Development,* 10.

[111] The project involves 16 partners, including CEA of France, Imperial College London, the Holst Centre, and a number of companies including Agfa and Solvay. "IMEC Launches Project X10D for Development of OPV Cells with Increased Efficiency at Lower Costs." *PVTech,* October 28, 2011; "EU Project Aims to Improve Organic Photovoltaics," *The Engineer,* October 31, 2011.

[112] IMEC Annual Report 2010, *<http://www.imec.be/Scientific Report/SR2010/2010/11159362.html>.*

[113] "New Funding for Promising Hasselt University Spin-offs Camargus and Lumoza," *LRM,* December 12, 2012.

[114] *<http://www.lumoza.be>.*

[115] The other consortium members were IMEC's affiliated Holst Centre, Envoi Industries AG (Germany), and Politic (Germany). "European Project Reaches Milestone Bidirectional Communication for Thin-Film RFIDs," *Flexible Substrate,* March 2012.

ether ketone (PEEK). IMEC researchers were experimenting with the placement of indium gallium zinc oxide diodes and transistors and organic transistors and meonaries in flexible foil at low temperature. These devices were then used to design circuits on foil, and the devices were used in backplanes for rollable active matrix organic light-emitting diode displays.[116]

In 2013, IMEC and Japan's Fujifilm Corporation reported that they had jointly developed a photoresist technology for organic semiconductors suitable for application in photolithographic patterning on large-size, flexible substrates. The new process reportedly will enable the realization of submission patterns as large, flexible organic substrates.[117]

The Holst Centre

The Holst Centre is an industry-government research organization founded by IMEC and TNO of the Netherlands in 2006. Located in the High Tech Campus in Eindhoven, Netherlands, the Centre's mission is to create generic technologies for ultra-low-power wireless sensors and large area flexible electronics.[118] Participating companies pay an "annual membership" as well as an "initial membership," which increases as the Centre's intellectual property (IP) portfolio grows. At the Centre, research partners augment their own exclusive R&D with "shared R&D," the results of which are shared between program partners on a nonexclusive basis through "customized agreements" that are "tuned to each partners needs and situation."[119] The Centre's research horizon aims for a commercial impact within 3 to 10 years. There are two basic categories of research programs:

- **Technology programs (TPs)** develop research roadmaps for particular technologies and deliver "fundamental understanding" and new "state-of-the-art concepts and demonstrators."

[116] "IMEC—Strategy: Large Area Electronics and Systems-in-Foil," *<http://www.imec.be/ScientificReport/SR2012/1116058.html>*.

[117] "Fujifilm and IMEC Develop New Technology for Organic Semiconductors Enabling Submicron Patterns," *Flexible Substrate,* October 2013.

[118] Holst Centre, *Executive Report 2012,* 8. The High Tech Campus was formerly Philips Research Laboratories. A strong proponent of open Innovation, Philips opened its facilities to other companies and research organizations in 1998, and in 2012, sold the Campus to a Dutch consortium of private investors. Philips remains a tenant. The Campus features extensive R&D infrastructure and facilities, including clean rooms, laboratories, and testing equipment, and houses more than 8,000 R&D staff. High Tech Campus Eindhoven, *<http://www.hightechcampus.com>*. The WTEC panel, which visited Holst in 2010, observed that "[r]esearch projects in the Centre take advantage of extensive clean room/microfabrication facilities and characterization instrumentation on campus, which are available on a fee-for-use basis. It was clear to the panel that the rapid buildup of the Holst Centre (for example, in terms of industry participation) since its founding in 2005 is attributable in large part to the excellent research facilities that are in place on the High Tech Campus." WTEC, *European Flexible Electronics,* 8.

[119] Ibid., 9.

- **Technology integration programs (TIPs)** integrate thematic applications to emerging technologies and prove the technologies through field trials and prototyping. These programs also seek to link emerging technologies to other and different applications. The Centre conducts applied research up to the demonstration level, after which industry partners assume responsibilities for independent prototyping and product development.[120]

Holst was formed at the initiative of Philips Research, which was seeking to establish a local open innovation research center involving participation by a significant number of companies as well as "independent knowledge institutes orchestrating open innovation." Philips asked TNO and IMEC to develop a business plan for a research center focusing on micro- and nanotechnologies, and they developed a proposal that won the support of the Dutch government. Philips was the Holst Centre's first industrial partner and the "participation of Philips gave Holst Centre a flying start."[121] Legally the Centre is a Dutch entity and part of TNO. IMEC participates through a separate legal entity, Stichting IMEC Nederland, established after difficulties were encountered establishing a transnational research institute.[122]

The Holst Centre prioritizes development of demonstrators of real-world applications of emerging technologies. Its approach encourages collaboration among various scientific and engineering disciplines and sharing of facilities, costs, and in some cases, personnel. In Holst's systems-in-foil division, for example,

> companies with know-how and IP in substrates and materials (DuPont Tejin Films, Agfa and Merck) can work along equipment suppliers and organic electronic manufacturers (Orbotech, ASML, Singuls Mastering and Plastic Electronic) and integrated device manufacturers (Philips), who understand the specs and system design required by the market. The whole value chain is represented.[123]

Funding

The Holst Centre's funding is derived from a variety of public and private sources in roughly the following order: public funding (45 percent), EU project funding (10 percent), and research contracts with companies (45 percent).

The Centre's public funding was provided by the Dutch Ministry of Economic Affairs during the startup period 2005-2012. It was determined that a total of €72 million in public funding would be required to sustain the Centre in

---

[120] Andrew Mina, David Connell, and Alan Hughes, "Models of Technology Development in Intermediate Research Organizations" (paper prepared for DR UID Summer Conference, "Opening up Innovation: Strategy, Organization, and Technology" June 16-18, 2010, Imperial College, London), 13.

[121] "Holst Centre Combines IMEC, TNO, and Industry's Capabilities," *Printed Electronics Now,* October 2010.

[122] Mina et al., "Intermediate Research Organisations," 13.

[123] Ibid., 14.

2013-2016, and this budget has been secured through joint contributions by governments and public organizations that include the Ministry of Economic Affairs, the province of Noord-Brabant, the Brainport Eindhoven region, TNO, IMEC, the Dutch Organization of Scientific Research (NOW), and a fiscal ruling issued by the Dutch government (known as TKI toeslag).[124] Over time public funding has declined as a proportion of the Centre's budget as payments from the private sector have increased, but public funding remains critically important.[125]

Flexible OLED Manufacturing Technology

At this writing some glass-based OLED lighting devices have begun to enter the commercial market, but the introduction of flexible OLED lighting has been hamstrung by the absence of reliable and efficient production processes. While R2R manufacturing is seen as the eventual solution, its introduction has been seen as not being able "to bring flexible OLEDs to the market within 10 years."[126] The Holst Centre is currently engaged in a number of collaborations to develop technologies that will enable the efficient production of flexible OLED lighting products for commercial use.[127] In June 2013, Georg Götz of the Holst Centre indicated that in collaboration with its academic and industrial parties, Holst would develop generic technologies for flexible OLED lighting architectures produced by R2R processes with emphasis on flexible and transparent encapsulation, solution processing of organic layer stacks, and replacement of indium tin oxide with a highly conductive polymer formulation (PEDOT:PSS).[128]

In early 2012, the Holst Centre and IMEC launched an effort to build on prior efforts in areas such as "organic and oxide transistors and flexible OLED lighting to develop an economically scalable route to high-volume manufacturing of large area flexible active-matrix OLED displays."[129] Technological challenges include the deposition of thin organic layers with a well-controlled homogeneous thickness combined with highly conductive and transparent electrodes; encapsulation

---

[124] EURIS, "Inventory of Good Practices on Open Innovation," Holst Centre, Eindhoven, The Netherlands; "Public and Industrial Agreements Enable Further Growth of Holst Centre," Holst Centre news release, April 5, 2012.

[125] EURIS, "Inventory of Good Practices," 8.

[126] Project Overview (for CORDIS Fact Sheet), *Flex-o-Fab*.

[127] In 2012, the Holst Centre and Solvay disclosed they had developed a bendable 69 square-centimeter OLED panel with layers deposited through a combination of vacuum deposition and solution processing. "Solvay and the Holst Centre Present an Efficient (30lm(w) large OLED lighting panel," Oled-info.com.

[128] PEDOT:PSS is a transparent, conductive polymer with high ductility. Its current applications include the coating of photographic films as an antistatic agent. "Large Area Flexible OLEDs: Solution Processing and R2R Technologies," *Flexible Substrate,* May 2013.

[129] "Holst Centre and IMEC Launch Research Program on Flexible OLED Displays," *IMEC news,* January 17, 2012. Holst and IMEC are also collaborating in the development of large area fully organic photodetector arrays that can be fabricated on flexible substrates. "IMEC and Holst Centre Unveil Fully Organic Images," *Flexible Substrate,* August 2013.

must be flexible, be transparent, and provide protection of the active layers for a long timeframe.[130] In September 2012, it was disclosed that Sumitomo Chemical, a major producer of high-end materials for polymer OLEDs, would join the Holst Centre's research program on Printed Organic Lighting and Signage, a move that was expected to "speed efforts to develop manufacturing processes for low-cost flexible organic light-emitting diodes."[131] In 2012, the Holst Centre and Solvay reported the development of a bendable 69 square centimeter OLED panel that enjoyed the same efficiency as Solvay's own rigid panels on glass.[132]

PragmatIC Collaboration

In 2011, the Holst Centre entered into a research collaboration agreement with PragmatIC Printing Ltd., a Cambridge, UK-based pioneer in printed logic circuits, for the further development and exploitation of flexible electronics. PragmatIC has developed a proprietary process that allows the fabrication of electronic circuits in a single layer of thin-film semiconductor, using an embossing process, and avoiding the multiple-layer structures required for conventional thin-film transistors. The Holst Centre and PragmatIC are jointly developing qualified processes and materials that will be licensed to Holst Centre program partners and PragmatIC licensees.[133]

Solliance

Solliance is a collaboration between Dutch, Belgian, and German research organizations and companies to develop thin-film photovoltaic technology. Participating research organizations include IMEC, Holst Centre, ECN (the largest Dutch energy research institution), TNO, and Forschungszentrum Jülich, a German institute for energy and climate research. Solliance has received financial support from the Dutch province of Noord Brabant, which contributed €38 million to create a new shared laboratory in Eindhoven with state-of-the-art equipment.[134] The Brabant Development Agency, an economic development entity co-funded by the Dutch government and Noord Brabant, is a member of Solliance.

Solliance R&D activities are conducted with industry partners at Solliance's own research center and at laboratories at IMEC and Forschungszentrum Jülich. Research themes include testing, characterization, and monitoring; laser

---

[130] "Large Area Flexible OLEDs: Solution Processing and R2R Technologies," *Flexible Substrate,* May 2013.

[131] "Sumitomo Chemical Joins Holst Centre OLED Research Program," Holst Centre Press Release, September 18, 2012.

[132] "Solvay and the Holst Centre Present an Efficient (30 lm/w) Large OLED Lighting Panel," *Flexible Substrate,* March 2012.

[133] "PragmatIC's Imprinted Planar Nano-Devices Included in Focus of Holst Centre Program," *Flexible Substrate,* February 2013.

[134] <*http://www.solliance.eu*>.

technologies; light management by mechanical texturization; transparent conductive layers; monolithic interconnection; thin-film deposition; new OPV device development, sheet-to-sheet processing, R2R processing, and in-line monitoring.[135] The basic goal of Solliance is to originate an efficient method of manufacturing OPV foils, which are less costly to produce than are conventional photovoltaic cells.[136] In 2013, Solliance reported production of the world's first photovoltaics made exclusively with inkjet printing processes, a technique that will reportedly enable rapid scale-up of manufacturing operations for flexible, lightweight, semi-transparent PV cells for integration into construction materials.[137]

Companies

Solvay S.A.

Solvay is a Belgium-based multinational chemical company that is pursuing a business strategy emphasizing high-value-added, technology-intensive products, with a particular focus on the North American market. Two fields that Solvay sees as "platforms for future growth based on radical innovation" are organic electronics and sustainable energy.[138] Solvay supplies specialty materials and inks for OLEDs and OPV devices, and its "Solvene EAP" materials have applications in printed memory devices.[139]

Solvay has engaged in significant collaborations with the Holst Centre and IMEC in developing flexible OLED lighting technology and organic photovoltaics.[140] Solvay has made substantial investments in Plextronics, a U.S. maker of printed electronics devices, with an equity stake of 47 percent as of January 2014, and when Plextronics entered Chapter 11 proceedings, Solvay made a buyout offer valued at $32.6 million.[141] Solvay acquired a

---

[135] "Solliance and Forschungszentrum Jülich Join Forces," *<http://www.brainport.nl/enhigh-tech-systems-materials/solliance-and-forschungszentrum-juelich-join-forces>*.

[136] "Advances and Improvements with Manufacturing," *Flexible Substrate*, March 2012.

[137] "Solliance Develops World's First All Inkjet Printed OPVs," *Flexible Substrate,* January 2014.

[138] Leopold Demiddeleer, Solvay General Manager of Future Businesses Competence Center, in "Plextronics Closes $14 Million Financing Round," *Printed Electronics Now,* August 20, 2009.

[139] OE-A, *Organic and Printed Electronics* (June 2013), 70.

[140] "European Collaboration Towards Efficient, Low-Cost Tandem Organic Solar Cells," *Printed Electronics Now,* October 28, 2011. In 2012, Solvay and the Holst Centre jointly demonstrated large area, flexible, high efficiency OLED lighting tiles with a surface area of 69 square centimeters. These devices have an energy efficiency which is 2-3 times higher than conventional incandescent bulbs. "Flexible OLED Lighting Reaches High Energy Efficiency Thanks to Shared Research Effort," *Printed Electronics Now,* March 8, 2012; "IMEC, Solvlay Announce World Record Efficiency for OPV Module," *Printed Electronics Now,* September 24, 2012.

[141] "Solvay Acquires Plextronics to Accelerate its OLED Display Development," *Flexible Substrate,* April 2014; "Plextronics Shorts Out," *Chemical and Engineering News,* January 27, 2014; "Solvay Commits $15 Million to Support Plextronics' Innovative Technology Development in OLED and

minority equity stake in Illinois-based Polyera, a producer of functional inks, in 2010.[142]

Philips N.V.

Philips, a Dutch multinational, is one of the world's leading producers of electronics and lighting products and systems. It was the first company to demonstrate RFID tags using organic transistors.[143] Philips has exited a number of electronics businesses during the past decade, including displays and semiconductors. A spinoff, Polymer Vision, pursued development of a foldable e-ink screen, went bankrupt in 2009, and, having been acquired by Taiwan's Wistran Corporation, was shut down in 2012.[144] Philips has continued to pursue research themes in flexible electronics, such as OLED lighting. Philips established an "open innovation" center at its primary R&D site in Eindhoven, the High Tech Campus, where the Holst Centre was established at Philip's initiative.[145] Collaboration between Philips and the Holst Centre has been close and extensive since the latter's inception.[146]

GERMANY

Germany has the potential to emerge as a major force in global competition in flexible electronics. Its chemical industry, which includes players like BASF, Merck, Evonik, and newcomers such as Novaled, is one of the most innovative and sophisticated in the world. Its machinery industry produces some of the world's best precision equipment and instruments, and its printing industry is a major asset in a field in which manufacturing processes are likely to be dominated by various forms of printing. Germany has a formidable array of excellent public research organizations, including the Fraunhofer Gesellschaft and the Max Planck Institutes backed by a powerful Ministry of Education and Research. Large-scale efforts to develop flexible electronics, including manufacturing capabilities, and the most advanced organic electronics cluster in Europe are located in Dresden.

OPV," *Printed Electronics World*, July 27, 2011; "Solvay Has Made a $10M Investment in Plastics" *Plastics Engineering*, October 2007.

[142] "Solvay Expands its Printed Electronics Development with Investment in Minority Stake in Polyera," *Hugin*, September 7, 2010.

[143] OE-A, *Printed, Organic & Flexible Electronics*, 260.

[144] "Wistron Folds Up Polymer Vision—Collapsible eReaders Are Once Again Science Fiction," *The Digital Reader*, December 2, 2012.

[145] WTECH, *European Research and Development*, 99–100. Philips sold the High Tech Campus in 2012 but remains a tenant on the site.

[146] As of late 2011, Philips Research was co-author of more than 20 Holst Centre patent filings and more than 100 technical notes. "Holst Centre, Philips Research Celebrate Successful Five-Year Collaboration with New Contract," *Printed Electronics Now*, October 24, 2011.

In Germany scientific research is funded according to a formula allocating fixed percentages of annual contribution by the federal and Land (state) governments. The German research system is characterized by dispersion of authority and funding of specific projects by multiple government entities. An implicit consensus exists at the state and federal level that most government funding should be directed toward "bridging the gap between knowledge creation and application."[147] In flexible electronics, as in other advanced technologies, the German research system emphasizes collaboration and coordination between research actors, the development of commercially relevant technologies, and the achievement of concrete short- and medium-term results as opposed to potentially game-changing "blue sky" discoveries.

Federal Support for Research

The principal German government entity funding research and development in the area of flexible electronics is the Ministry of Education and Research (BMBF), formerly known as the Federal Ministry for Scientific Research (BMwF). BMBF provides direct financial assistance to research and development projects that involve collaborations between universities and public research institutes, on the one hand, and the private sector on the other hand. BMBF funding in flexible electronics has typically been directed toward projects of short and medium duration (1-3 years) and involving major industrial participants such as Merck, BASF, Siemens, and Philips (Table 5-11). BMBF provides additional indirect support for the sector through its funding of the Fraunhofer and Max Planck institutes, which are extensively engaged in the field.

So-Light

The So-Light project, a 2009-2012 effort funded by BMBF with €8 million, was designed to address the complete supply chain from primary materials, manufacturing processes, and optical components to OLED lighting applications. Pronounced an "outstanding success," the project has reportedly enabled consortium partners to develop prototypes for OLED signage, including specialty applications in the automotive and architectural lighting markets.[148] (See Table 5-12.)

---

[147] Markus Winnes and Uwe Schimack, *National Report: Federal Republic of Germany,* Institute for the Study of Societies, TSER Project No. SOE1-CT96-1036 (May 1999); Jakob Elder and Stefan Kühlmann, "Coordination within Fragmentation: Governance in Knowledge Policy in the German Federal System," *Science and Public Policy,* 2008, 267.

[148] "German OLED Project 'Exceeds Expectations.' " *Flexible Substrate,* February 2013.

TABLE 5-11 BMFB Funding of Research Consortia—Flexible Electronics

Project	Theme	Timeframe	Project Cost (Millions of Euros)	BMBF Funding (Millions of Euros)	Industry Partners	Universities/Research Institutions
Polytos	Printed organic switches/ chips	2009-2012	13.8	7.2	BASF, Merck, Peppesl & Fuchs, Bosch, SAP	Darmstadt Heidelberg Mannheim
Polytos 2	Printed organic switches/ chips	2012-2013	1-6	4.9	BASF, Merck, Bosch, SAP PolyIC, Heidelberger, Druckmaschinen	Darmstadt Heidelberg Mannheim
NanoPEP2	Nanostructures and plastic electronics print platform	2012-2013	6.8	3.0	BASF, Heidelberger, Druckmaschinen	Darmstadt
Gluco Sens	Organic electronic novel glucose sensors	2010-2013	4.6	2.3	Rosch, BASF	Freudenberg Heidelberg
Cobalt	Cost efficient OLEDs for lighting applications	2012-2015	42.9	17.2	Philips, BASF, Aixtron	
Print OLED	Application of vapor deposition processes emitter/matrix systems to printing processes for OLEDs	2009-2013	12.2	5.9	Merck, BASF, Philips, Osram	Karlsruhe It, Darmstadt, Brownschweig
HOP-X	Hybrid organic photodetectors for radiography	2012-2015	3.8	1.9	Siemens, Merck, Plastic Logic	Leibniz

continued

TABLE 5-11 Continued

Project	Theme	Timeframe	Project Cost (Millions of Euros)	BMBF Funding (Millions of Euros)	Industry Partners	Universities/Research Institutions
R2Flex	Roll-to-roll production of organic components on flexible substrates	2010-2012	10.7	6.2	Novaled, Heliatek, Ledon OLED, 3D Micromac, VON ARDENNE, ALANOD, Crephys, Tridonic Dresden	FHG, IPMS, FHG, FEP
So-light	OLED applications in lighting and signage	2009-2012	14.6	8.0	Novaled, AEG, Siteco, Sensiat Imaging, Aixtron, Symoled, Fresnel, Hella	FHG, IMPS Munster
KoSIF	Flexible autonomous sensor systems on films	2013-2017	6.0	3.8	Würth, FESTO, HSG-IMAT	Stuttgart Media U, Stuttgart IGM, Max Planck, MPI, IMS Chips
LightinLine (liLi)	OLED manufacturing for lighting application	2009-2012	7.5	3.3	Applied Materials , Merck	Braunschoeig
cyCESH	Efficient production of OLED devices	2013-2016	6.1		Novaled, Cynora	Regensburg
Popup	Materials for organic PV	2013-2016	16.0	8.2	Merck, PolyIC, Siemens, Centrosolar Glas, Leonard Kurz Stiftung	Ctr for Applied Energy Syhstems; Karlsruhe Inst. Tech; Ctr for Solar Energy/Hydrogen Research Ctr for Applied Energy Syhstems; Karlsruhe Inst. Tech; Ctr for Solar Energy/Hydrogen Research

SOURCE: BMBF, *Computer-automation.de.*

TABLE 5-12 Impact of Germany's So-Light Project

Research partner(s)	Achievement
Novaled	Progress toward fully air-stable electron transport layer
Sensient	New host materials for OLED emitter layers
Nonaled/Sensient	New p-doped hole transport system with lower absorption, lower cost scalability, to be commercialized by Novaled
AIXTRON/Fraunhofer COMEDD	Optimized OVPD process on a Gen2 substrate size
LEDON OLED Lighting	Developed efficient electrical controlling technology enabling greater system efficiency
Fresnel Optics	Successfully processed external flat primary optics directly on rear surface of OLED panel
HELLA	Made design studies for automotive indoor lighting and car rear lights with red OLEDs
BMB MIS	Developed thin OLED backlight for LCD signage application
AIXTRON	Demonstrated new high deposition rate process
Siteco	Manufactured a suspension luminaire containing OLEDs for building façade integration

SOURCE: "So-Light Project Successfully Concluded," *Flexible Substrate,* February 2013.

OLED 2015

The OLED 2015 initiative, launched in 2006, involved research into OLED technology for lighting and displays by a consortium comprised of 33 entities. BMBF contributed €100 million to the project and industry partners another €500 million. At the time this effort was by far Germany's largest commitment to the promotion of a nation/capability in organic electronics. OLED 2015 was comprised of a number of subprojects that in some cases led to follow-on projects. Project OPAL 2008 developed improvements in the efficiency, service life, and surface of OLEDs, technology which was transferred to OSRAM, an industry partner, during the course of the project and permitted small-scale commercial use of the new OLED technology.[149] OPAL 2008 was succeeded by TOPAS 2012, a project involving Philips, OSRAM, AIXTRON, and BASF to develop OLEDs for future lighting systems.[150] TOPAS has itself been succeeded by GENESIS, another government-funded project to scale down to manufacture-compliant

[149] acetech, *Organic Electronics in Germany* (2011) op. cit., 43.

[150] "OLED Lighting Research Backed by German Government," *Electronics Weekly,* January 13, 2010. OSRAM has utilized the technology developed in TOPAS to introduce a new luminaire lighting product, the Rollercoaster, featuring transparent OLED panels. Production will begin in 2014. "OS-RAM Reports Advances in Transparent OLED Development, To Start Production in 2014," *Flexible Substrate,* January 2013.

processes and substrate sizes.[151] Project OPEG, another OLED 2012 subproject, increased the efficiency of organic solar cells from 5 percent before the project to 8.3 percent. The two projects OPAL and OPEG also provided the bases for nearly 50 Ph.D. theses.[152]

Innovation Alliance Organic Photovoltaics (OPV)

BMBF launched the OPV initiative to complement OLED 2015 by developing technology for the application of OPV.[153] BMBF contributed €60 million to this project and expected industry contributions of €300 million.[154] The principal industry participants in this effort were Merck, BASF, Bosch, and Schott. BASF and Bosch invested in a startup company, Heliatek GmbH, to engage in the development of organic solar cells, in 2006.[155] Heliatek, based in Dresden, developed a succession of increasingly efficient organic solar cells and in 2012 set a new world efficiency record with organic solar cells that achieved 10.7 percent cell efficiency.[156] In March 2012, Heliatek inaugurated a proof-of-concept production line for the manufacture of flexible solar panels for building integrated PV applications.[157]

Organic Electronics Cluster of Excellence

In 2007, BMBF committed €40 million to the establishment of a cluster of excellence, "Forum Electronics in the Rhine-Neckar," in the vicinity of Heidelberg. In 2009, the cluster was launched, comprising 16 large and medium-sized companies (including BASF and Heidelberger Druckmaschinen Ag, a world leader in printing technology) and 11 research institutes and universities, including Darmstadt Technical University's Institute for Printing Presses and Printing Methods IDD. One of the BMBF-funded collaborations undertaken in the cluster was NanoPEP, a project involving BASF, the University of Heidelberg, and the Technical University of Darmstadt in an effort to develop nanobased functional materials and printing techniques for processing these materials.[158] NanoPEP

[151] "OSRAM Reports Advances in Transparent OLED Development."

[152] acetech, *Organic Electronics in Germany,* 43.

[153] "Innovation Alliance Organic Photovoltaics," *<http://www.fona.de/en/10007>.*

[154] acetech, *Organic Electronics in Germany,* 19.

[155] Heliatek was spun off from the Technical University of Dresden and the University of Ulm. It has received funding from BMBF, the EU, the Free State of Saxony, and the German Federal Ministry of Economics and Technology (BMWi). Other investors include Wellington Partners and RWf.

[156] "Heliatek's Organic Tandem solar Cell Verified at 10.7% Record Conversion Efficiency," *PV Tech,* April 27, 2012.

[157] "Heliatek Opens Groundbreaking Production Facility for the Manufacture of Organic Solar Films," *M2Presswire,* March 13, 2012.

[158] "Printed Electronics Research Project Moves Into Next Phase," *Package Print Worldwide,* August 2, 2012.

produced initial functional elements under laboratory conditions in the cluster's clean room using modified printing methods. In order to transfer these processes to an industrial scale within a 2-year timeframe, NanoPEP2 was launched in 2012. The follow-on effort will use practical demonstrations to show the functionality of the printed components, which can take the form of flexible OLEDs or solar cells produced in the cluster's clean room.[159]

New Materials for OLEDs from Solutions (NEMO)

The NEMO project (2009-2012) developed new solution-processable materials for OLEDs suitable for integration into large-area OLED components with applications in signage, televisions, and illumination for objects or rooms.[160] The project was co-funded by BMBF that contributed €32 million.[161] Merck supervised the research, which developed and tested new phosphorescent materials for red, green, and blue.

The Fraunhofer Gesellschaft

The Fraunhofer Gesellschaft is a distributed network of public research institutes with the core mission of the pursuit of knowledge with practical utility. The Fraunhofer's mission is to serve as a bridge between Germany's science base and industry, and most of its activities involve applied research with specific industrial and commercial objectives.[162] Each Fraunhofer Institute is linked with one or more German research universities with curricula and faculty that are relevant to the institute's competencies. The institutes utilize basic research developed in the universities and other German research organizations and "generate relevant application-oriented knowledge themselves on demand from (industrial) clients. While this may often be strongly linked to research in universities . . . it nevertheless

---

[159] "Joint Research Project of BASF, Heidelberg and TU Darmstadt for Printed Electronics Centers the Next Phase," *Chemical Business NewsBase,* August 2, 2012. The printing press "serves as the platform for modified or new coating units and functions as the integrator for the newly-developed processes. The printed layers are only a few nanometers thick and must be highly homogeneous and defect-free." Transferring such highly complex printing processes to industrial scale requires a deep understanding of the printing process itself. Accordingly, the Institute for Printing Presses and Printing Methods at the Technical University of Darmstadt is developing a model defining the production parameters and examining the specific physical causes of inhomogeneities in the printed layers which can lead to failure of the device (ibid.).

[160] "BMBF Project NEMO on New OLED Materials under Merck Leadership Successfully Concluded," *Flexible Substrate,* September 2012.

[161] "OLED Project Focuses on Lighting, Signage," *EE Times,* February 11, 2009.

[162] Rebecca Harding, "Resilience in German Technology Policy: Innovation Through Institutional Symbiotic Tension," *Industry and Innovation,* December 2000, 228; Anton Heuberger, "Applied Research: The Fraunhofer Method," *Industries et Techniques,* February 23, 1988, JPRS-ELS-88-006; "Joseph von Fraunhofer and Max Planck Can Feel Satisfied," *Handelsblatt,* August 9, 1991, JPRS-EST-91-015.

constitutes a knowledge-creation sub-system of its own."[163] The Fraunhofers are active in research in more than 250 business fields and areas of competency representing virtually every subject relevant to a modern industrial economy.[164]

The Fraunhofer Gesellschaft is a registered association under private law (eigentragener verein) and technically independent of government direction. However, its research agenda is broadly consistent with those of the German government and the European Union, and occasionally the Fraunhofer carries out special missions at the behest of the German government.[165] Roughly one-third of its annual income is derived from "core" funding from the German federal government (Bund) and from the states (Länder) at a 90:10 ratio, respectively. Roughly another one-third of its income consists of payments by German government entities and other public organizations for contract research on various themes deemed to be in the public interest, such as environmental, energy, water quality, and defense/security-related research. Roughly the final one-third of the Fraunhofer's revenue is derived from industry through the performance of contract research with commercial applications. Because industry contract research payments are frequently partially comprised of subsidies from state, federal, and EU governments, the cumulative proportion of public funding of the Fraunhofer's operating budget may range as high as 80 percent.[166] In addition, the separate capital budget for expenditures on new research organizations and institutes has included large contributions from the Bund, the Länder, and the European Regional Development Fund (ERDF).[167]

The Fraunhofer Gesellschaft enjoys an outstanding global reputation and is widely credited as a key factor underlying German manufacturing competitiveness and the success of German goods in export markets.[168] Fraunhofer institutes,

---

[163] Fraunhofer Gesellschaft Annual Report, 2010, 15.

[164] For a comprehensive analysis of the Fraunhofer-Gesellschaft and its place in the German system of innovation, see National Research Council, *21st Century Manufacturing: The Role of the Manufacturing Extension Partnership Program,* ed. Charles W. Wessner (Washington, DC: The National Academies Press, 2013).

[165] Early in the reunification of Germany, the German research ministry tasked the Fraunhofer with reorganizing and restructuring the applied research institutes of the former German Democratic Republic. Dieter Thierbach, *Deutsche Enheit in Forschung und Technologie* (BMFT, 1991). The ministry has also assigned the Fraunhofer specific roles in developing key technologies such as new materials and information technology. *Materialforschung* (BMFT, 1986), JPRS-EST-86-036; *Forschung und Technik zum Whole der Menschen: Jahresbericht,* 1984, JPRS-WST-86-015; "Expose Chips with X-Rays," *Frankfurter Allgmeine Zeitung,* October 24, 1985, JPRS-WST-86-018.

[166] Diego Comin, Gunnar Trumbull, and Kerry Yang, *Fraunhofer: Innovation in Germany* (Harvard Business School Monograph 9-711-022, January 6, 2012).

[167] Fraunhofer Annual Report, 2011, 20. The ERDF is a fund administered by the EU to counter regional economic imbalances in Europe.

[168] "Germany—The New Mini-Superpower," *Christian Science Monitor,* January 30, 2011; "What's Behind the Success of German Manufacturing Industry?" *Xinhua,* February 23, 2012; "How the German Economy Became a Model," *Spiegel Online,* March 21, 2012; House of Commons, Committee on Science and Technology, *Second Report: Technology and Innovation Centres,* February 9, 2011, 41.

with an average staff size of 400, possess deep technological competencies and are normally extremely well equipped.[169] The Fraunhofer holds a huge patent portfolio that can be made available to industrial clients seeking to license advanced technologies.[170] The Fraunhofer develops product prototypes and industrial processes on behalf of its clients and demonstrates them in in-house simulation platforms and pilot manufacturing lines. The Fraunhofer's services are particularly important for Germany's small- and medium-sized enterprises, frequently regarded as the key to German export competitiveness, but which could not afford to invest in expensive research infrastructure and commonly have no organic R&D capability.[171] These firms are able to contract with the Fraunhofer on highly favorable terms to develop sophisticated technological solutions for the competitive challenges that they confront.[172] "[T]he research facilities of the Fraunhofer serve as external, very well equipped research departments" of small- and medium-sized German firms.[173]

At least 10 Fraunhofer Institutes and organizations are deeply involved in applied research activities relevant to flexible electronics. After the mid-1980s, in response to the direction of the German research ministry, the Fraunhofer devoted substantial resources to the development of advanced materials with high-technology applications.[174] Today the institutes are engaged in a broad range of research activities aimed at fostering new materials with flexible electronics applications. (See Table 5-13.)

---

[169] The Fraunhofer benefits from the "power and generosity" of the German machine tool industry, which loans equipment to the institutes on generous terms. The machine builders benefit from these arrangements because Fraunhofer develops improvements in the equipment, tests the machines in manufacturing environment and supplies data to the companies, and introduces the machines to potential customers. "Applied Research: The Fraunhofer Method," *Industries et Techniques,* October 20, 1987, JPRS-ELS-88-006.

[170] At the end of 2011, the Fraunhofer held 6,131 patents, 673 of which were registered in 2011. Normally Fraunhofer owns the intellectual property rights (IPRs) developed in the course of its research projects. Industry partners may also receive an exclusive license, but this is limited to the given application that was the subject of the research effort. The Fraunhofer remains free to license the technology to other users for different applications. Interviews with Fraunhofer Institute for Process Engineering and Packaging IVV, Friesing, Germany, June 13, 2012; Institute for Interfacial Engineering and Biotechnology IGB, Stuttgart, Germany, June 14, 2012; and Institute for Manufacturing Engineering and Automation IPA, Stuttgart, Germany, June 14, 2012.

[171] Christian Hamburg, *Structure and Dynamic of the German Mittelstand* (Heidelberg and New York: Physica-Verlag, 1999), 58–59; "Germany—The New Mini-Superpower," *Christian Science Monitor.*

[172] Fraunhofer R&D is cost competitive in that the institute's workforce is comprised, in part, of students, and the fact that the contract price does not include the sunk cost already incurred by the institutes to develop the skills relevant to the project. In addition, many industrial R&D contracts are partially subsidized by the national and/or regional and local governments and the European Union. See generally "Fraunhofer-Gesellschaft: The German Model of Applied Research," in National Research Council, *21st Century Manufacturing,* 224–284.

[173] Hamburg, *Structure and Dynamic of the German Mittelstand,* 58–59.

[174] BMFT, *Materialforschung,* March 5, 1986, JPRS-EST-86-036.

TABLE 5-13 Fraunhofer Institutes—Materials Competencies and Services in Flexible Electronics

Institute	E-Paper	E-Plastic	E-Textiles	Organic Conductor	P.-type Organic Semiconductor	N-type Organic Semiconductor	Dielectric	Carbon Nanotubes
POLO		xx	xx				x	
COMEDD		xx	xx	xx	xx	xx		
EMFT (IZE)	x	x	xx	x	x	x	x	
ENAS (IZE)	xx	xx	x	x			xx	x
IAP		x	x	x	xx	xx	xx	
ILT			xx	xx				
IPA								x
ISC			xx	xx			xx	
ISE		x	x	x	x	x		
ISIT			xx	xx				

SOURCE: OE-A, *Organic and Printed Electronics: Applications, Technologies and Suppliers,* June 2013.
NOTE: xx = products and services; x = competence.

Similarly, with respect to manufacturing processes and equipment for flexible electronics, the Fraunhofer Institutes offer a broad range of products, services, and competencies. (See Table 5-14.)

Fraunhofer Institute for Organic Materials and Electronic Devices Dresden

In 2012 the Fraunhofer IPMS Center for Organic Materials and Electronic Devices Dresden (COMEDD) was given the status of a new standalone Fraunhofer Institute. Fraunhofer COMEDD's mission is to conduct applied research in optoelectronic microsystems and surface modules, such as OLED lighting and OPV. A WTEC team that visited COMEDD in 2010 characterized its facilities as "vast and impressive."[175] Fraunhofer COMEDD features a clean room with the following equipment:

- Pilot line for the fabrication of OLEDs on 370 × 470 mm^2 substrates;
- Two pilot lines for 200 mm wafers for OLED integration on silicon substrates;
- Research line for R2R fabrication on flexible substrates.[176]

The initial total investment in COMEDD during the 2007-2009 timeframe was €24.6 million, of which the ERDF contributed 60 percent, the Land Government of Saxony 20 percent, and the federal government 20 percent.[177]

Fraunhofer COMEDD's research infrastructure enables it to offer industrial users product development services that begin with R&D and system concept through pilot fabrication. Its recent industrial partners include Universal Display Corporation, a major U.S.-based developer of OLED technology; VON ARDENNE Anlagentechnic GmbH, a major German producer of coating equipment; Heliateck GmbH, a German startup focusing on organic vacuum deposited solar cells; and AIXTRON SE, a German maker of metalorganic chemical vapor deposition equipment for the semiconductor industry.[178] Fraunhofer COMEDD and Tridonic Dresden (formerly LEDON OLED Lighting), participants in R2Flex, a BMBF-funded project to develop OLED lighting components, reported development of a desk luminaire with flexible OLED that can potentially be manufactured with an R2R process.[179]

---

[175] WTEC, *European Research and Development,* 11.

[176] "New Trends in OLED Lighting," *Flexible Substrate,* January 2014; "COMEDD Now a New Research Institute of the Fraunhofer," *Printed Electronics World,* July 12, 2012.

[177] Fraunhofer COMEDD website, <*http://comedd.fraunhofer.de*>.

[178] "Public Funded OLED Project So-Light Successfully Concluded," *Printed Electronics World,* February 6, 2013; "Universal Display and Fraunhofer Agreement for White OLED Lighting," *Printed Electronics World,* May 3, 2012; Organic Solar Cells Reach 6% Confirmed Efficiency," *Printed Electronics World,* September 1, 2009; "OLED Brings Out the Shine," *Printed Electronics World,* May 28, 2013.

[179] "EU R2R Program Creates Flexible OLED Desk Lamp," *Flexible Substrate,* October 2013.

TABLE 5-14 Fraunhofer Institutes—Equipment/Process Competencies and Services in Flexible Electronics

Institute	Polymer Film Encapsulation	Inkjet Printing	Photo-Lithography	Laser Ablation	Evaporation	Sputtering	Spin Coating	Clean Room
POLO	xx				xx	xx		
COMEDD	xx		x	xx	xx	xx	xx	xx
EMFT (IZE)	xx		xx	xx	xx	xx	xx	xx
ENAS (IZE)	x	xx		xx			xx	
IAP	x	xx	xx	xx	xx	xx	x	xx
ILT	xx	x	x	xx	xx			
IPA		x	x					x
ISC	xx	xx	xx	xx				xx
ISE	x	x	x		x	x	x	x
ISIT	xx	xx	xx		xx	xx	xx	xx

SOURCE: OE-A, *Organic and Printed Electronics: Applications, Technologies and Suppliers,* June 2013.
NOTE: xx = products and services; x = competence.

Fraunhofer COMEDD played an important role in IMAGE, a consortium that conducted research for 3 years, concluding in 2014, to develop printable electrode materials for high-performance lighting devices and organic solar cells. The project was co-funded by BMBF and France's Agence Nationale de la Recherche. Fraunhofer COMEDD's project partners included Carnot MIB, based in Bordeaux (project lead) and the companies Arkema and Tridonic (advisory support).[180] Dr. Olaf Hild, the business unit manager at Fraunhofer COMEDD, commented that in IMAGE "we were able to construct the electrodes very thin, transparent and flexible and to integrate them in our processes. Thus Fraunhofer COMEDD is now in a position to manufacture flexible organic devices such as OLED lighting films, organic solar cells or sensors as film according to customer requirements." Fraunhofer COMEDD and Carnot MIB were reportedly seeking industry partners to exploit these developments.[181]

Fraunhofer Institute for Photonic Microsystems (IPMS)

Fraunhofer IPMS conducts applied research in electronic, mechanical, and optical components and their integration into intelligent systems and devices.[182] Fraunhofer IPMS created COMEDD as an internal research center in 2008. When COMEDD was spun off to create a new institute, Fraunhofer IPMS continued some research in the field of flexible electronics such as process technology for printing OLED-based screens for signage and decorative lighting applications.[183]

Fraunhofer Institute for Applied Polymer Research (IAP)

Fraunhofer IAP, based in Potsdam, specializes in polymers and their applications, including synthesis, processing, and testing. One of the institute's principal themes has been developing new active polymer materials for organic electronic components such as OLEDs and adapting them to solution-based processes.[184] Fraunhofer IAP is currently collaborating with the Karlsruhe Institute of Technology in a 2012-2016 R&D project to develop printable organic solar cells with a

---

[180] The institutes in France are applied research organizations that have successfully applied for and been granted the "Carnot" designation by the government, which also provides some funding to the institutes correlated to the amount of contract research each institute performs for French industry. See "The Carnot Initiative in France," in National Research Council, *21st Century Manufacturing,* 368–389.

[181] "EU Completes Project on Printed Transparent Electrodes for Flexible OLEDs and Solar Cells," *Flexible Substrate,* March 2014; "Project IMAGE Yields Printable Transparent and Flexible Electrodes for OLEDs and Solar Cells," *EETimes Europe,* January 15, 2014.

[182] *<http://www.ipms.fraunhofer,de/en.html>*.

[183] "Transparent OLEDs for Signage and Decorative Lighting Applications," *Flexible Substrate,* January 2013; "New Developments in Highly Efficient OLEDs," *Flexible Substrate,* May 2013.

[184] OE-A, *Organiz Electronics 2nd Edition,* 2007, 52; Fraunhofer IAP, "Functional Polymer Systems," *<http://www.iap.fraunhofer.de/en/Forschmgsbereiche/Functionale-Polymerssysteme.html>*.

conversion efficiency exceeding 10 percent.[185] In 2012, Fraunhofer IAP successfully concluded a 3-year €29 million R&D project, "NEMO" (New Materials for OLEDs from Solutions) involving Merck and 10 other partners from industry on academia to develop phosphorescent materials for red, green, and blue applications in large surface OLED components.[186] Fraunhofer IAP is currently developing OLED, OPV, and OTFT technologies for applications in flexible signage, security, and energy harvesting. An area of focus is the combination of different types of organic electronic devices, such as OTFT-driven OLEDs and OPV-powered OLEDs.[187] In 2013, Fraunhofer IAP opened a new pilot line for producing newly developed OLEDs on a near-industrial scale.[188]

Fraunhofer Institute for Silicate Research (ISC)

Fraunhofer ISC develops inorganic-organic hybrid polymers (ORMOCER®) with emphasis on applications in polymer electronics systems. A Fraunhofer ISC team worked for nearly 20 years to develop a coating technology based on ORMOCER® that effectively protects flexible electronics surfaces from the deleterious effects of oxygen and water. The team developed a barrier lacquer that was combined with silicon dioxide. Dr. Sabine Amberg-Schwab, who headed the research team, commented that

> [t]he results were astounding. A barrier effect that is far better than could be expected from adding only the two layers. The reasons for this are special effects that are generated between the two materials.[189]

---

[185] This project is supported with a grant of €4.25 million from the Federal Ministry of Education and Research (BMBF). "Solar Power from Plastic Foils," *Printed Electronics World,* July 5, 2012.

[186] Dr. Udo Heider, the Head of the OLED unit at Merck, commented on NEDO that "the success of the project is an enormous and important step for printable material systems with very good performance data. We are enabling our customers to use cost-efficient manufacturing processes, which thanks to their low material losses in production will ultimately benefit the environment." The project was co-funded by BMBF. Other partners included the universities of Potsdam, Regensburg, and Tübingen and the Humboldt University of Berlin; Heraeus Precious Metals GmbH & Co. KG; Enthone GmbH; and Delo Industrie Klebstoffe. "BMFB Project NEMO; Research on New OLED Materials," *Printed Electronics World,* September 18, 2012.

[187] In a large clean room environment, the institute offers a number of processing techniques, including spin coating for material evaluation in laboratory devices, inkjet printing, and high-precision slot die coating on a robot-controlled manufacturing line on which devices can be fabricated on a pilot scale in sizes up to 150 sw. mm. OE-A, *Organic and Printed Electronics,* 86.

[188] The new line was developed in cooperation with the plant manufacturer MBRAUN and was supported by funding from the federal research ministry. "New Pilot Line for Organic Electronics," *Printed Electronics World,* February 11, 2013. "Fraunhofer IAP Opens Pilot Line for Organic Electronics," *Flexible Substrate,* February 2013.

[189] "Flexible Films for Photovoltaics," *Printed Electronics World,* May 31, 2011; "Ultra-High Basics for Encapsulation of Flexible Organic Electronic Devices," *Flexible Substrate,* March 2014.

OROCER®s are lacquers with properties that can be adapted to a variety of substrate types according to specific customer requirements.[190] Fraunhofer IAP is participating in the ENAB-SPOLED project, co-sponsored at €4 million by government agencies in Germany, Austria, and the UK, and will use its polymer expertise to develop charge transport polymers for solution processing for OLED lighting devices.[191]

Fraunhofer Institute for Electron Beam and Plasma Technology (FEP)

Fraunhofer FEP develops efficient vacuum coating methods (including coating of flexible products) and electron beam technologies. In 2013, Fraunhofer FEP is presenting at trade shows a novel R2R manufacturing process for high barriers (coatings for protection against moisture and other agents) and functional films for flexible displays. This technology is intended to address a major obstacle to the efficient production of large area flexible displays, the lack of cost-efficient and reliable methods for encapsulating those displays. The institute states that it has developed R2R technology that on a pilot scale can apply multiple layers of protective coatings (zinc tin oxide or aluminum oxide) to surfaces of 400 mm width and up to 500 m long with the lowest water-vapor permeability properties in the world.[192]

Fraunhofer Institute of Reliability and Microintegration (IZM)

Fraunhofer IZM specializes in integration of electronic, optical, actuator, and sensor functions. Fraunhofer IZM has spun off several of its internal departments to create new Fraunhofer institutes pursuing flexible electronics themes, including the Fraunhofer Institute for Electronic Nanosystems (ENAS; 2008) and the Fraunhofer Institute for Molecular Solid State Technology (EMFT). The institute develops multifunctional systems for applications on foils, textiles, and other flexible substrates. The institute's Laboratory for Textile-Integrated Electronics (TexLab) develops interconnection technologies for stretchable and textile substrates, with applications in fields such as medical engineering, security and logistics, fashion, lighting, and construction. (See Table 5-15.)

Fraunhofer Institute for Electronic Nanosystems (ENAS)

Fraunhofer ENAS' Printed Functionalities department conducts R&D in the field of flexible large area organic electronics and printed electronics. Its main

---

[190] Fraunhofer Polymer Surfaces Alliances POLO, "Transparent High Barrier Film for Organic Electronics: Roll-to-Roll Pilot Production," 2013.

[191] "ENAB-SPOLED Project Targets Solution-Processed OLEDs for Lighting," *Flexible Substrate,* October 2013.

[192] Fraunhofer FEP, "Functional Films for the Displays of the Future," May 31, 2013, <*http://www.fep.fraunhofer.de/en/press_and_media/Pressemittilungen/06_213.html*>.

TABLE 5-15 FhG IZM TexLab Electronic Textile Applications

Project	Product
ConText	Textile ECG and EMG sensors for monitoring heart and muscle activity in sports clothing
AlarmTextil	Large area fabric with integrated sensors for alarm systems
InsiTex, Place-IT	Seat occupancy sensor and interior lighting for vehicles
TextraLog	GTextile RFID transponders for logistics
Pocket Lock Backpack	Antitheft antifraud protection embedded in clothing and textile accessories
Place-IT, LumoLED, DesignMesh	Lighting and displays on/in fabric
Sinetra, Textees	Sensors integrated into clothing for personal safety applications

SOURCE: Fraunhofer IZM, "Textile-Integrated Electronic Systems."

fields of concentration are digital fabrication, printed functionalities, and hybrid R2R printing applications.[193] The institute works closely with the Chair of Digital Printing and Technology at the Chemnitz University of Technology. Its infrastructure includes R2R printing machines for inkjet, flexo, and gravure printing. The institute is focusing on applications that include RFIDs, printed radio frequency identification antennae, and printed batteries.[194]

Fraunhofer Institute for Modular Solid State Technology (EMFT)

Fraunhofer EMFT develops sensors and actuators, including devices for flexible applications. One of the institute's current main topics is the integration of various foil components, including organic circuits, printed batteries, sensors, ultrathin ICs, and photovoltaic cells, into smart flexible systems.[195] The institute offers its industrial partners a research infrastructure and expertise in R2R fabrication and testing.[196] In 2002, the institute established the Bavarian Polytronic Demonstration Center (BDP) to support its Flexible Systems business unit. The BDP features state-of-the-art production equipment for the microfabrication of product demonstrators on foils.[197] In 2013, Fraunhofer EMFT disclosed that it

[193] OE-A, *Organic and Printed Electronics,* 87.

[194] Fraunhofer ENAS, "Printing Technologies for Functional Layers and Components," *<http://www.evias.fraunhofermde/en /core_competencies/printing_technologiesforfunctionallayersandcompnents.html>*.

[195] Fraunhofer EMFT was spun off as a stand-alone institute from the Fraunhofer Institute for Reliability and Microintegration IZM in 2010. "Fraunhofer EMFT Becomes an Independent Institute," *DeviceMed,* July 15, 2010.

[196] OE-A, *Organic and Printed Electronics,* 86.

[197] Fraunhofer EMFT Annual Report, 2012, 20.

had developed a glove with sensors that identify toxic substances and indicate their presence by changing colors.[198]

Fraunhofer Institute for Laser Technology (ILT)

Fraunhofer ILT pursues R&D themes associated with the industrial uses of lasers. Among other things the institute has developed high-speed plastic welding techniques for flexible materials substrates and processes for the high-speed, high-resolution patterning of thin films.[199]

Fraunhofer Institute for Integrated Circuits (IIS)

Fraunhofer IIS develops microelectronics systems, devices, and associated software. It has recently developed shirts with electronic sensors and measurement systems embedded in textiles for medical and fitness applications.[200]

Fraunhofer Institute for Solar Energy Systems (ISE)

Fraunhofer ISE develops solar energy technology. The institute is pursuing the development of more efficient organic solar cells that do not require indium tin oxide.[201]

Fraunhofer Institute for Silicon Technology (ISIT)

Fraunhofer ISIT specializes in silicon-based microelectronics. It is currently engaged in research on the combination of conventional silicon circuits with organic electronic components that exploits the advantages associated with each technology, such as bendable displays with integrated memory functions. The potential use of inkjet printing to less expensive substrates (e.g., paper, PET foils)

---

198 "Fraunhofer Develops Color-Changing Glove That Warns of Toxic Substances," *Flexible Substrate,* May 2013.

199 OE-A, *Organic and Printed Electronics,* 88; Fraunhofer ILT, "Laser Ablation for Thin Film Structuring," <*http://www.ilt.fraunhofer.de/en/publication-and-press/brochures/borchure_laser_ablation_for-thin_film_technology.html*>; "Comparison of Laser Ablation of Transparent Conductive Materials on Flexible and Rigid Substrates," *Flexible Substrate,* February 2013.

200 The Fraunhofer IIS FitnessSHIRT is a T-shirt with measuring systems for ECG and respiration recording, providing continuous monitoring of cardiac and respiratory functioning. Fraunhofer IIS, "FitnessSHIRT: Improving Safety Through Telemonitoring," <*http://www.iis.fraunhofer.de/en/bf/med/mss/fitnesshirt.html*>. The Fraunhofer IIS RespiSHIRT is a T-shirt suitable for normal activities. It includes an embedded respiratory measurement system that transmits data via wireless signal to a Smartphone or PDA, where it is analyzed. Fraunhofer IIS, "RespiSHIRT," <*http://www/iis/fraunhofer.de/en/bh/med/mss/respishirt,html*>.

201 OE-A, *Organic and Printed Electronics,* 89.

and the integration of flexible batteries into the production process are expected to conserve energy and reduce the cost of electronic modules.[202]

Fraunhofer Institute for Manufacturing Engineering and Automation (IPA)

Fraunhofer IPA develops production processes that are conducted under clean room conditions using digital 2-D and 3-D printing and additive technologies. The institute develops solutions for the industrial handling, separation, transport, and storage of foil substrates; high-precision assembly processes for small devices mounted into multifunctional plastic film and foil-based systems; and processes for selective coating applications utilizing inkjet and electrophotographic printing and dispensing technologies.[203]

Fraunhofer Institute for Material and Beam Technology (IWS)

Fraunhofer IWS specializes in research on laser and surface technology. Researchers at this institute have developed a new manufacturing process for producing thermoelectric generators (TEGs) through a printing process on large, flexible surfaces consisting of environmentally friendly materials. Embedded devices have the potential to produce electricity generated by changes in temperature. The technology may be employed to produce electricity from waste heat in platforms such as automobiles, cooling towers, large computer centers, sewage systems, and industrial production lines.[204]

Fraunhofer Institute for Chemical Technology (ICT)

Fraunhofer ICT in Karlsruhe has been working with industrial partners to develop a tool for characterizing polymer nanocomposites, onBOX, which can be used during the production process itself. onBOX is mounted to the exit nozzle of the conveyor to analyze the polymer compound while it is in the mixing plant, using a combination of ultrasound, microwaves, and spectroscopy to assess the composition of the compound. A computer uses the data so generated to fix precise mixing ratios needed to produce the desired material and to identify the precise manufacturing process required, feeding the information to the machine's control system.[205]

---

[202] OE-A, *Organic and Printed Electronics*, 89.

[203] OE-A, *Organic and Printed Electronics,* 88.

[204] "Fraunhofer Printed Thermoelectric Generators Could Capture Energy from Waste Heat," *Flexible Substrate,* May 2013.

[205] "Testing Nanomaterial Smart Plastics in Real Time at Fraunhofer ICT," *Flexible Substrate,* November 2013.

Max Planck Gesellschaft

The Max Planck institutes are public research organizations responsible for basic research. Like the Fraunhofer, the Max Planck society is an independent nongovernmental nonprofit research association supported by funding from the Bund and the Länder. Max Planck comprises roughly 80 thematic research institutes and is regarded as one of the finest research organizations in the world. Max Planck and its institutional predecessor, the Kaiser Wilhelm-Gesellschaft (KwG), can count more than 30 Nobel laureates among their scientists, including Albert Einstein (1921).

Max Planck Institute for Solid State Research

The Organics Electronics Group at the Max Planck Institute for Solid State Research in Stuttgart is pursuing research themes applicable to flexible electronics. A particular area of interest is organic transistors. The group is studying ways to reduce the operating voltage of such transistors, improve the stability of p-channel and n-channel transistors, design organic transistors with channel lengths down to 100 nm, and improve the high-frequency performance of organic transistors. The group is also investigating manufacturing techniques for organic transistors including inkjet and microcontact printing. In addition to its work on transistors the group is exploring various organic and hybrid nanostructures that take advantage of unique properties of certain materials such as carbon nanotubes and organic/inorganic radial superlatives.[206] In 2012, MPI-P and BASF opened a joint research and development center, the Carbon Materials Innovation Center (CMIC), at BASF's site in Ludwigshafen. CMIC will investigate the potential of graphene and other carbon materials for applications including touchscreens and solar cells.[207]

Max Planck Institute for Polymer Research (MPI-P)

The MPI-P in Mainz performs research on polymers for applications in a variety of fields including flexible electronics. In a joint research project with Japan's National Institute of Materials Science (NIMS) the institute developed the world's first supramolecular thiophene nanosheet, a 2-D material with a thickness of 3.5 nanometers that has potential application in organic electric devices without the expense and energy consumption associated with vacuum vapor deposition processes.[208]

---

[206] Max Planck Institute for Solid State Research, <*http://www.fkf.mpg.de/65473/Research*>.

[207] "BASF, Max Planck Institute for Polymer Research Inaugurate Joint Graphene Research Lab," *Flexible Substrate,* November 2012.

[208] "German Researchers Synthesize World's First Supramolecular Thiophene Nanosheets," *Flexible Substrate,* May 2013.

Max Planck Institute of Colloids and Interfaces

The Max Planck Institute of Colloids and Interfaces in Potsdam-Golm disclosed in 2013 that it has developed a technique that utilizes a conventional inkjet printer to create conductive structures on paper. The printer prints a catalyst in a specific pattern on paper, after which heat is applied to convert the printed pattern into graphite (conductive) while the adjacent areas on the paper are converted into amorphous carbon (nonconductive).[209]

Companies

BASF

Germany's BASF is the world's largest chemicals company. A BASF subsidiary, BASF New Business GmbH, which is tasked with starting up new businesses in promising technology areas, oversees BASF's operations in organic and printed electronics. The company is developing printable material systems for thin-film transistor applications and phosphorescent materials for OLEDs.[210] BASF has developed a significant number of major research collaborations with universities and public research organizations in Europe and abroad with themes relevant to flexible, printed, and organic electronics.[211] (See Table 5-16.)

Heliatek

Heliatek, an OPV maker, was spun off from the Technical University of Dresden and the University of Ulm in 2006. It has continued to work with these academic institutions in developing OPV technology. A number of large industrial companies have invested in Heliatek, including Bosch, BASF, and RWE. In 2012, Heliatek started up its first manufacturing facility for flexible organic solar panels in Dresden, utilizing an R2R process with vacuum deposition at low temperatures.[212] In 2013, Heliatek disclosed that in collaboration with the two

---

[209] "Max Planck Institute Researchers Use an Inkjet Printer to Create Electrically Conductive Paper," *Flexible Substrate,* October 2013.

[210] OE-A, *Organic and Printed Electronics*, 44.

[211] "National University of Singapore, BASF Embark on Joint Graphene Research," *Printed Electronics Now,* January 20, 2014; "BASF, Top American Universities to Research New Functional Materials," *Printed Electronics Now,* March 14, 2013; "BASF, Max Planck Institute for Polymer Research Inaugurate Joint Research Laboratory for Graphene," *Printed Electronics Now,* September 24, 2012; "BASF, Three Top European Universities Team Up on Functional Materials Research," *Printed Electronics Now,* July 28, 2009; "BASF, Heidelberg and TU Darmstadt Collaboration Shows Promise," *Printed Electronics Now,* August 2012; "BASF to Set Up Electronic Materials R&D Center Asia Pacific at Sungkyunkwan University in South Korea," *Printed Electronics World,* November 7, 2013.

[212] "Heliatek Opens Groundbreaking Production Facility for the Manufacture of Organic Solar Films," *Printed Electronics Now,* March 12, 2012.

TABLE 5-16 BASF—Academic Research Collaborations for Flexible Electronics

Country	Institutions	Themes
Singapore	National University of Singapore	Graphenes for organic electronic devices
United States	Harvard MIT U Mass Amherst	New materials for automotive, building, and energy industries
Germany	Max Planck Inst. for Polymer Research	Applications of innovative carbonized materials (graphene)
Switzerland, France, Germany	U. Strasbourg, Freiburg, and ETH Zurich	Functional materials with new properties
Belgium	IMEC	Process chemicals for semiconductors
Germany	T.U. Darmstadt	Intelligent printing processes with applications in flexible components
Korea	Sungkyunkwan U	Electronics materials

universities, it had succeeded in pushing the conversion efficiency of OPV cells to an unprecedented 12 percent.[213]

Novaled Ag

Novaled was established in 2001 as a spinoff from the Technical University of Dresden with the support of the Fraunhofer IPMS.[214] It is a leading developer of OLED technology and holds or has pending more than 500 patents in the field.[215] It also supplies proprietary materials for use in the manufacture of OLED and PV devices. In 2013, Korea's Samsung Group concluded an agreement to acquire Novaled.[216] In early 2014, Novaled and Plastic Logic jointly demonstrated a flexible, plastic, fully organic AMOLED display, which is expected to accelerate the commercialization bendable and wearable displays.[217]

[213] "Heliatek Achieves 12% Organic Solar Cell Efficiency," *Optics.org,* January 23, 2013.

[214] WTECH, *European Research and Development,* 95.

[215] OE-A, *Organic and Printed Electronics,* 64.

[216] At the time Samsung Venture Investment held 10 percent of Novaled. Pursuant to the agreement, Samsung affiliate Cheil Industries will acquire roughly a 50 percent stake, and Samsung Electronics Co. Ltd. will acquire 40 percent. "Samsung Agrees to Buy German Screen Lighting Firm Novaled," *Bloomberg,* August 9, 2013.

[217] "Plastic Logic, Novaled Partner to Demonstrate a World First for Displays," *Printed Electronics Now,* February 7, 2014.

AIXTRON

AIXTRON is a German producer of metalorganic chemical vapor deposition equipment for the semiconductor industry. The company was spun out of RWTH Aachen, one of Germany's leading research universities, in 1985, and it currently generates most of its revenues from sales to manufacturing clients in Asia. As of late 2013, AIXTRON had sold more than 3,000 deposition systems globally and had grown to more than 800 employees. AIXTRON is currently engaged in R&D for process technologies for OLED displays and lighting, organic material large area deposition, and applications using carbon nanostructures (graphene and carbon nanowires and nanotubes).[218] AIXTRON is currently leading the Production Work Package of the EU's Graphene Flagship project and is working to develop large-scale equipment for water-based graphene and continuous production of foil-based graphene for transistors and transparent conductive films.[219]

Merck Chemicals

Merck Chemicals, based in Darmstadt, Germany, holds one of the world's largest portfolios of organic semiconductor patents, including technologies applicable in printed and flexible electronics.[220] Merck operates a chemicals research site in Southampton, UK, the Chilworth Technical Centre, and its specialty materials have been incorporated in Plastic Logic's display products pursuant to joint development, test, and commercialization arrangements with that company.[221] In 2013, Merck was selected by BMBF to head the POPUP consortium (2013-2016), a collaboration to develop materials with OPV applications.[222] In 2013, Merck won the first place Solar Energy Award (category: PV Materials Enabling Award) at the European Photovoltaic Solar Energy Conference for its SolarEtch structuring posters, which enable selective etching of antireflective coatings and passivization layers on solar cells and transparent conductive materials.[223]

---

[218] "AIXTRON Celebrates 30th Anniversary." *Printed Electronics Now,* December 5, 2013.

[219] "AIXTRON Plays Key Roles in Two-Dimensional Nanomaterial Projects," *Printed Electronics Now,* October 31, 2013. In early 2014, AIXTRON and Germany's Manz AG disclosed a strategic cooperation agreement to demonstrate efficient organic layer deposition up to a substrate size of 2,300 × 2,500 mm based on AIXTRON's propriety OVPD process technology. The new process is expected to enable efficient production of OLEDs for displays and lighting applications. "AIXTRON and Manz AG Agree to Strategic Cooperation for OLED Manufacturing," *Flexible Substrate,* January 2014.

[220] IDTechEx, *Printed, Organic & Flexible Electronics* (2011) op. cit., 255. Merck is a market and technology leader in liquid crystal mixtures used in all display applications. "Merck Exhibits Innovative Display Materials at IMID, FPD International," *Printed Electronics Now,* October 11, 2011.

[221] "Merck KGaA and Plastic Logic Jointly Develop New Generation Organic Semiconductors," *Printed Electronics Now,* April 14, 2010.

[222] "POPUP: Novel Organic Solar Cells," *Science Daily,* December 13, 2013.

[223] "Merck KGaA Receives Solar Industry Award 2013," *Printed Electronics Now,* October 11, 2011.

PolyIC

PolyIC Gmbh was established in 2003 as a joint venture between Germany's Leonhard Kurz Stiftung & Co. KG, a maker of stamping tools and foils (51 percent share) and Siemens AG (49 percent share). PolyIC develops technologies for touch sensors, passive devices, and pointed displays, notably flexible RFIDs.[224] PolyIC, one of the first firms to introduce printed flexible RFIDs, specializes in making thin, flexible circuits that are inexpensive and disposable.[225] In 2013, PolyIC and majority owner Kurz jointly demonstrated a number of applications that combined functional films and decorative elements, including an automotive center console utilizing touch sensor film and decorative film permitting touch control of keys, climate control, and entertainment systems.[226]

Evonik Industries AG

Evonik, based in Essen, is one of the world's leading producers of specialty chemicals. It has been a collaborator with AU Optronics of Taiwan in a joint venture, Evonik Forhouse Optical Polymers (EFOP), which operates a plant producing acrylic polymers for the TFT-LCD industry.[227] Evonik participated in the EU FP7 project ORICLA (2010-2012), along with IMEC, Holst Centre, and PolyIC, in the development of thin-film ultra-high-frequency RFID tags.[228] In 2013, Evonik's collaboration with the Holst Centre was expanded when the company joined Holst's research program on organic/oxide semiconductors, devices which can be printed on thin films.[229] In 2012, Evonik introduced FLEXOSKIN, a barrier film that can be used to cover flexible photovoltaic devices that is transparent but blocks moisture and harmful ultraviolet (UV) radiation.[230]

FINLAND

Although Finland has a population only slightly more than 5 million, it has had a disproportionately large impact on global innovation, and has been rated as Europe's most innovative business environment.[231] The Academy of Finland,

---

[224] OE-A, *Organic and Printed Electronics*.

[225] "A New Industry Shapes the Future of Printing," *Printed Electronics Now*, December 2008.

[226] "PolyIC, Kurz Present Touch Applications for Automotive Industry at IC 2013," *Printed Electronics Now*, November 13, 2013.

[227] "Evonik and AU Optronics Corp. Conclude Strategic Partnership," *Printed Electronics Now*, December 2, 2010.

[228] "IMEC Paves Way for Intelligent Item-Level RFID Tagging to Replace Bar Codes," *Printed Electronics Now*, December 13, 2012.

[229] "Evonik, Holst Centre Partner to Extend Thin Film Electronics," *Printed Electronics Now*, February 12, 2013.

[230] "Evonik Offers FLEXOSKIN Barrier Film for Protection of Flexible PV," *Printed Electronics Now*, January 11, 2012.

[231] Lisbon Council and Allianz Dresdner Economic Research, "The Lisbon Review 2008."

TABLE 5-17 Finnish SMEs and Flexible Electronics Technologies

Company	Technology
Beneg	Roll-to-roll atomic layer deposition equipment providing barrier coating for flexible electronics and PVs
Canatu	Flexible, highly transparent conductive carbon nanomaterial-based thin films for customized touch sensors
Iscent	Printable holographic-like light-scattering films for smart packaging

a governmental funding body, finances basic scientific research activities in Finland, while Tekes, an agency of the Ministry of Trade and Industry, funds applied research.[232] Tekes provides about 30 percent of total Finnish public funding of R&D; in 2010 it channeled €633 million in R&D support to Finnish universities, research organizations, companies, and public organizations.[233] Tekes' funding prioritizes collaborative projects ("programmes") bringing together universities, research organizations, and companies, including EU projects.[234] Between 2007 and 2013, Tekes implemented the Functional Materials Programme, a major R&D project to develop new applications and competitive advantage for Finnish industry through materials technology. Tekes provided €70 million of the project's €140 million budget. The project included themes applicable to flexible electronics, and three participating Finnish SME's developed R2R materials, equipment, and manufacturing processes that were transferrable to industrial applications.[235] (See Table 5-17.)

Valtion Teknillinen Tutkimuskeskus (VTT, State Technical Research Center) is a nonprofit government-owned research organization subordinated to the Ministry of Employment and the Economy that provides applied research services in a broad range of technologies. It receives "basic" funding from the government of Finland and additional revenue from Tekes, the EU, and other Finnish and foreign governmental entities. (See Table 5-18.)

All told VTT derives greater than 73 percent of its turnover revenue from public funding.[236]

[232] Tekes, *Tekes Review* 289/2012, 20. The Academy of Finland is supervised by the Ministry of Education and Culture, Tekes by the Ministry of Employment and the Economy. Ibid.

[233] *Tekes Review,* 25. In 2010, 61 percent of Tekes' R&D funding was directed to small- and medium-sized enterprises and 70 percent was provided to companies with less than 500 employees (ibid., 28).

[234] *Tekes Review,* 32.

[235] "Dr. Markku Heimo, Spinverse Ltd., and Dr. Markku Lumra, Teches," *Research Europe,* April 15, 2013. Beneq reports that its R2R atomic layer deposition process (ALD) developed in the course of this project "is a true paradigm shift enabling high-throughput PV and OLAE applications." "Roll-to-Roll Atomic Layer Deposition Technology for Producing Single Layer Ultra Barrier Films," *Flexible Substrate,* February 2013.

[236] VTT, *VTT Review,* 2012, 40.

TABLE 5-18 VTT Sources of Revenue (2012)

Source	Amount (Millions of Euros)
Basic government funding	94.0
Tekes	57.5
Other domestic public sector	25.7
EU	29.8
Other foreign public sector	4.4
Private sector (Finland)	58.2
Private sector (foreign)	16.9
Total Turnover	**286.5**

VTT began investing in research in "printed intelligence" in the late 1990s[237] with an emphasis on commercialization, including the employment of business development specialists.[238] This effort enjoyed some successes, such as the development with the packaging company Mreal of a process for printing RFID patterns directly onto packaging without using a silicon chip.[239] In 2006 VTT established the Center for Printed Intelligence (VTT/CPI), which has grown into a research center with more than 100 employees and is equipped with pilot R2R manufacturing facilities. The Center collaborated with Ciba to develop printable functionalities in high-volume packaging and diagnostics.[240] However, as VTT observed in 2010 "despite such efforts, the number of end products developed for/with our customers has been rather limited," a phenomenon attributed to "expectations and excitement" that "often heightens to levels that current technological capabilities are not yet able to meet."[241]

[237] "Printed intelligence" refers to components and systems that extend the applications of printing beyond traditional text and graphics and perform actions as part of functional products or information systems. It includes printed electronics components and systems on flexible substrates. PrintoCent, "Printed Intelligence," <*http://www.printocent.net/intelligence.htm*>.

[238] VTT, *Research, Development and Commercialization Activities in Printed Intelligence,* 2010, 4–5. Printo, a 3-year project that began in 2001, was funded by Tekes to develop methods for fabricating passive and active electronics and optical and optoelectronic elements using R2R fabrication processes. Printo led to the establishment of a pilot R2R production line at a VTT site in Oulu with a clean room and the capability to process paper, plastic, and other flexible substrates up to 20 cm in diameter. The same facility also included two gravure printers with thermal, infrared, and ultraviolet curing units, an R2R lamination unit, and an R2R hot-embossing machine. "Northern Lights," *Plastic Electronics,* April-May 2008.

[239] "Radio Barcodes Printed on Consumer Packaging," *Printed Electronics World,* June 23, 2005.

[240] "Ciba and VTT Technical Research Centre of Finland Expand Collaboration in Printed Electronics," *Nanotechwire,* June 8, 2008.

[241] VTT, *Printed Intelligence,* 4.

In 2009, VTT, the University of Oulu, the Oulu University of Applied Sciences, the City of Oulu, and Oulu Innovation Ltd. founded PrintoCent, an organization tasked with creating business from emerging printed intelligence technologies through collaborations with large companies and demonstrator and piloting projects with smaller companies.[242] PrintoCent launched a €10 million effort (2009-2011) to construct an applications design and pilot manufacturing environment for printed electronics and diagnostics in Oulu, an effort supported by public funding from the ERDF, the city of Oulu, and the State Provincial Office of Oulu's Education Department.[243] PrintoCent was expected to invest €15 million annually in R&D projects and to help create a skilled local workforce in the field of printed intelligence.[244] By the end of 2013 PrintoCent employed more than 200 professionals. As of the end of 2013 PrintoCent had spawned 18 startup companies.[245] In late 2013, the Chief Business Development officer of Ynvisible, a consortium member specializing in electrochromic displays using proprietary inks, summarized the benefits of the consortium to his company:

> As a small company, we aim to do much of our experimental work in research and pilot facilities. This reduces our risks in making capital investments into equipment that we may not ultimately need or have very little use for. The PrintoCent facilities have evolved over more than 10 years, and based on earlier experiences. This same expertise accumulated within PrintoCent can help our company save time and money, as we don't have to reinvent the wheel and repeat all of the earlier mistakes. Many people involved in PrintoCent have several years of experience in taking production of novel printed systems from lab into volume production of final products.[246]

In March 2012, the PrintoCent Roll-to-Roll and Hybrid Integration pilot factory was inaugurated, featuring six pilot lines offering scaling-up manufacturing, demonstration, and piloting services.[247] According to a VTT spokesman the new facility was "the most advanced industrialization capability and service, being at least 2 or 3 years ahead of others.[248] In 2012, VTT won an award at the IDTechEx

---

[242] PrintoCent is formally the Printed Electronics and Optical Measurements Innovation Centre.

[243] "Converting State-of-the-Art Research Results Into Significant Business," *Printed Electronics World,* June 11, 2009.

[244] VTT, *Research and Development Activities in Printed Intelligence,* 2009, 6.

[245] "PrintoCent Consortium Seeks to Develop, Commercialize Printed Intelligence," *Printed Electronics Now,* December 2012.

[246] One of the PrintoCent spinoffs is TactoTek Oy, which develops formable optical touch panels for applications in mobile phones, tablet computers, and other consumer electronics and industrial application ranging from coffeemakers to construction equipment. VTT, "Introducing Printed Electronics into Mass-Produced Articles," *<http://www.vtt.fi/references/introducint_printed_electronics_into_mass_produced_articles.jsp?lang=en>*.

[247] Harri Kopola, "Printed Intelligence Technology at VTT" (ERATO Seminar sponsored by JST ERATO Someya Bio-Harmonized Electronics Project, October 31, 2012).

[248] "World's First Pilot Factory for Printed Intelligence Industrialization Opens at VTT," *EurekAlert,* March 13, 2012. In September 2012, the facility was upgraded to R2R assembly as well as R2R

Printed Electronics USA forum for "Best Technical Development Manufacturing" for its hybrid pilot manufacturing facility. The judges agreed that "the combination of manufacturing capability and expertise in one location gives unique opportunity to develop new products quickly and effectively from prototypes to proof-of-production level piloting."[249]

Companies

Beneq

Beneq is a Finnish maker of thin-film coatings and coating equipment. Established in 2005 the company has grown rapidly through acquisitions. Financial support from Tekes was critical to the company's initial growth, according to co-founder Tommi Vaino: "Thanks to the support of Tekes, we were able to set up a research environment in our facilities in Vataa. Without that support, we'd have been dead in the water."[250] In 2013, Beneq sold three of the world's first scaled-up R2R atomic layer deposition systems, technology that can be employed for encapsulation of OLED and flexible photovoltaic devices.[251]

Enfucell Oy

Enfucell was founded in 2002 to commercialize the results of a decade of work at the Helsinki University of Technology led by Dr. Zhang Xiachang developing power sources for low-power applications. Enfucell's Soft Battery is a printed flexible battery for use in disposable and short-use products such as RF sensors, RFID tags, cosmetics, drug delivery patches, and functional packaging. Soft Battery is comprised of environmentally friendly materials that can be disposed of with normal household waste. The company, which has now a number of awards, received early-stage funding from Tekes.[252]

---

printing. "VTT Upgrades PrintoCent Pilot Factory," *Printed Circuit Design & Fab,* September 20, 2012; "R2R Manufacturing of Organic PV Using Gravure and Rotary Screen Printing Technique," *Flexible Substrate,* May 2013.

[249] "Printed Electronics USA 2012 Awards Recognize New Developments," *Flexible Substrate,* January 2013.

[250] Tekes, "Beneq: Coating Manufacturer Ready to Grow," 2013, *<http://www.tekes.fi/en/tekes/results-and-impact/cases1/2013/beneq-coating-manufacturer-ready-to-grow>*.

[251] "Beneq Wins Tekes Commercialization Breakthrough Award," *Printed Electronics Now,* December 19, 2013; "The Status and Outlook of R2R Atomic Layer Deposition Technology," *Flexible Substrate,* January 2014.

[252] "Great Potential Seen in Soft Batteries," *Helsingin Sanomat,* March 20, 2007; "Enfcell Brings Expertise to the Printed Battery Market," *Printed Electronics Now,* September 2009; http://enfucell.com.

6

East Asia

Japan, South Korea, and Taiwan are well positioned to enter the field of flexible electronics and to dominate some of the emerging commercial applications. They have numerous large industrial groups with extensive manufacturing capabilities in sectors directly relevant to the production of flexible electronic devices, including microelectronics, optoelectronics, printing, electronic materials, photovoltaics, and displays. All three have developed increasingly strong research capabilities relevant to flexible electronics in government laboratories, universities, and company research and development (R&D) centers, and they are demonstrating that technological gaps can readily be filled by acquisition of technology abroad. With close government support, companies based in these countries have become adept at moving new technologies from the laboratory to the market quickly and efficiently. Moreover, China has developed a robust indigenous liquid crystal display (LCD) industry which it is using to leverage market entry in organic light-emitting diode (OLED) displays, albeit with a limited domestic supply chain and research base.

Past promotional efforts in the East Asian countries have established a strong foundation for the development of flexible electronics. Large-scale government-backed R&D efforts in semiconductors, optoelectronics, photovoltaics, new materials, and nanotechnology have ensured that with respect to the "convergence technology" of flexible electronics, Korea, Taiwan, and Japan have highly developed capabilities with respect to the major converging fields.

SOUTH KOREA

South Korea enjoys an established industrial base, supply chain, and extensive know-how and intellectual property associated with the manufacture of displays, representing a major advantage in the emerging field of flexible displays. The Samsung and LG groups are world leaders in the production of displays and both companies have already commercialized OLEDs, which will play a significant role in many flexible electronics applications. Samsung is also one of the world's leading manufacturers of semiconductors. A 2013 analysis of the emerging flexible display market concluded that South Korean companies

> appear better placed to benefit from flexible display than their global competitors, at least to begin with. They are at an advanced stage of development in most of the processes needed to make this technology a reality and should also be the first to enjoy the advantages of a vertically integrated supply chain. [LG and Samsung] are leading the way in developing this new technology . . . and have the capability to manufacture and release differentiated products that will change the market.[1]

The Korean giants have proven nimble competitors, displacing Japanese firms as leaders in displays. Korean manufacturers are sometimes faulted for "lacking unique or original ideas" but are credited with "faster decision-making and a greater willingness to take risks with unproven ideas."[2]

The sheer scale of recent commercial successes by Korean electronics groups such as Samsung and LG sometimes obscure the fact that South Korea has some areas of potential weakness as it seeks to develop into a leader in flexible electronics. It is comparatively weak in basic and fundamental research.[3] The country's science establishment is criticized for being overly bureaucratic, stressful, and unrewarding for researchers, and too focused on short-term results.[4] Small- and medium-sized enterprises, the principal source of innovation, are widely seen as making an inadequate contribution to the Korean economy, which may or may not reflect

---

[1] HSBC Global Research, *Flexible Display: Fantastic Plastic—A Shape-Shifting Game Changer,* April 2013, 3.

[2] "South Korean Electronics Companies Are Beating Japanese Competitors to the Punch in Getting Cutting Edge Technologies Developed by Japanese Researchers into Their Product Ranges," *Asahi Shimbun,* July 9, 2010; "CES Reveals Korea's IT Firms Are Lagging Their Rivals," *Chosun Ilbo Online,* January 12, 2010.

[3] Changhee Lee, Seoul National University, "Flexible and Printed Electronics—A Korean Initiative," September 24, 2010.

[4] In 2012, a survey of 293 scientists in Korea and 226 scientists of Korean descent living in the United States found that 72 percent of the Korea-based scientists indicated they would leave Korea if the opportunity arose. Sixty-six percent of Korean scientists living abroad said they wanted to go back to Korea but "low pay and a poor research environment keep them abroad." When asked "what do you think of Korean science and technology policies?" 79 percent of Korean and 67 percent of Korean-American scientists responded "poorly" or "very poorly." The most prevalent complaints were "an environment to where it is difficult to advance in research, unrealistic research fee regulations, bureaucracy meddling in research and a rise in temporary research positions as compared to permanent positions." "For Scientists in Korea, Careers of Stress, Insecurity," *JoongAng Daily Online,* September 24, 2012; A Call for Scientific Reinvention," *JoongAng Daily Online,* April 21, 2012.

the dominance of the chaebol.[5] Materials and machinery for flexible electronics applications are heavily sourced from abroad.[6] Although flexible electronics is a quintessential "convergence" technology, Korean government support and regulatory oversight of industry has historically been rigidly segmented by sector, leading to obstacles and delays in the introduction of overlapping new technologies.[7] South Korea's university system is said to rank "far below Korea's economic status," and the country has had difficulty in attracting researchers from abroad. Barriers between academic disciplines have impeded the development of new technologies.[8]

The Korean government's strategy in flexible electronics has been to support research on the core technologies in the field (OLEDs, LCDs, e-paper, touch panels, flexible PCBs, RFIDs and organic photovoltaic cells [OPVs]). Support has been extended to equipment and materials suppliers, including small- and medium-sized enterprises. Weaknesses in fundamental research are being addressed through international collaborations.[9] In March 2011, Korea enacted the Industrial Convergence Promotion Act, effective in September 2011, to speed regulatory approvals for new products based on convergence technologies.[10]

Convergence Technology Initiative

In early 2011, South Korea's Ministry of Knowledge Economy announced a $1.4 billion plan to propel Korea into the "league of top five international

[5] "Non-Chaebol Firms Losing Ground to Chaebol in S. Korea," *Yonhap,* July 3, 2013. Korean SMEs are said to suffer from the "Peter Pan Syndrome," unwilling to grow beyond a certain size and forfeiting an estimated total of 160 benefits, including tax credits and deductions and exemptions from regulations available only to SMEs. Mid-sized businesses accounted for only 0.04 of Korea's 3.12 million enterprises in 2010. "Peter Pan SMEs Loath to Grow UP," *JoongAng Daily Online,* January 2, 2013; "Peter Pan Syndrome," *The Korea Herald Online*, January 5, 2013.

[6] "Printed Electronics in Korea," *Printed Electronics World,* February 15, 2011.

[7] In 2004, LG Electronics developed a mobile phone capable of measuring blood sugar level and managing the administration of medication. The company had to abandon plans for commercialization, however, because government regulations categorized the phones as medical devices, giving rise to a prohibitively burdensome approval process. Overlapping regulations and standards have been blamed for delays in the commercialization of Internet TV and Internet telephones (VoIP) in Korea. "Regulation Hindering Technological Advances," *Dong-A Ilbo Online,* March 27, 2010. "A Technological Hermit Kingdom," *JoongAng Daily Online,* March 23, 2010.

[8] Comments of Choi-Yang-hee, Dean of Graduate School of Convergence Science and Technology at Seoul National University, in "Korea Needs Change for Convergence," *JoongAng Daily Online,* May 23, 2009.

[9] Professor Changhee Lee, Seoul National University, "Flexible and Printed electronics—A Korean Initiative," September 24, 2010.

[10] "Korea Needs Change for Convergence," *JoonAng Daily Online,* May 23, 2009. In 2011, Seoul National University unveiled 55 interdisciplinary fields in an effort to "break down traditional boundaries between academic disciplines." "Interdisciplinary Studies," *Korea Times Online,* July 8, 2011. In 2010, Yonsei University was selected by the Ministry of Knowledge Economy as the site of a state-funded program to train IT scientists modeled on MIT's Media Lab, a "world class media convergence technology research center." *Dong-A Ilbo Online,* August 26, 2010.

technology power houses" by 2020. The centerpiece of this plan is the promotion of six so-called "convergence technologies" that are expected to generate sales of $330 billion by 2025, or about one-third of the country's gross domestic product (GDP) in 2010. Four of these are directly applicable to flexible electronics:

- Transparent flexible displays and related application products;
- Neuro tools that fuse information technology with neurology and nerve medical devices;
- Graphene materials and parts; and
- Super-fine print electronics manufacturing systems.[11]

The plan was developed by a team headed by Hwang Chang-gyu, former President of Samsung Electronics, and Minister of Knowledge Economy Choi Joong-Kyung.[12] The government reportedly plans to launch a "huge" printed electronics development fund with some matching investments by industry.[13]

The government is arranging extensive financial support for private companies that enter high-technology convergence industries and other designated sectors with $6 billion worth of loans, guarantees, and other forms of financial support:

- The Ministry of Trade, Industry and Energy (MOTIE) announced in 2013 that it would invest $42 million over the next 6 years for commercializing graphene applications in the information technology (IT) sector.[14]
- The government Ministry of Knowledge Economy (now MOTIE) committed in 2010 to provide $875 million, to be matched by private-sector funds, to develop 10 "World Premier Materials" for industrial use, including substrates for flexible displays.[15]
- The Korean Printed Electronics Association told a visiting foreign delegation in 2012 that "the South Korean government wishes to invest $48 million in printed electronics over the next six years," primarily for R&D.[16]
- In 2013, the Chairman of South Korea's Financial Services Commission said that the government, "public financial support organizations,"

[11] The other two convergence technologies to be fostered are offshore plants to industrialize deep sea resources and compact multipurpose module nuclear reactors.

[12] "Korea Hones in on Growth Engines," *JoongAng Daily Online,* March 22, 2011.

[13] "Printed Electronics in Korea," *Printed Electronics World,* February 15, 2011.

[14] "S. Korea to Spend 40 Mln USD on Graphene Development," *Xinhua,* May 21, 2013.

[15] "S. Korea Launches Project to Develop New Materials," *Yonhap,* September 30, 2010; "Samsung, LG, Hyundai Motor Join Hands for Next Generation Flexible Display Development," *MK English News Online,* July 5, 2010.

[16] "IDTechEx Visits South Korea," *Printed Electronics World,* April 18, 2012.

TABLE 6-1 Korean Flexible Electronics Development Programs

Entity	Area of Concentration
Electronics and Telecommunications Research Institute (ETRI)	Flexible transistors, OLEDs
Korea Electronics Technology Institute (KETI)	OLED lighting
Korea Institute of Machinery and Materials (KIMM)	Machinery for printed electronics
Korea Research Institute for Chemical Technology (KRICT)	Inks for printed electronics; printing technology for solar cells
Kongkuk University	Flexible displays
Pohang University of Science and Technology (POSTECH)	Nano-ink and substrates for flexible displays
Sungkyunkwan University	Graphene, flexible power sources
Korea Advanced Institute of Science and Technology (KAIST)	Nanotube fabrication, flexible substrates, wiring for flexible displays
Samsung Group	Flexible displays
LG Group	E-paper, flexible displays

> and companies would collaborate to establish the $5.3 million growth Ladder Fund to make investments supporting entrepreneurship.[17]

The new promotional effort builds on flexible electronics development programs that are well under way in Korean government laboratories and a number of companies. (See Table 6-1.)

Government Entities Supporting R&D

Historically, responsibility for Korea's science and innovation policy has been divided between a number of competing ministries in a manner that was "sometimes unhelpful to Korea's drive to improve its S&T and innovation performance."[18] Recent consolidations have been implemented that may address this problem, with most key functions now located within the Ministry of Trade, Industry and Energy (MOTIE) and the Ministry of Education, Science and Technology (until 2008 known as MOST).

---

[17] "$5 Bn to Pour Into SMEs in Next 3 Yrs," *MK English Online,* May 23, 2013.

[18] Organisation for Economic Co-operation and Development (OECD), *OECD Reviews of Innovation Policy: Korea 2009,* 2009, 180.

Ministry of Trade, Industry and Energy (MOTIE)

In 2013, the Ministry of Knowledge Economy (MKE) was renamed the Ministry of Trade, Industry and Energy (MOTIE), with jurisdiction over industrial and energy policy and foreign trade and investment.[19] In 2009, MKE unveiled an $870 million plan to promote Korean development of "10 World Premier Materials (WPM)" by 2010, helping Korea develop into a "global parts and components powerhouse, increasing the value of parts/components exports from the 2008 level of $180 billion to $500 billion by 2018. In addition to R&D support the WPM effort financially supports mergers and acquisitions between Korean and foreign materials companies.[20] One of the 10 materials being developed is substrates for flexible electronics displays.[21] MOTIE is also responsible for developing and implementing the massive R&D effort involving six convergence technologies that was announced in the spring of 2011.

MOTIE is concerned about Korean companies' heavy dependence on imported production equipment—Samsung and LG, for example, procure most of their semiconductor manufacturing equipment abroad, particularly from Japan.[22] In 2009, MOTIE started a program to provide R&D support to Korean companies to develop indigenous equipment, including machinery for "growth engine" technologies with implications for flexible electronics, such as the production of light-emitting diodes.[23]

Ministry of Education, Science and Technology (until recently, MOST)

The Ministry of Education, Science and Technology (until 2008 the Ministry of Science and Technology) coordinates Korea's science and technology activities. It provides support for university-based research on flexible electronics themes.

Public Research Institutes

During the developmental phase of Korea's economy in the 1960s and 1970s, the country lacked the infrastructure for innovation with industrial and commercial applications. In response the government established the Korea Institute of

---

[19] "Seoul's Commerce Ministry to Change Name With Added Role," *Yonhap,* March 22, 2013.

[20] "Govt. to Invest $1 Trillion Won to Develop Premier Materials," As*irang News,* November 16, 2009.

[21] Gye-man Kang and Hyung-deuk Jean, "Samsung, LG, Hyundai Motor Join Hands for Next Generation Flexible Display Development," *MK English News Online,* July 5, 2010.

[22] "Korea can aptly be described as a semiconductor powerhouse. In producing the world's best products in the sector, however, the country imports most of the manufacturing equipment needed for semiconductors." "Korean Company Arrives to Compete in Extremely Touch Sector," *Dong-A Ilbo Online,* March 8, 2013.

[23] "Equipment Industry is Crucial to Growth," *The Korea Herald Online,* August 17, 2009.

Science and Technology (KIST) in 1966 as a technical center to assist Korean companies to adopt and utilize foreign technologies. The government continued to open new specialized public research institutes relevant to key industries the government was seeking to foster, including electronics, telecommunications, chemicals, and energy. These centers played a key role in adapting and disseminating foreign technologies to local industry, as well as functioning as training centers for research personnel who migrate to jobs in industry. In the 1980s these institutes were consolidated and began to engage industry in collaborative research projects relevant to national economic development. The public research institutes are funded by government core grants that cover roughly half their budgets, with the remainder of their revenue derived from contract research for industry, government ministries, and local governments.[24]

Korea Electronics Technology Institute (KETI)

KETI was established in 1991 as a government-funded research institute under the supervision of the Ministry of Commerce, Industry and Energy (MOCIE), now the Ministry of Knowledge Economy. Seventy percent of KETI's funding is provided by the government and 30 percent by the private sector. KETI promotes the development of new technologies in electronics, telecommunications, and information technology. KETI provides support to small- and medium-sized enterprises and operates a business incubator. KETI's research divisions and centers are pursuing a number of research themes with potential applications in the field of flexible electronics. (See Table 6-2.)

KETI administers the Korean Printed Electronics Center (KPEC), established in Jeonbok in 2009 to promote development of nanolevel printed/flexible electronics. KPEC represents an investment of $75 million, with $21 million invested in state-of-the-art production equipment. KPEC is concentrating on providing process, equipment, and characterization services for R&D by industry, government, academia, and industrialization. It is pursuing printed/flexible electronics themes that are expected to have applications in organic lighting, digital signage, automotive sensors, solar cells, and intelligent windows.

Korea Institute of Science and Technology (KIST)

KIST was established in 1966 jointly by the governments of South Korea and the United States to perform R&D to support Korea's economic growth, by developing technologies with industrial applications. Today, KIST is Korea's foremost science and technology institution. KIST has been promoting development manufacturing technology for flexible transistors for a number of years.[25]

---

[24] OECD, *OECD Reviews of Innovation Policy*.

[25] In 2007, in a collaboration with MIT, a team of KIST engineers succeeded in making a "low-voltage flexible transistor that could be worn like clothing or used to make wristwatch-style mobile

TABLE 6-2 KETI Research Themes

KETI Organization	Themes
Energy Nano Materials Research Center	• LEDs, thin-film solar cells
Flexible Display Research Center	• Flexible displays (LCD, OLED, e-paper) • Printed electronic devices • Smart windows • Flexible device and display driving circuits
Display Components and Materials Research Center	• Nano-inks and particles • Printed electronics materials and processes • OLED lighting
Printed Electronics Research Center	• Printed electronics
Electronic Materials and Device Research Center	• Stretchable semiconductor circuit materials • Electronic ink and pasted materials
Convergence Communication Components Research Center	• Flexible RF component modules • Flexible wireless transmission platform
Reliability Physics Research Center	• Lifetime prediction for flexible embedded module
Convergence Sensor Research Center	• Touch and tactile sensors

A United Kingdom (UK) trade and investment mission visiting KIST in 2012 reported that it was setting up a roll-to-roll (R2R) photo-curing system that had been developed in house.[26] The system has been used to demonstrate printing of RFID antennas. KIST was also reportedly researching process technologies for silver nanoparticles, organic photovoltaics, and inkjet printing of thin-film transistors.[27]

Electronics and Telecommunications Research Institute (ETRI)

ETRI, a government entity, is Korea's largest research institute and traditionally played a central role in the development of the Korean semiconductor industry.[28] In 2010, ETRI researchers developed a flexible nonvolatile memory based

phones. The new transistor was developed by making a paper-thin polyethylene terephthalate plastic circuit board, which is bendable, inexpensive to produce, energy efficient and safe for the user." "Transistor Could Lead to Wearable Computers," *Joong Ang Ilbo,* October 9, 2007.

[26] Photo-curing systems have been utilized by a number of companies engaged in printed electronics. The process is used to cure metallic inks (e.g., silver, copper, copper oxide). The advantage of the roll-to-roll process is that it is fast and can be used with low temperature substrates such as paper. "IDTechEx Visits South Korea," *Printed Electronics World,* April 18, 2012.

[27] "IDTechEx Visits South Korea," *Printed Electronics World,* April 18, 2012.

[28] ETRI was a key member of the government-industry consortia that developed Korea's 256 megabit and 1 gigabit dynamic random access memories. "Taedok to Become Mecca for Venture Firms," *Chonja Sinmun,* April 10, 1998.

on memristors, promising electronic circuit elements discovered in 2008 that hold the promise of new types of dense, inexpensive, low-power memory devices that could replace transistors in future computers in a much smaller space.[29]

National Research Foundation of Korea (NRFK)

NRFK is a quasi-governmental organization established by statute in 2009 that specializes in research funding and management.[30] Based in Daejon, it was formed through the merger of the Korea Science and Engineering Foundation, the Korea Research Foundation, and the Korea Foundation for International Cooperation and Science and Technology. It operates under the direction of the Ministry of Education, Science and Technology. Its basic mission is to provide comprehensive support for research in the natural sciences and the humanities, and it funds research at domestic universities with flexible electronics themes.

Korea Institute of Machinery and Materials (KIMM)

KIMM is a government research institute that develops and disseminates source technologies in mechanical engineering and conducts reliability tests. The government provides $54 million of its annual budget of $138 million, with the remaining $84 million derived from KIMMs operating revenue.[31] KIMM's Nano-Mechanical Systems Research Division is pursuing a number of research themes directly relevant to flexible and printed electronics:

- Flexible low-cost printed electronics devices technology,
- Flexible low-cost solar cell technology,
- Self-cleaning surface fabrication technique using nature-inspired technology,
- 3-D plotting system for tissue engineering scaffold,
- Ultra-precision machining and forming process for micro/nano features,

---

[29] "Memristors change their resistance depending on the direction and amount of voltage applied, and they 'remember' the resistance when the voltage is removed." Most memory devices store data as a charge, but memristors would enable a "resistive RAM, a nonvolatile memory that stores data as resistance rather than charge." Memristors were postulated by Professor Leon Chun at the University of California in 1971. Hewlett-Packard subsequently demonstrated that they exist and developed an understanding of how they operate. To date, memristors have been made using metal oxide thin films. A team led by Sung-Youl Choi, the leader of ETRI's flexible device research, used thin graphene oxide films to make a memristors-based flexible nonvolatile memory. The graphene oxide devices could be printed on rolls of plastic sheets and used in wearable electronics and plastic radio-frequency identification (RFID) tags. "Flexible Graphene Memristors," *Printed Electronics World,* December 9, 2010.

[30] *Law of the Korea Research Foundation*, Proclaimed March 25, 2009, Law No. 9518 (June 26, 2009).

[31] *<http://www.kimm.ks/english/htm/about_01_t3.php>*.

- Micro-/nanoscale machining and fabrication process based on femtosecond laser,
- Fabrication and process development for advanced MEMS,
- Nano-imprinting equipment technology,
- E-beam lithography and nanostamp fabrication technologies,
- Nanoprinting and nanodevices fabrication using nano-imprint lithography,
- E-beam lithography and nanostamp fabrication technologies,
- Nanoscale measurement and analysis,
- Development of thermoelectric device and module,
- Separation and transparent heater of carbon nanotube,
- Mass production and application of quantum dot, and
- Structural analysis and reliability evaluation of nano/micro structures.

KIMM sells technologies that it develops to private companies for a flat fee or for a percentage of revenues generated.[32]

Korea Research Institute for Chemical Technology (KRICT)

KRICT is a government-supported research institute for chemical technology. The government directly provides $33.6 million of KRICT's budget of $75.4 million. KRICT derives $41.8 million of its budget from income from R&D projects, including another $35.7 million in competitive government grants, $4.8 million from the private sector, and $1.3 million from royalties. KRICT's Device Materials Research Center is pursuing a number of research themes with potential application in flexible electronics:

- Ink materials and processing technology for printable electronics;
- Low-cost/large area printing technologies for organic solar cells;
- Preparation of thin films, coating materials, and nanostructured materials.

In 2010, a team of KRICT scientists reported that they had succeeded in developing a process for using inkjet printing for the deposition of an active layer of polymer-based organic photovoltaic cells (OPVs). The technology could have applications printing OPVs on flexible substrates. The research was funded by a grant from the Ministry of Knowledge Economy's Fundamental R&D Program for Core Technology of Materials.[33]

---

[32] "Lucrative State-Run Lab Deal Called Model Sale," *JoongAng Ilbo,* June 8, 2007. In 2010, it sold nano-imprinting technology for sensors with medical applications to U.S.-based NanoLambda for $80,000 plus 1.6 of all sales of image sensors created with the new imprinting technology. "State-Run Laboratory Export Advanced Nanoimprinting Technology," *Yonhap,* March 15, 2010.

[33] "Highly Efficient Inkjet Printed Organic Photovoltaic Cells," *Japanese Journal of Applied Physics,* 2010.

Korea Electrotechnology Research Institute (KERI)

KERI is a nonprofit government-funded research institute based in Changwon specializing in electrotechnology and electric power. In 2008, a KERI research team headed by Lee Geon-woang reported that it had developed a new transparent electrode that is bendable and can be applied to thin films to create flexible displays and touch screens. The electrodes were formed by combining carbon nanotubes, binders, stabilization compounds, and other chemicals. The liquid one-component solution can be applied in an R2R process in the form of paste. Lee said that the new manufacturing process can replace the conventional indium tin oxide (ITO) process "and revolutionize designs for touch screens, flexible displays, solar cells and various sensor devices."[34]

Industry/Government Consortia

Flexible Display Consortium

In July 2010, three of South Korea's major industrial conglomerates, Samsung, Hyundai Automotive Group, and LG (Lucky-Goldstar) announced formation of a consortium to develop substrates for flexible displays as one of the World Premier Materials projects sponsored by the Ministry of Knowledge Economy. The consortium's proposal was submitted to the Korea Evaluation Institute of Industrial Technology (KEIT) for approval. The project involves government funding of $81.8 million. The companies would contribute funds for equipment and facilities. Other participating companies include LG Chem, LG Display, Kolon Industries, Inktec, Samsung Mobile Display, Cheil Industries, and Samsung Electronics.[35]

Thin-Film Solar Cells

In May 2011, it was announced that Samsung Electronics, LG Electronics, and Dongjin Semichem would participate in a government-industry joint R&D project to "develop next-generation thin film solar cells that can be placed in glass metal or polymer (flexible) substrates."[36] According to the Ministry of Knowledge Economy this effort is one of five major R&D projects that collectively will receive 350 billion won ($328 million) from the government.

---

[34] "Engineers Create New Transparent Electrodes for Display Screens" *Yonhap,* March 20, 2008.

[35] "Samsung, LG, Hyundai Motor Join Hands for Next Generation Flexible Display Development," *MK English News Online,* July 5, 2010.

[36] "Samsung, LG to Join Development of Thin Film Solar Cells," *Yonhap,* May 31, 2011.

60-Inch Flexible OLED Consortium

In 2012, the Korean government chose LG Display to lead a consortium to develop 60-inch transparent and flexible OLED displays by 2017 as part of the government's Future Flagship Program.[37] The consortium will include Avaco, a maker of equipment for producing displays that reportedly received a $6.7 million order to supply LG with tools and parts for making flexible OLED panels. LG reportedly envisions applications for flexible OLED displays that include home consumer products, windows displaying information at bus stations, and curved displays for retail settings in which displays wrap around columns or dangle from the ceiling.[38]

Printed Electronics Project

In 2012, the Korea Printed Electronics Association (KoPEA) disclosed plans to launch a national Printed Electronics Project, a $48 million effort running from 2012 to 2017. Funding will reportedly be provided by MOTIE. Tentative technology themes have been identified including OLED lighting, RFID touch-screen displays, digital signage, and flexible printed circuit boards. The project will be led by large companies such as Samsung Electronics, LG Electronics, and POSCO and will include small- and medium-sized companies. Foreign participation is welcome.[39]

Key Companies

LG Display

LG Display is the world's largest maker of LCD panels. It was formed through a joint venture between South Korea's LG Electronics and Royal Philips Electronics NV of the Netherlands and was known as LG Philips LCD Co. Philips sold its stake in 2008, and the company's name was changed to LG Display.

Samsung Electronics

Samsung Electronics is a major South Korean producer of IT equipment, semiconductors, solar cells, and displays. Samsung Electronics Vice President Hung Wan-pyo said in July 2010 that "Samsung will introduce a smart phone that can be folded just like paper before the year 2015."[40] Samsung withdrew from the

[37] "LG Display Chosen by Korean Government to Lead OLED Project," *Printed Electronics World,* August 8, 2012.

[38] "LG Convinces South Korea to Fund Development of Flexible 60-Inch OLED Displays," *Extremetech,* July 17, 2012.

[39] "UK Opportunity: Korean Project 2012-17," *Plastic and Printed Electronics,* May 1, 2012.

[40] "Samsung Aims to Introduce Flexible Smartphone Before 2015," *MK English News Online,* July 22, 2010.

production of e-paper in 2010, citing cost issues.[41] Samsung was established in 2012 through consolidation of Samsung's various displays business units.

Cheil Industries

Cheil Industries is a Samsung subsidiary that was established as a textile company in 1954 and that diversified into chemicals (1989) and electronics materials (1994). Annual revenues are about $5 billion.[42] Cheil is a major supplier of electronic materials to Samsung group companies including materials with flexible electronics applications.[43] Cheil entered the OLED business in 2009 and invested $180 million in 2011 in a factory for OLED material production to support Samsung's Galaxy S4 smartphone.[44] Cheil has reportedly made progress in developing polyimide plastic substrate and is expected to play a significant role in Samsung Display's Commercialization of plastic display products.[45] In April 2013, Cheil was reportedly considering the acquisition of Novaled AG, a German maker of proprietary OLED products.[46]

Samsung Electro-Mechanics (SEMCO)

SEMCO, a member of South Korea's Samsung Group, developed industrial inkjet printheads and copper ink in 2008. SEMCO was the first Korean company to enter the industrial inkjet printhead market, and its nano-copper ink for printed circuit boards was the first of its kind in the world. The inkjet printhead can be used to print conducting inks onto flexible substrates such as films and textiles.[47]

Korea Kumho Petrochemical Co.

Korea Kumho Petrochemicals is a South Korean manufacturer of electronic chemicals and synthetic resins and rubber. In 2009, it announced plans to build a plant to manufacture carbon nanotubes, which are being used as conductors in

---

[41] "Samsung Electronics Halts E-Paper Production," *Yonhap,* August 23, 2010.

[42] Cheil's CEO, Park Jang-woo, previously headed the Digital Media Division at Samsung Electronics and Samsung Electro-Mechanics. Cheil's Vice President Lee Seo-hyun is the daughter of Samsung Electronics Chairman Yi Kun-hee. "Cheil Is Into Electronics, Chemicals, Fashion," *Joong Ang Daily Online,* March 28, 2012.

[43] Cheil makes "electronic or plastic parts that are crucial in producing the major finished products by Samsung Electronics and other Samsung-affiliated companies." Choi-Ji-wan, NH Research Center, cited in "Cheil Is Into Electronics, Chemicals, Fashion," *Joong Ang Daily Online,* March 28, 2012.

[44] "Cheil Industries Celebrates Shipment of Amoled Displays," *Joong Ang Daily Online,* April 3, 2013.

[45] HSBC, *Flexible Display.*

[46] "Samsung Group Unit Cheil Ind Considering Buying Novaled AG," *Financial Express,* April 2, 2013.

[47] "Samsung Electro-Mechanics Launching Industrial Inkjet Business," *Printed Electronics World,* April 1, 2009.

a number of flexible electronics technologies. The plant, to be located in Jeonju, is expected to achieve an output of 300 tons by 2013.[48]

Hanwha Chemical

Hanwha Chemicals, a member of the Hanwha Group, is a South Korean petrochemical enterprise that produces polyethylene products, PVC resins, solar cells, anodic materials for batteries, and pharmaceuticals (including biosimilars). In 2008, the company indicated that it had started mass production of carbon nanotubes at a plant in Inchon. "The plant has the capacity to produce 100kg of single-walled carbon nanotubes and four tons of multi-walled carbon nanotubes."[49] The company indicated it would invest $78 million in carbon nanotubes for 2013. Hanwha acquired Iljin Nanotech Co., a Korean carbon nanotube producer, in May 2008. In 2011, Hanwha acquired a 19 percent equity stake in U.S.-based XG Sciences, a U.S. manufacturer of graphenes.

University R&D

South Korea's university system has been the subject of considerable criticism within the country. Many of the country's universities suffer from under-enrollment, poor management, parochialism, and resistance to change.[50] In 2009, the Ministry of Education, Science and Technology compiled a list of badly run Korean universities and told each to "clean up its act."[51] Non-Korean professors comprise less than 5 percent of the faculty in Korean universities, and as of 2008, 39 institutions had no foreign faculty members, a significant weakness in a globalized economy.[52] A recent Organisation of Economic Co-operation and Development (OECD) survey ranked South Korea last among member countries in terms of college education environment.[53] A recent Korean editorial commented that

> [c]onsidering how much intellectual assets and national wealth an American university creates, Koreans have no choice but to worry about the future of their universities and the nation.[54]

[48] "Korea Kumho Petrochemical to Build Carbon Nanotube Plant," *Yonhap,* September 30, 2009.

[49] "Hanwha Chemical Begins Mass Carbon Nanotube Production," *Yonhap,* December 22, 2008.

[50] "Whither Our National Universities?" *JoongAng Daily Online,* June 22, 2011; "How Bad Universities Fill Their Seats," *JoongAng Daily Online,* June 17, 2011.

[51] "Universities Get a Failing Grade," *JoongAng Daily Online,* June 17, 2011.

[52] "Foreign Professors Are Rare Breed," *The Korea Herald,* October 6, 2008.

[53] Korean Universities have 32.7 students per professor, on average, or more than double the OECD average of 15.8. The figure is 15 in the US, 11.5 in Germany and 10.4 in Japan. "Korean Students are Badly Served by Their Universities," *Chosun Ilbo Online,* June 8, 2011.

[54] "MIT and Korean Universities," *Dong-A Ilbo Online*, May 20, 2011.

The Korean government is implementing sweeping reforms in the university system in an effort to "raise their competitiveness both at home and abroad."[55] The universities are making a concerted effort to add foreign professors to their faculties.[56] In order to "globalize" Pohang University of Science and Technology (POSTECH) became an English language–only campus in 2010 and announced that it would invest $44.2 million for 3 years to draw 10 Nobel Prize and Fields Medal laureates to its faculty.[57] In 2008, the government indicated it would limit funding for research to 17 "World Class Universities," effectively cutting off 35 institutions from further government research support.[58]

Notwithstanding the problems of Korea's universities generally, a handful of the country's best institutions appear to be improving relative to their foreign counterparts and are engaged in significant research in the sciences and engineering.[59] In 2010, Korean universities ranked 11th in global standing for publications in major international scholarly journals, up from 19th in 1999.[60] In 2013, a global ranking of Korean engineering faculties by Quacquarelli Symonds placed Seoul National University (SNU) at 17 and Korea Advanced Institute of Science and Technology (KAIST) at 20th in chemical engineering, whereas in 2012 SNU was 38th and KAIST was not in the top 50.[61] During the 5-year period 2006-2010, 12 Korean universities each registered more than 1,000 patents.[62] (See Table 6-3.)

Several of these leading universities are making significant research contributions in the field of flexible electronics, most notably KAIST, UNIST, POSTECH, SNU, Kongkuk, Sungkyunkwan, and Kyung Hee universities.

Korea Advanced Institute of Science and Technology

KAIST is a graduate school specializing in science and engineering education and research. It was established in 1971 pursuant to special legislation and is supported with government funding. KAIST is Korea's leading center for medium- to long-term strategic R&D projects of national importance.

In 2006, Suh Nam-pyo became president of KAIST and launched an ambitious effort to transform the institution into one of the top 10 universities in the

---

[55] "National Universities Face Sweeping Reforms," *Yonhap,* September 28, 2010.

[56] "Korean Universities Hire More Foreign Faculty," *The Korea Herald Online,* August 2, 2010.

[57] "POSTECH Becomes All-English Campus," *JoongAng Daily Online,* March 13, 2010.

[58] "Handful of Top Universities to Get State Research Funds," *The Korea Times Online,* December 1, 2008.

[59] In 2010, five Korean universities ranked among the top 200 universities in the world in the World University Rankings by the Times and Quaquarelli Symonds, a British university rating agency, compared with only two in 2007. KAIST moved from 95th in 2008 to 79th in 2010. Seoul National University received the highest ranking at 50th. "5 Korean Universities Rank Among Global Top 20," *Chosun Ilbo Online,* September 8, 2010.

[60] "Korea Ranks 11th in Papers for Scholarly Journals," *Chosun Ilbo Online,* December 15, 2010.

[61] "Korean Engineering Faculties Jump in World Ranking," *Chosun Ilbo Online,* May 8, 2013.

[62] "KAIST Tops Patent List Among Domestic Universities," *Chosun Ilbo Online*, July 5, 2011.

TABLE 6-3 Patents Registered by Korean Universities

Institution	Number of Patents Registered
KAIST	4,403
Seoul National University (SNU)	3,536
Yonsei University	2,610
Korea University	2,516
Hanyang University	2,247
Sungkyunkwan University	1,880
Inha University	1,374
Kyung Hec University	1,274
Pohang University of Science and Technology (POSTECH)	1,223
Konkuk University	1,080
Kyungpook National University	1,074
Pusan National University	1,006

world. KAIST sent shockwaves through Korean academia in 2007 by denying tenure to 15 out of 35 applicants, and by increasing tuition levels for students with mediocre grades.[63] In 2009, KAIST president Suh became the first Asian winner of the ASME Medal, an annual award bestowed by the American Society of Mechanical Engineering for distinguished engineering achievement that is regarded as the Nobel Prize of engineering.[64] KAIST is pursuing a number of research themes with application to flexible electronics. (See Table 6-4.)

Ulsan National Institute of Science and Technology (UNIST)

UNIST is a recently established university offering undergraduate and graduate curricula in mathematics, the sciences, engineering, and technology. "Many consider it to be the MIT of South Korea."[65] It has begun to report research discoveries relevant to the field of flexible electronics. In 2013, a group of Korean scientists led by UNIST professor Lee Sang-young announced development of the world's first imprintable and bendable lithium-ion battery, based on nanomaterials, which was expected to enhance the development of flexible mobile

[63] "How KAIST and POSTECH Reforms Can Save the Country," *Chosun Ilbo,* October 8, 2007.

[64] "KAIST President Wins Int'l Mechanical Engineering Prize," *Chosun Ilbo Online,* July 3, 2009. Suh's campaign to raise KAIST's standing has proven controversial, and he has been blamed for fostering a culture of competition that has led to a string of suicides among faculty and students. "KAIST Case Should Send a Warning to Elite Science Education in Korea," *Kyunghyang Shinmon Online,* April 12, 2011.

[65] "YSU Expands Research Role with S. Korean Institute," *Vindicator,* April 3, 2013.

TABLE 6-4 KAIST Research Achievements—Flexible Electronics

Year	Team Leader	Result	Government Support	Potential Application
2012	Jeon Seok-woo	Stretchable new material (3 × original length) that retains conductivity		Flexible displays, solar panels
2012	Lee Keon-jae	Flexible battery with high performance		Flexible electronic gadgets
2011		Metal wires for applications on polymer substrates		Flexible displays
2011	Byeong Soo Bae	Plastic and glass cloths with limited thermal expansion	NRFK MOST	Flexible displays, solar panels
2010	Byeong Soo Bae	Rollable transparent glass-fabric reinforced composite film (GFRH ybrime)	NRFK	Flexible displays

SOURCES: "KAIST Paves the Way to Commercialize Flexible Display Screens," *Printed Electronics World,* February 28, 2011; "Korean Researchers Develop New Flexible Metal Wire Manufacturing," *Plastic Electronics,* May 27, 2011; "S. Korean Scientists Develop New Flexible Stretchable Display," *Yonhap,* July 3, 2012; "Korean Team Claims to Create World's First Flexible Batter," *Don-A Ilbo,* August 28, 2012; "Bendable Inorganic Thin Film Battery for Fully Flexible Electronic Systems," *Nano Letters,* July 30, 2012.

devices.[66] In 2013, UNIST researchers "demonstrated high-performance polymer solar cells (PSCs) with the highest power conversion efficiency reported to date (8.92 percent) for plasmonic PSC using metal nanoparticles."[67] PSCs are bendable, inexpensive to fabricate, lightweight, and have a relatively low potential for negative environmental effects, and the discovery moved the technology much closer to the 10 percent conversion efficiency seen as the level necessary for commercialization. In May 2013, UNIST scientists disclosed that they had combined silver nanowires with grapheme to create a thin, transparent, and stretchable electrode that shows almost no change in its electrical resistance when bent or folded.[68]

Pohang University of Science and Technology

POSTECH is a science and engineering university that supports R&D in basic and applied sciences, usually in government/industry commissioned

[66] The research was partially funded by the Ministry of Education, Science, and Technology. "S. Korea Develops World's 1st Imprintable, Flexible Battery," *Yonhap,* January 15, 2013.

[67] "A Giant Leap to Commercialization of Polymer Solar Cell," *Printed Electronics World,* May 10, 2013; "Multipositional Silica-Coated Silver Nanoparticles for High-Performance Polymer Solar Cells," *Nano Letters,* April 22, 2013.

[68] "Korean Researchers Demonstrate New Transparent, Stretchable Electrodes," *Printed Electronics World,* May 31, 2013.

TABLE 6-5 Sungkyonkwan University Research Achievements—Flexible Electronics

Year	Team Leader	Result	Known Government Support	Potential Application
2010		Touch-sensitive screen		Self-powering touch sensors
2009	Lee Hyo-oung	Graphene processing at low temperature	MOST NRFK	Manufacturing
2010	Ahu Jong-hyun	R2R processing of transparent electrodes		Replace ITO-based circuits
2010		Screen printing of large area OLEDs		OLED-based lighting
2009		Flexible tactile displays		Intelligent cloth, gloves, Braille displays

SOURCES: "Large-Area OLED Lighting Fabricated by Screed Printing," *Flexible Substrate* (November 2010); "Scientists Develop Transparent Electrode Manufacturing Process," *Yonhap* (June 20, 2010); "Breakthrough Made in Next-Generation Chip Material," *Dong A.com* (January 15, 2009); "Self-Powered Flexible Electronics," *MIT Technology Review* (April 30, 2010); "Sungkyonkwan University and the University of Nevada Develop Flexible Tactile Displays," *Flexible Substrate,* June 2009.

projects. One of its research organizations is South Korea's foremost nanotechnology research center, the National Center for Nanomaterials Technology (NCNT). The Korean government provided $40 million of the $110 million that was invested to establish the center, and the private sector provided $40 million. One of NCNT's three "priority support areas" is the development of flexible display core materials, including nano ink, substrate, and electronic materials. NCNT is also reportedly initiating R&D into graphene-based electronics.[69]

Sungkyonkwan University

Sungkyunkwan, located in Seoul, is one of Korea's foremost universities. Researchers at the university are active in developing materials and technologies with potential applications for flexible electronics. (See Table 6-5.)

Seoul National University

SNU is a highly regarded national research university in Seoul whose graduates dominate Korea's government, academia, and business communities. SNU

[69] "Korea's POSTECH Explore Nanotechnology Frontiers," *Nano Science Works,* January 20, 2010.

administers the Inter-University Semiconductor Research Center (ISRC) conducting basic research on semiconductor technology. ISRC has 550 researchers based in schools and businesses around Korea and 1,200 annual trainees. ISRC's key feature is its state-of-the-art machinery rather than individual research projects.

Konkuk University

Konkuk University, located in Seoul and Chungju, specializes in science and technology. Konkuk administers the Flexible Display Research Centre (FDRC). Since 2007, FDRC has been collaborating with the Korean office of Finland's VTT Technical Research Centre to establish a joint laboratory for printed electronics based on R2R process technology. The project is funded by the Korea Foundation for International Cooperation of Science and Technology (KICOS) through a grant provided by the Ministry of Education, Science and Technology (MEST). In June 2011, an international research team headed by Konkuk professor Park Bae-ho announced findings confirming the ripple structure of graphenes, thus identifying a technological hurdle that must be overcome before graphenes can find widespread applications in electronics.[70]

Kyung Hee University

Kyung Hee University, a private university located in Seoul and Suwan, is regarded as one of the best institutions of higher learning in Asia. Kyung Hee operates as Advanced Display Research Center (ADRC) in conjunction with thin-film transistor (TFT)-LCD National Laboratory, with facilities to produce and characterize TFT-LCD, OLED, and Field Emission Display (FED) technologies. Kyung Hee is the only university in the world with this capability.[71]

Sunchan National University

Sunchan is a national research university. In 2010, in collaboration with Rice University in the United States, it developed a printable carbon nanotube RFID transmitter that, when embedded in packaging, "would allow a customer to walk a car full of groceries or other goods past a scanner on the car, the scanner would read all the items in the card at once, total them up and charge the customer's account while adjusting the store's inventory."[72]

[70] "Int'l Research Team Confirm Ripple Structure of Graphenes," *Yonhap,* June 30, 2011.

[71] "Advanced Display Research Center (ADRC) at Kyung Hee University," *Korea Times,* June 14, 2012.

[72] "Nano-based RFID Tag, You're It," *Rice University News & Media,* March 18, 2010.

International Collaboration

Korean observers of the country's strengths and weaknesses in flexible electronics regard as major weaknesses the country's reliance on foreign sources for critical materials and equipment and its lagging position in basic research. International collaborations, overseas R&D centers, acquisitions and technology acquisitions, and exchanges are seen as important mechanisms for addressing these concerns.

Samsung/Liquavista

In January 2011, Samsung Electronics acquired Liquavista BV, a Dutch display technology company. The acquisition allowed Samsung to utilize a new type of "electronic display technology called electrowetting that enables display production with low power consumption for electronic book readers and mobile devices. Samsung plans to use the technology to develop e-paper and transparent displays."[73] Amazon reportedly bought Liquavista from Samsung in 2013.[74]

Inktec/Thin Film

Inktec Co. Ltd. is a leading Korean manufacturer of commercial and electronic inks. In 2008, Inktec disclosed that pursuant to an R&D collaboration with Thin Film Electronics ASA, a Norwegian developer of functional polymer materials for nonvolatile memory applications, the two firms had succeeded in creating fully functional printed memory products in an R2R high-volume printing process. Yields achieved were in the 96-97 percent range. The printed memory film was extremely thin (200 nm).[75]

Yissum/Vaxan

In February 2011, Korea's Vaxan Steel and the Yissum Research Development Company, the technology transfer arm of the Hebrew University of Jerusalem, signed a licensing on research agreement for the development of silver nanoparticles and silver-coated copper nanoparticles for use in conductive inks.[76]

---

[73] "Samsung Acquires Dutch Display Tech Firm," *Yonhap,* January 20, 2011.

[74] "Amazon Buys Liquavista from Samsung," *Flexible Substrate,* May 2013.

[75] The two companies produced working 512 MB chip modules in 0.25 nm technology. In 2009, this achievement won the IDTechEx Technical Development Manufacturing Award, which is given for the most significant manufacturing device, process, or production plant in the printed electronics industry in the preceding 24 months. "Inktec and Thin Film Electronics Win Prestigious Award," *Printed Electronics World,* April 14, 2009.

[76] The silver-based conductive inks were developed by Professor Schlomo Magdassi, Dr. Alexander Kamyshmy, and Michael Grouchko from the Institute of Chemistry at Hebrew University. Inks based on these nanoparticles can be used in printed electronics applications on many types of surface,

Under the terms of the agreement Yissum grants Vaxan a license to commercialize the new inks exclusively in Asia, excluding Israel and the former Soviet Union, and Vaxan will pay Yissum research fees and royalties from future sales.[77]

KIMM/Singapore

In 2009, the government-run Korea Institute of Machinery and Materials (KIMM) concluded an agreement with Singapore's Institute of Materials Research (IMRE) and Singapore Institute of Manufacturing Technology (SIMTECH) to develop micro- and nanotechnologies with medical, electronic, automotive, aerospace, and environmental applications. Research themes include R2R printed electronics, nano-imprint lithography, and nanostructured functional coatings. The collaboration will draw on IMRE's expertise in materials science and engineering, SIMTECH's process and measurement technologies, and KIMM's systems and materials technologies.[78]

Korea-China Flexible Solar Cells Project

In 2010, a Korea-China joint research program was undertaken to develop advanced flexible plastic solar cells and new photovoltaic polymers. The participating institutions are Korea's Pusan National University Department of Nanomaterials Engineering and the Qingdao Institute of Bioenergy and Bioprocess Technology (QIBEBT), Chinese Academy of Sciences. The Korea-China joint research program is co-funded by Korea's Ministry of Education, Science and Technology and China's Ministry of Science and Technology.[79]

LG Display/PVI

In 2010, it was disclosed that Korea's LG Display would collaborate with Hydis Technologies Co. Ltd., a subsidiary of Taiwan's Prime View International, to seek leadership in the global e-paper display market. The parties are cross licensing technology and will mutually support production with respect to e-paper

including thin plastic films and paper. Silver nanoparticles are seen as a particularly promising material for inkjet printed electronics because silver is the most conductive metal, and, in contrast to other metals, oxidation does not harm the conductivity of the film on which silver-based circuits as a superior printing method to lithography and screen printing techniques because it is faster and less expensive.

[77] "Yissum and Vaxan Collaborate and Printed Electronics Ink," *Printed Electronics World,* February 2, 2011; "Korea, Israel Collaborate on Nanoparticles," *Korea Herald Online,* April 10, 2011.

[78] "Korea S'pore Institutes in R&D Tie-up," *The Business Times,* March 20, 2009.

[79] "New Korea-China Research Program Focuses on Nanomaterials for Flexible Solar Cells," *MIT Technology Review,* April 19, 2010.

and high-end TFT-LCDs. The two companies will collaborate on materials sourcing and technology development.[80]

KPF/Plextronics

In 2008, Korea Parts & Fasteners (KPF), a producer of forgings and bearings, entered into a memorandum of understanding with U.S.-based Plextronics, Inc., a maker of printed solar and lighting products to establish a production line for organic photovoltaic panels in Cheongju, Korea, and an R&D center. KPF committed $12 million and Plextronics, $10 million, and the collaborators will hold 51/49 percent shares, respectively. The new joint venture, KNP Energy Co. Ltd., "will receive government cash grants tax breaks and support in the building of its factory and research center."[81]

Hanwha Chemical/XG Sciences

In January 2011, Hanwha Chemical Corp., a chemicals-producing firm in South Korea's Hanwha Group, acquired a 19 percent stake in XG Sciences, a U.S.-based manufacturer of graphenes, materials that have promising potential for flexible electronics, for $3 million. Hanwha also secured rights to sell XG Sciences' graphenes to India and China.[82]

POSCO/XG Sciences

In June 2011, POSCO, South Korea's largest steelmaker, concluded a deal to acquire a 20 percent stake in U.S.-based XG Sciences Inc. The deal makes POSCO XG Sciences' largest shareholder. POSCO will manufacture graphenes under a licensing agreement with XG Sciences and hopes to establish a graphene manufacturing plant in China in 2012.[83]

CHINA

China presents a curious counterpoint to the United States and Europe in flexible electronics. In contrast to these regions it lacks most of the equipment and materials infrastructure to support manufacture of flexible displays, as well

[80] "PVI, Hyds Technologies and LG Display Launch Co-operation," *Printed Electronics World,* January 4, 2010.

[81] "Korean, U.S. Firms to Set up Solar Panel Joint Venture," *Yonhap,* May 29, 2008. Under the terms of the MOU, Plextronics supplies baseline solar device and process technology to KPF, and KPF develops advanced process technology using Plextronics' inks, device, and process technology. "Plextronics Signs MOU to Supply Solar Technology to Korean Company," *Printed Electronics World,* June 9, 2008.

[82] "Hanwha Chemical Buys Into U.S. Graphene Maker," *Yonhap,* January 19, 2011.

[83] "POSCO Buys into U.S. Graphene Manufacturer," *Yonhap,* June 8, 2011.

as a mature basic science research base. However, also in contrast to the United States and the European Union (EU)—where no major manufacturer of flexible displays has emerged—China is currently experiencing an investment rush by enterprises seeking to enter the production of AMOLED displays. The apparent anomaly reflects, in significant part, the potential "market pull" of Chinese consumer demand, as well as the demonstrated willingness of government authorities to finance large-scale investments in production capacity and, if necessary, to subsidize major operating losses.

China is a late starter in the field of flexible electronics. However, it is likely to exercise significant influence on the evolution of global competition in this industry. Given that a key impediment to the development of a commercial flexible electronics industry is the comparatively low consumer "demand pull," China's market is likely to play a major role in creating high-volume demand enabling the improvement of production efficiency and yields.[84] In addition, large-scale financial backing by local governments is enabling Chinese manufacturers to mount a challenge to Korea's supremacy in flexible consumer displays. While to date China's central government has not placed a high priority on development of an indigenous flexible electronics industry, it has provided incentives and encouraged local government developmental initiatives. Moreover, the government has targeted other closely related sectors (microelectronics, photovoltaics), a fact which is likely to contribute to the eventual build-out of a domestic flexible electronics industry chain.

China's huge domestic market, augmented by targeted government procurement programs, has generated demand for products closely associated with flexible electronics, inducing investments in local manufacturing operations by domestic and foreign enterprises. Between 2007 and 2013, the Chinese government implemented a rural home appliance discount plan that provided subsidies to rural households to purchase low-end appliances, including LCD TVs, providing a strong demand boost for indigenous production of flat-panel displays (FPDs). The existence of an FPD industrial base is enabling China's entry into AMOLED flexible displays.[85] In addition, China has emerged as the world's largest market for RFIDs, driven by government procurement of 1 billion national ID cards and increasing use by public entities of RFID tags for transportation logistics and contactless reusable tickets.[86]

---

[84] "Potential Threat to Samsung: Chinese AMOLED Panels to Pass Into Market," *Business Korea,* May 2, 2014.

[85] "Rural Home Appliance Subsidy Fruitful," *China Daily,* November 17, 2013; "China's Rural Home Appliance Subsidy Boosts Sales," *China Daily,* November 12, 2012. The subsidy policy promoted the purchase of low-end LCD TVs up to 47 inches, which favored domestic producers relative to Japanese and Korean makers of higher priced LCD TVs. In 2008, the subsidy was paid only when a product priced at 2,000 yuan ($324) or less was purchased. In 2007, the subsidy was expanded to products costing as much as 7,000 yuan ($1,134).

[86] "RFID, PE Industries Enjoy Growth in China," *Printed Electronics Now,* September 2010; "China's PE, RFID Markets Show Strong Growth," *Printed Electronics Now,* November 2009; "ID

Not all Chinese government policies have been consistent with a development of an indigenous capability in flexible electronics. In an industry in which, in global terms, small- and medium-sized enterprises are often at the cutting edge of innovation, Chinese central government policies in electronics have favored large enterprises, with the result that SMEs have encountered difficulty raising capital.[87] The central government's introduction of measures promoting "indigenous innovation," introduced during and after 2006, arguably induced Chinese display makers to utilize inferior domestically made inputs and acted as a drag on industry integration.[88]

Displays

As Korea and Taiwan are demonstrating, the existence of an established industrial base for the production of rigid consumer displays can leverage entry into flexible displays, most dramatically through the conversion of production lines for glass panels to enable manufacturing of flexible displays.[89] As recently as 2010, China accounted for less than 4 percent of global production capacity for flat-panel displays. However, reflecting massive and sustained capital investments—capacity has grown at a compound annual growth rate of 51 percent since 2010—China is forecast to account for greater than 21 percent of global capacity by 2015. While to date most spending has been concentrated on amorphous silicon TFT-LCDs, Chinese producers of displays are rapidly diversifying into high-end subsectors such as low-temperature polysilicon TFT-LCDs and AMOLED

Cards Drive RFID Demand in China," *Infowars.com,* August 30, 2005. In 2009, the Hong Kong Airport Authority introduced RFID technology to facilitate baggage tracking. The Hong Kong government has sponsored pilot projects to encourage the use of electronic product codes (EPCs) in RFID identification tags to improve supply chain efficiency. "Product Codes Help Synchronize Flow of Goods," *South China Morning Post,* October 10, 2008; "Motorola Deal to Improve Airport Luggage Handling," *South China Morning Post,* May 12, 2009.

[87] In the 1990s, Chinese industrial policy was "grasp the large, let go to the small." The Ninth Five Year National Development Plan provided with respect to the electronics industry the objective of creating 2-3 microcomputer makers with an annual production capacity of more than $1 billion. YungKai Yang, "The Taiwanese Notebook Computer Production Network in China: Implication for Upgrading of the Chinese Electronics Industry" (University of California at Irvine, Personal Computing Industry Center, February 2006), 19. A significant exception to this general pattern was the Torch Program, initiated in 1986, which directed government R&D support to SMEs. Heilman et al. (2013) op. cit., "China's Torch Programme."

[88] On this point see Chen and Ku (2014) op. cit., "Indigenous Innovation vs. Teng-Long Huan-Niao," op. cit., the gist of the Indigenous Innovation policies was to create incentives to enable Chinese enterprises to acquire, absorb, and modify foreign technologies to create new technologies with Chinese-owned intellectual property. *Provisional Measures for Accreditation of National Innovation Products*, Ministry of Finance, Ministry of Science and Technology and National Development and Reform Commission, December 2006.

[89] "Q&A," Dr. Janglin Chen of ITRI (Part 1)," *The Emitter,* July 2, 2013; ITRI, "Flexible Electronics Packing a Punch," March 31, 2011.

displays for smartphone and tablet PC applications.[90] Massive Chinese investments in AMOLED displays now appear to represent a direct challenge to the market leader, Samsung.[91]

Emergence of China's Flat-Panel Display Industry

China's come-from-behind effort to establish a domestic FPD industry is relevant to the subject of flexible electronics because the comparative lack of a domestic FPD industrial capability is seen as a major impediment to U.S. and European prospects in flexible electronics. China has demonstrated that with the requisite level of investment and acquisition of foreign technology, an FPD industry can be established in a relatively short timeframe in the face of strong competition. The experience of BOE Technology Group, China's market leader in FPDs, which suffered massive operating losses in the wake of the global financial crisis, underscores the risks facing Chinese firms in this sector.[92] However, central and local government subsidies ultimately offset these losses.[93] Sustained by government organizations, the newly established FPD industrial base is enabling China's current bid to challenge Korean dominance in flexible displays.

China's nascent LCD industry benefited substantially from supportive government policies at the national level.[94] However, the key government actors in

[90] "China to Account for More Than 70% of Flat Panel Display Equipment Spending," *CTimes,* February 7, 2014.

[91] "Potential Threat to Samsung: Chinese AMOLED Panels to Pour Into the Market," *Business Korea,* May 2, 2014; "Can China Break Samsung's AMOLED Grasp?" *China Focus,* April 8, 2014; "China Firms Invest in AMOLED Lives Largely," *SinoCast,* March 12, 2014.

[92] "BOE admitted that hurt by recession in the global consumer electronics market in the second half of this year, it saw the price of major LCD panels continued to drop and due to this, its loss incurred from operations this year would be huge." "China Panels Makers Expand Amid Fiercer Global Competition," *SimoCast,* December 12, 2011; "BOE Loss-Making Operation Expected to Continue in 2011," *SimoCast,* February 3, 2012.

[93] *SinoCast*, an online information website covering Chinese business, observed in 2012, "According to industry observers, it is inevitable for both BOE and Huaxing Photoelectronic to suffer an operational loss in 2011, thus various kinds of government subsidies, export rebates, and discount loans will play a significant role in helping them turn the loss into a profit on their financial sheets. . . . The Chinese government had provided subsidies and support to the nation's LCD panel makers for a period of time and in 2010, TCL received more than CNY 1.5 billion [$243 million] from the municipal governments of Shenzhen and Huizhou as well as from the provincial government of Guangdang. The figure it achieved in the first three quarters of 2011 hit CNY 340 million [$55 million]." "Panel Makers Depend Much on Government Subsidies," *SinoCast,* January 4, 2012.

[94] China's 11th Five Year Plan (2006-2010) included a subplan for development of the information industry, including promotion of the digital TV sector, which used flat panels to display images. Beginning in 2006, tax rebates on value-added tax (VAT) paid on intermediate components and subassemblies used in flat panel TV receivers were increased for 13 percent to 17 percent upon export of finished products. The industry enjoyed tariff and VAT concessions on imported machinery and materials used to produce LCDs. In 2009, the State Council issued a plan calling for the elimination of "bottlenecks" in the flat-panel display industry, including central government subsidies of up to $32 million. In 2012, the government raised the tariff rate on large-screen LCD panels from 3 percent

the growth of this industry were local governments.[95] Municipal governments have been alternately encouraged and restrained by central government authorities concerned over the potential for excessive investments.[96] Local authorities utilized their State Assets Supervision and Administration Commissions (SASACs), high-tech development zones (HTZs), and economic development authorities to support LCD investment projects, provide loans, and periodically to purchase additional equity in LCD enterprises.[97] (See Table 6-6.)

The ownership structure of the Chinese LCD makers typically provides for mixed public–private ownership, which enables periodic rounds of public–private equity financing.[98] In the case of BOE Technology, China's largest LCD maker, the government of Beijing wields "actual control" over the company through the wholly government-owned Beijing Electronics Holdings Co., Ltd., despite the fact that the latter's equity interest is extremely modest and diluted by other equity interests.[99] (See Figure 6-1.)

---

to 5 percent to protect domestic producers. Tain-Jy Chen and Ying-Hau Ku, "Indigenous Innovation vs. Ten-Long-Huan-Niao: Policy Conflicts in the Development of China's Flat Panel Industry," *Industrial and Corporate Change,* February 2014.

[95] Professor Jang Jin of Korea's Kyung Hee University, who chaired the International Meeting on Information Display in 2013, observed that "China's local governments have made huge investments in the display industry. China's LCD display industry is currently facing a 1-2 year technology gap with Korea's, while the technology gap in OLED technology extends further to five year's difference. However, within ten years, China will begin to dominate the display industry in its entirety. They are hiring talented engineers from Taiwan and paying them high salaries." "Display Technology Is Emerging in China," *Korea IT Times,* June 26, 2013.

[96] In 2009 the State Council urged local governments to provide financial support for strategic investment projects in the flat panel display industry, committing to augment this support with central government funds. The result was a "plethora" of investment initiatives "mostly orchestrated by the local governments," and in 2010 the Ministry of Industry and Information issued an ordinance revoking the authority of local governments to approve LCD panel projects, requiring central government approval of each project. Subsequently the NRDC has screened LCD investment proposals favoring in particular proposals featuring comparatively advanced technology levels. Chen and Ku, "China's Flat Panel Industry," 11–12.

[97] At the national level, central SASAC was established in 2003 to own and manage state-owned assets "on behalf of the State Council," China's supreme executive authority. A network of local-level SASACs was subsequently established. The local SASACs operate under the administrative control of local governments, which make appointments and manage budgets of the enterprises that they wholly or partially own. Central SASAC retains business authority over local SASAC offices and "guides and supervises" the work of state-owned assets supervision and administration of the government at the local level." *Interim Regulations on Supervision and Management of State-Owned Assets of Enterprises,* Decree of the State Council of the People's Republic of China, No. 378, Premier Wen Jiabao, May 27, 2003.

[98] In 2011, China's central SASAC called for engagement of non-state-owned capital, including private and foreign capital, in the state holdings of state-owned enterprises. "Equity diversification is crucial for SOEs to form a modern corporate management structure [to enable SOEs to] compete with other multinational companies in the global arena," said SASAC Chairman Wang Yang. "Chinese State-Owned Enterprises to Use Overseas Funds," *Asia Pulse,* January 9, 2011.

[99] Beijing Electronics Holdings Co. Ltd.'s direct 2.04 percent equity interest in BOE Technology is smaller than the holdings of five other entities: Beijing e-TOWN International Investment & Devel-

TABLE 6-6 Ownership of Leading Chinese TFT-LCD Firms (2010)

Company	Ownership
BOE Technology Group Co. Ltd	Beijing government, Beijing Development Zone, Hefei government, private investors
Shanghai Tianma Micro-electronics Co. Ltd.	Shanghai government organizations (Shanghai Guoyou Zichau, Shanghai Changjiang Group), Tianma, private investors
InfoVision Optoelectronics (Kunshan) Co. Ltd.	Kunshan government (51%) IVO Holding (Taiwan) (49%)
Shanghai SVA NEC Liquid Crystal Display Co. Ltd.	Shanghai government (SVA Group), 75%; NEC (Japan), 25%
Shenchao	Shenzhen government, private investors

SOURCE: Rho, "A Study on the Chinese Flat Panel Display Industry" (2010) op. cit., 21.

A key aspect of local government support for Chinese LCD producers has been HTZs, many of them originally established pursuant to a central government initiative but characterized by "bottom up" local government direction and featuring a wide variety of promotional tools from zone to zone.[100] The Kunshan New & Hi-Tech Industrial Development Zone (KSND), for example, established in 1994, is the site of China's first large-scale indigenous OLED production line, established by Visionox in 2008.[101] The KSND, like similar organizations in other areas, exists as an administrative entity capable of setting its own operating rules and industrial incentives. The KSND offers a variety of incentive packages to high-technology companies, which may include grants, training, and concessional rents and utilities. High-technology companies qualify for an income tax rate of 15 percent, as

opment Co. Ltd (10.98 percent, state-owned); Beijing BOE Investment & Development Co. Ltd., (6.37 percent, state-owned); Beijing Economic-Technological Investment & Development Corp. (6.27 percent, state-owned); Hefei Rangke Project Investment Co. Ltd. (4.99 percent, state-owned); and Beijing BDA Technological Investment Development Co., Ltd (4.44 percent, private). BOE Technology Annual Report, 2013, 41.

[100] The so-called Torch Plan, launched by the State Council in 1988, sought to establish "Silicon Valleys" across China in the form of HTZs. Despite its genesis as a central government initiative, between 1988 and 2005 only about 1 percent of Torch Plan finances derived from the national budget, with the remainder attributable to local sources including local enterprises and banks. Although the HTZs were originally intended as incubation areas for innovative domestic enterprises, many of them came to be dominated by foreign-owned entities and large export-oriented companies. "Progress of Torch Plan Reviewed," *Keji Ribao,* April 11, 1990; Sebastian Heilman, Lea Shih, and Andreas Hofem, "National Planning and Local Technology Zones: Experimental Governance in China's Torch Programme," *The China Quarterly,* December 2013.

[101] Market Intelligence & Consulting Institute, "China's Visinox Begins OLED Production in Kunsham," October 17, 2008.

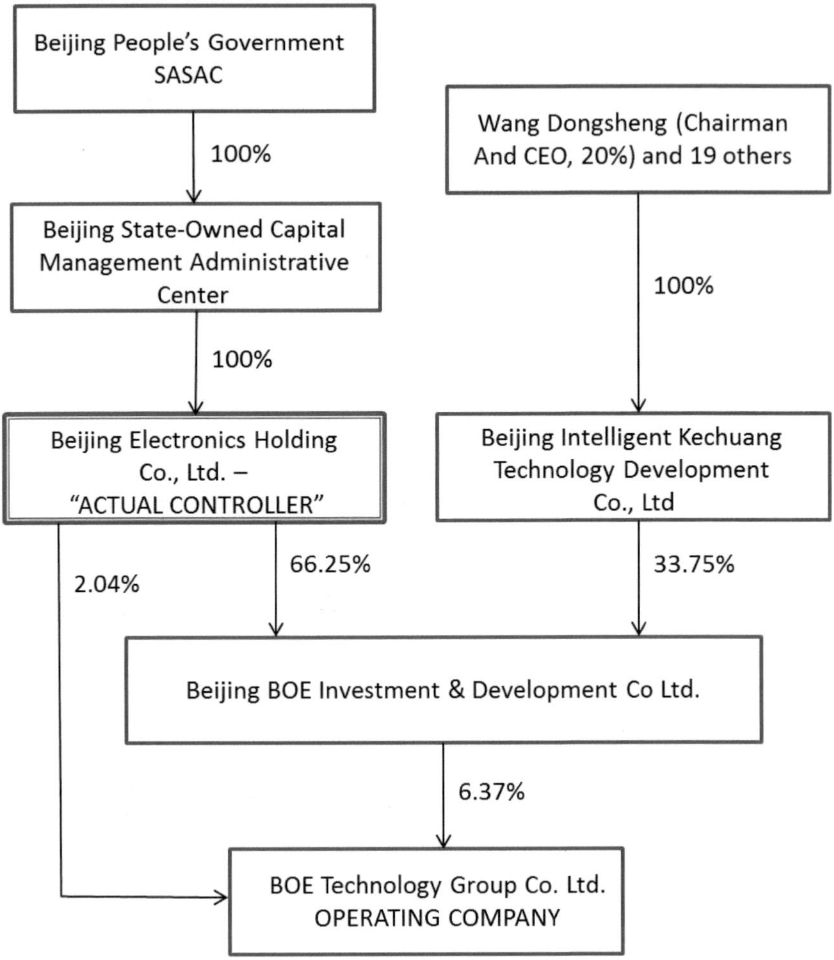

FIGURE 6-1 Schematic showing control arrangement for BOE Technology Group.
SOURCE: BOE Technology Group Co., Ltd, Annual Report 2013, 45.

opposed to the standard rate of 25 percent.[102] KSND is focusing its developmental effort on six "pillar" sectors, one of which is OLEDs, an area in which it collaborates with Tsinghua University, which has established a science park in Kunshan.[103]

---

[102] Kunshan Expat Association Website, <*http://www.kunshanexpat.com/business-update/41-the-ksnd-business-promotion-bureau*>.

[103] "Special Supplement: KSND, Rising Star in the Yangtze River Delta," *China Daily,* September 7, 2007.

Local governments have found creative ways to induce investment in displays, including flexible displays. In August 2011, BOE Technology reportedly reached an agreement with local authorities in Ordos, Inner Mongolia, to invest $3.5 billion to set up AMOLED display production lines in Ordos, in return for which the BOE group received exploitation rights for 1 billion tons of coal from the city of Ordos. The exploitation rights were held by a BOE subsidiary, BOE Energy Technology, and later in 2011 BOE Technology agreed to sell 60 percent of its stake in BOE Energy to Beijing Industry Development & Investment Management Co., Ltd., a Beijing government-owned entity that held a 3.27 percent equity stake in BOE Technology, and another 20 percent of its equity stake in the state-owned coal producer Beijing Haohua Energy Resource Co. Ltd.[104] Through the stock sales BOE Technology received $566.9 million, which it used to offset losses and to finance the new AMOLED lines in Ordos.[105] In effect, Ordos used exploitation rights for coal to enable BOE Technology to raise capital from its existing public shareholders for a local investment in AMOLED manufacturing.

AMOLED Displays

In 2011, when Korea's Samsung held a near monopoly in the production of AMOLED displays (92 percent share), Chinese LCD makers were reportedly preparing to enter the AMOLED display market and challenge the Korean market leaders.[106] Market analyst Vinita Jakhanwal from the consultancy iSuppli commented that

> in turning its attention to AMOLED, China plans to be a significant supplier and gain greater influence in the overall mobile display industry . . . and assurances of capital investment from both Beijing and at local government levels, China could well become an important player in the AMOLED space.[107]

The major Chinese market entrants were expected to be established LCD makers, and a number of these companies declared their intention to enter AMOLED production, including BOE Technology, Tianma, Visionox, Shenzhen ChinaStar Optoelectronics, and IRICO.[108]

---

[104] BOE Technology Annual Report, 2011, p. 53; "BOE Transfers Mining Shares for OLED Investment," *WantChinaTimes.com,* October 29, 2011.

[105] "BOE Tech Group to Sell 80% of energy Unit," *SinoCast,* November 15, 2011; "BOE Gains 3.6 Billion Yuan Transfer Proceeds, Wholly Put an AMOLED," *GG-LED.com,* February 29, 2012, <*http://www.gg-led.com/asdisp3-G56095fb-159-html*>. "BOE Transfer Mining Shares for OLED Investment in Ordos," *WatchChinaTimes.com*, October 29, 2011.

[106] "China Aims for Lead in AMOLED Sector," *Asia Pulse,* November 18, 2011.

[107] "Suppli Sees China Targeting AMOLED Market," *Optics.org,* November 27, 2011; "China Aims to Be Major Player in AMOLED Industry," *Yonhap,* November 18, 2011.

[108] In addition to these players, the Dutch-Belgian research organization IMEC was reportedly planning to build a 3.6 Gen AMOLED display fab in Nanjing in collaboration with JGroup, and Blue Excited Technology planned to make small AMOLED displays based on technology developed

In 2013 and 2014 numerous reports spoke of a major, multifirm buildup of indigenous Chinese production capacity for small AMOLED displays for handsets. A newly formed company, Shanghai-based Everdisplay Optronics, demonstrated prototype 5.5-inch OLED displays in 2013 and was said to be investing in production capacity for 15,000 units per month to commence in late 2014.[109] BOE and Visionox were reportedly considering investment in production lines for AMOLED panels smaller than 10 inches, with Visionox indicating it was hiring experts from the United States, Korea, and Taiwan to assist in this effort.[110] At least three other Chinese makers were reportedly considering market entry, raising concerns of overcapacity:

> Concentrating on the low to medium-price handsets from Chinese producers is attractive for now, but if domestic consumers choose . . . more technically sophisticated foreign modes in the future, there is a risk that too much investment will lead to an oversupply as demand for the devices shrinks. This would be similar to the glut, created by aggressive Chinese moves into LED lighting manufacturing at the beginning of the decade.[111]

According to a 2014 report by the consultancy NPD Display Search, which specializes in the display industry, Chinese companies' spending on FPD production equipment, which represented 22 percent of global spending as recently as 2010, is expected to exceed 70 percent in 2014 and 2015, representing investments in both low-temperature polysilicon and AMOLED display applications.[112]

Despite broad and deep government financial support, Chinese firms entering the market for AMOLED displays against entrenched Korean players face daunting challenges. A 2011 report by the consultancy CCID Consulting summarized the situation facing Chinese entrants into OLED displays:

> So far, China has not yet formed its OLED industry chain, without any domestic full-set OLED manufacturing equipment producers and with key equipment and full-set equipment technologies dominated by the Japanese, South Korean and European enterprises. It lacks raw materials including indium tin oxide (ITO), glass photoresist, dessicant and UV curing adhesives for packaging. China's development of AMOLED techniques is facing great difficulty due

at Xinyang Normal University. "AMOLED Production: Entering a New Era?" *Information Display,* March/April 2013.

[109] "China Looking to Break into Small Screen AMOLED Market 2015," *Plastic Electronics,* May 7, 2014.

[110] "Potential Threat to Samsung: Chinese AMOLED Panels to Pour into the Market," *Business Korea,* May 2, 2014.

[111] "China Looking to Break into Small Screen AMOLED Market 2015," *Plastic Electronics,* May 7, 2014.

[112] "China to Account for More than 70% of Flat Panel Display Equipment Spending," *CTimes,* February 7, 2014.

to the backward TFT technologies, inferior panel technologies and inadequate management.[113]

The technological and supply chain obstacles to China's success in AMOLED displays are being addressed, in part, by offshore procurement and arrangements and local investments by western materials and equipment suppliers.[114] In addition, Chinese enterprises have reportedly hired significant numbers of engineers with relevant competencies from Taiwan, Korea, and other advanced countries.[115]

Key Companies

Beijing Orient Electronics (BOE) Technology Group Co. Ltd.

The foremost Chinese challenger of Korea's position in AMOLED displays is BOE Technology, a partially state-owned enterprise that is China's largest manufacturer of TFT-LCDs. The company's roots go back to the 1950s to a Beijing-based maker of cathode ray tubes that gradually accumulated technology, production experience, and human skills and resources and which began seeking entry into TFT-LCD markets in the mid-1990s.[116] In 2002, pursuant to a South Korean government effort to rescue the financially troubled Korean semiconductor producer Hynix, BOE Technology Group acquired Hydis, Hynix's flat-panel display unit, for $380 million. To facilitate the sale of Hydis to BOE, three banks that were majority owned by the Korean government provided a total of $180 million in loans to BOE.[117] Through the Hydis purchase BOE became

[113] "Large OLEDs Top Lists in China," *Solid State Technology,* April 7, 2011. BOE's investment in AMOLED display manufacturing in Ordos, Inner Mongolia, prompted the observation that "it will be a significant challenge to operate an AMOLED-display fab in such a remote location without any supply chain or plentiful water supply, and for BOE to make the leap into GEN 5.5, the leading edge of AMOLED display production." Paul Semanza, "OLEDs in Transition," *Information Display,* October 2011, 16.

[114] "Merck KGaA Inaugurates Liquid Crystals Center in Shanghai," *Printed Electronics Now,* December 3, 2013; "Qingshan L: Appointed as New Lab Director for AIXTRON China Ltd," *Printed Electronics Now,* March 21, 2013; "BOE Technology Group Teams with Applied Materials to Deliver Leading Edge Display Technologies for Next-Generation Televisions and Mobile Displays," *Printed Electronics Now,* July 2013: "Universal Display's State-of-the-Art Chemistry Laboratory in Hong Kong," *Printed Electronics World,* April 29, 2011.

[115] "Display Technology is Emerging in China," *Korea IT Times*, June 26, 2013. "China makers are heavily reliant on recruiting talent from Taiwan to develop AMOLED panels, which poses challenges for Taiwan's panel industry, as top engineers are continuing to transfer to companies in China." "China Handset Vendors to Rely on Local AMOLED Panel Production from 2016," *Digitimes,* February 26, 2014.

[116] *Understanding China's Manufacturing Value Chain: Opportunities for UK Enterprises in China* (University of Cambridge, 2008), 29.

[117] Congressional Research Service, *The Semiconductor Industry and South Korea's Hynix Corporation,* May 27, 2003, CR5-12. The banks were the Korea Exchange Bank, the Korea Development Bank, and Woori Bank (ibid.). The transaction was part of a larger effort by Hynix to sell nearly

the first Chinese firm to possess core TFT-LCD technology, which the company planned to use to become the world's third largest maker of TFT-LCDs. BOE CEO Wang Dongsheng commented in 2003 that "the purpose of the acquisition was to obtain the core technology. This core, for crystal flat-display technology, was in the hands of South Korean and Japanese companies. An acquisition was the only way we could get ahold of it."[118]

BOE began volume production of TFT-LCDs in Beijing in early 2005, but reflecting its limited technological capability it suffered operating losses for a number of years. However, it continually succeeded in raising funds from municipal governments in Beijing and Hefei (in Anhui Province) as well as from a number of state-owned enterprises—and as a result, it was able to sustain an aggressive investment effort.[119] In 2010, the company reported it was being granted a government subsidy of $122 million.[120] In 2012, in the wake of "huge" operating losses, BOE Technology reported that two of its subsidiaries, Hefei BOE Optoelectronics Technology Co., Ltd. and Beijing BOE Display Technology Group Co. Ltd., each received discounted loans from the municipal governments of Hefei and Beijing totaling 223 million yuan ($36 million) and that its LCD production lines qualified for return of "overpaid" value-added tax (VAT) totaling 1.39 billion yuan ($225 million).[121] In 2013, BOE Technology disclosed that it had signed an agreement with the government China Development Bank giving it total credit from the bank of 20 billion yuan ($3.2 billion) for 2013 and 2014.[122] In mid-2013 BOE disclosed that the government

---

$1 billion in non-core assets. "Rivals Furious As Hynix Set to Sell Dollars 1bn of Non-core Assets," *Financial Times,* January 31, 2003.

[118] "China Imports Hi-Tech through Hydis Purchase: BOE Says an Acquisition Is the Only Way to Get Ahold of the Core Technology," *South China Morning Post,* February 20, 2003. The aftermath for Hydis itself was less sanguine:

> "[Hydis] worker acrimony increased when BOE used technology transferred from Hydis to build a new display factory in Beijing. When Hydis later ran into financial trouble, BOE did not pump in more money, leaving it to file for bankruptcy protection in 2006, according to Hydis employees. Now sold to a Taiwanese company, Hydis is a shell of its former self. 'BOE got the technology they wanted,' said Hwang Pil Sang, a Hydis worker. 'All we got was layoffs.'" "Soured Deal Embitters Shanghai and Seoul: A Warning for Those Seeking China Cash?" *International Herald Tribune,* February 25, 2009.

[119] Sungho Rho, "Technological Discontinuity and Industrial Catch-up: A Study on the Chinese Flat Panel Display Industry" (presented at Globelics 2010 8th International Conference, University of Malaya, Kuala Lumpur, Malaysia, November 1-3, 2010), 22. "BOE Tech Sells Shares to Raise 1,96 Yuan," *South China Morning Post,* October 13, 2006; "BOE Technology Wins Government Aids," *SinoCast*, November 11, 2008.

[120] "China's BOE Technology Group May Turn Around, Buoyed by Government Subsidy," *China Business News,* January 20, 2010.

[121] "Panel Makers Depend Much on Government Subsidies," *SinoCast*, January 4, 2012.

[122] "BOE Tech. Signs Financial Cooperation Agreement with CDB," *SinoCast,* January 8, 2013.

Export-Import Bank of China had extended it a line of credit of 30 billion yuan ($4.9 billion) for the next 3 years.[123]

Beijing Visionox Technology Co. Ltd.

Visionox was established in 2001 to commercialize OLED technology being developed at Beijing's Tsinghua University. In 2002, pursuant to China's 863 Program, or State-Hi Tech Development Plan, Visionox and Tsinghua studied the application of OLED technology to flat-panel displays under the Tenth and Eleventh Five-Year Plans.[124] This project reportedly successfully overcame "technological difficulties in terms of OLED material, manufacturing technology and product implementation, completed the technical study of pilot-scale experiment and realized small batch production."[125] Visionox's commercial strategy emphasizes development of its own proprietary technology, and in the first decade after its founding it applied for and/or secured 288 patents. Visionox's R&D was supported by numerous government organizations, including the Ministry of Science and Technology, the Ministry of Industry and Information Technology, the Economic Development and Reform Commission, and the National Natural Science Foundation Committee.[126]

In 2011, Visionox declared its intention to become a manufacturer of AMOLED displays.[127] Visionox indicated that it was building a 5.5 Gen AMOLED manufacturing line that was expected to become operational in mid-2014.[128] In 2013, Visionox announced that it had developed the Chinese mainland's first flexible AMOLED display, a 3.5-inch display with a bending radius of nearly 10 cm that could withstand repeated bending.[129] A year later Visionox announced that it had

[123] "China's BOE Technology Signs Agreement With Export-Import Bank of China," *China Business News,* June 20, 2013.

[124] The 863 program (so named because of its inception in March 1986) was established on the initiative of four of China's leading weapons scientists to diversity China's R&D programs from strictly military themes to dual use and civilian technologies. The 863 Program concentrated on the fields of information technology, lasers, space technology, automation, energy, new materials, and biology. "Innovation Can Solve China's Energy Predicaments," *Keji Ribao,* July 13, 2006; "Call for More Investment in China's High Tech Sector," *Asia Pulse,* February 2, 2006; "Fair, Dedicated, Realistic, Coordinated, Innovative," *Keji Ribao,* March 25, 1990.

[125] Ministry of Science and Technology, "The 1st Large-Scale OLED Production Line of Mainland China Settled in Kunshan," Press Release, November 23, 2005.

[126] *<http://www.visionox.com/en/about.aspx>*.

[127] "Visionox Announces Its Commitment to Gen 4.5 AMOLED Production," *OSADirect,* August 8, 2011. Visionox also produces OLED lighting panels, which are comparatively inexpensive. "Updates on Visionox's OLED Lighting Program: New Panels and Lamps Unveiled," *OLED-Info.com,* April 22, 2012.

[128] "Visionox Starts Constructing a 5.5-Gen AMOLED Line, Develops a Flexible 3.5' Monochrome AMOLED," *OLED-Info,* August 2, 2013.

[129] "Kunsham OLED Industry Makes a Breakthrough," *Kunshan Today,* August 5, 2013.

developed a flexible full-color 3.5-inch AMOLED display, albeit with a smaller bending radius of under 5 mm.[130]

Truly-Huizhou AMOLED JV

In December 2013, Hong Kong–based Truly International Holdings Ltd., an industrial group making LCD modules, touch panel products, and camera modules, concluded an agreement with the municipal government of Huizhou, in Guangdong Province, to launch a joint venture to produce AMOLED displays on a 4.5 GEN production line beginning in the second half of 2015, with a planned monthly output of 15,000 units.[131] The AMOLED line will be located in the Huizhou Zhankai High-Tech Industrial Development Zone, and the management of the Zone is taking a 27.12 percent stake in the venture, along with another 19.88 percent share by the SASAC owned by the Huizhou Municipal Government. (See Figure 6-2.)

The joint venture's total investment will be 6 billion yuan ($971 million) to be funded "by a combination of registered capital contributions, bank loans and government subsidies." The AMOLED joint venture has "received great attention [sic] from Guangong provincial government and Huizhou municipal government as well as industrial subsidy support."[132]

Tianma Micro-Electronics Co.

Tianma is a maker of LCDs founded in Shenzhen in 1983 and has established LCD operations in Shanghai, Shenzhen, Chengdu, Wuhan, and Xiamen.[133] In 2010 Tianma's Shanghai-based subsidiary, Shanghai Tianma Microelectronics Co., Ltd., announced plans to build a 4.5 GEN AMOLED production line in Shanghai. Shanghai would reportedly contribute 210 million yuan ($34 million) to this project, with another 281.6 million yuan ($46 million) derived from "government grants and subsidies."[134] In 2014, Tianma was reportedly preparing to invest 1 billion yuan ($162 million) in a 4.5 GEN AMOLED production line,

---

[130] "Visionox Developed Full-Color 3.5" Flexible AMOLED," *OLED-Info.com,* March 27, 2014.

[131] Truly is incorporated in the Cayman Islands and operates out of Hong Kong. Founded in Hong Kong in 1978, it has become a major force in the electronics industry in Guangdong Province, with a 1 million square meter local production base. <*http://www.trulymid.com/en/newsInfo.asp? Pid?pid24*>; Truly International Holdings Annual Report, 2013.

[132] Truly International Holdings Limited, *Shareholders Agreement on Formation of a Joint Venture Company,* December 17, 2013, 5. Truly International's 2013 annual report (p. 112) states that in that year the company received "incentive subsidies" of about $13 million from the Hong Kong government "to encourage the operation of a PRC subsidiary for the development on export sales and advance technology."

[133] <*http://www.tianma.cn/english*>.

[134] "Shanghai Tianma to Invest in GEN 4.5 AMOLED Pilot Line," *OSA Direct,* August 30, 2010.

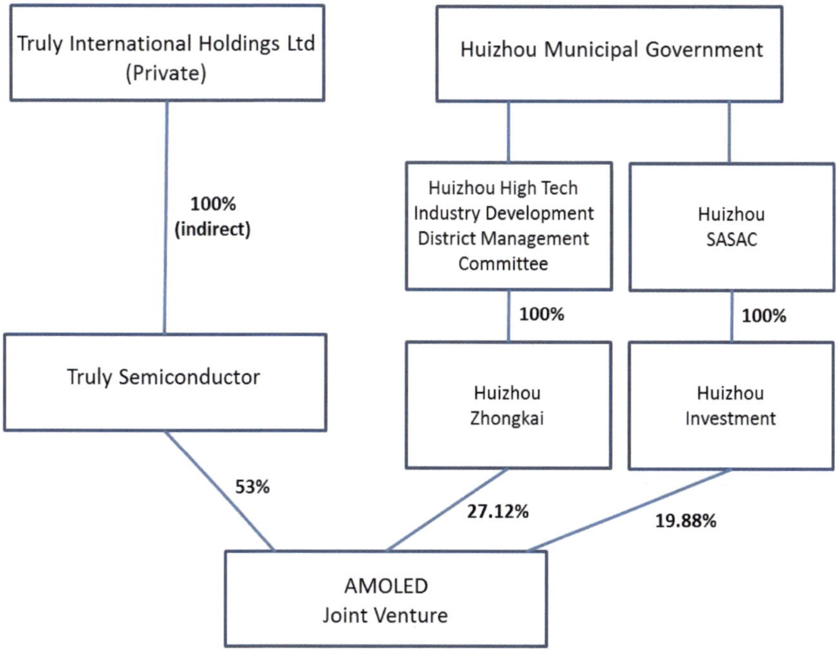

FIGURE 6-2 Schematic showing control arrangement for Truly/Huizhau AMOLED Joint Venture.
SOURCE: Truly International Holdings Limited, *Shareholder Agreement on Formation of a Joint Venture Company* (December 17, 2013).

primarily to serve domestic companies such as Huawei, Lenovo Coolpad, and ZTE Corp.[135]

Hehui Optoelectronics

Shanghai Jinshan Hehui Optoelectronics Works is reportedly establishing a 4.5 GEN AMOLED manufacturing line in Shanghai's Jinshan Industrial Zone.[136] The line is reportedly being established "largely through government funding." Hehui has reportedly succeeded in recruiting more than 70 engineers from Taiwan. The line is reportedly producing small volumes of handset panels

---

[135] "China Firms Invest in AMOLED Lines Largely," *SinoCast,* March 12, 2014.

[136] "Hehui Optoelectronics President Visits Jobsite of SBC-MCC Group in Shanghai," *MCCChina,* Press Release, July 8, 2013.

but is expected to begin mass production of AMOLED panels in 2014 and is "also likely to produce flexible displays."[137]

IRICO

IRICO Group Corporation is a state-owned enterprise with extensive electronics manufacturing operations whose shares are held by the central government SASAC, and according to its website, "for many years, the Party Central Committee and the State Council have always been concerning and supporting the development of IRICO."[138] IRICO is a major producer of LCDs and for a number of years has stated its intention to enter the manufacture of OLED displays.[139] In 2011, the company was reportedly planning an issue of nonpublic shares to finance establishment of a 4.5 GEN AMOLED production line.[140] As of mid-2013 IRICO was reportedly preparing to build an AMOLED production line "in the next two years."[141]

Ascent Solar-Suqian JV

In 2013, Colorado-based Ascent Solar Technologies concluded an agreement with the municipal government of Suqian, in China's Jiangsu province, to establish a joint venture to manufacture Ascent's proprietary thin-film copper/indium/gallium/selenium (CIGS) photovoltaic modules on flexible polyimide film.[142] The joint venture will be located in the Suqian Economic and Industrial Development Science Park. The municipal government has committed to inject $32 million into the joint venture, and Ascent will contribute its proprietary technology and intellectual property as well as equipment from its Colorado plant. Suqian will also provide a 5-year corporate tax holiday followed by a 5-year 50 percent tax rebate; full rebate of the value-added tax for 2 years and a 50 percent rebate over the subsequent 5 years; and free accommodations for up to 3 years for "key scientists, engineers and management personnel of the JV."[143]

---

[137] "Hehui to Mass Produce Small-to-Medium-Size AMOLED Panels in 2014," <*http://www.ctxme.com/latestnews_info.php?id=208*>.

[138] <*http://www.ch.com.cn/english/txt.jsp?uritype+tree.TreeTempUri&wbtreeid=1459*>.

[139] "Highlights from Printed Electronics Asia 2012," *Printed Electronics World,* October 10, 2012.

[140] "IRICO to Raise Funds to Establish Generation 4.5 AMOLED Production Lines," *GG-LED,* April 7, 2011.

[141] "DisplaySearch: AMOLED Manufacturing Capacity Forecast to Nearly Triple in 2012," *Green Street Journal,* July 29, 2013.

[142] "Ascent Solar to Build New Manufacturing Plant in China," *Printed Electronics World,* July 16, 2013.

[143] "Ascent Solar Signs Definitive Agreement to Build New Manufacturing Plant in Suqian of Jiangsu Province, China," *Printed Electronics Now,* January 3, 2014.

Kunshan Printed Electronics Ltd.

Kunshan Printed Electronics Ltd. (KSFPE) was established in Suzhou in 2011 by Dr. Zhang Xiachang, the inventor of SoftBattery and co-founder of Finland's Enfucell. KSFPE was funded by local investors in cooperation with a local machinery operator. The company has established an R2R production line for printed batteries with a capacity of 40 million cells per year.[144] Applications of the batteries include RFID labels, smartcards, and ultrathin calculators. The company's collaborators include Finland's government research organization, VTT, and the Finnish printed battery company affiliated with Dr. Zhang, Enfucell.[145]

China Star Optoelectronics Technology (CSOT)

China Star Optoelectronics Technology (CSOT) was formed as a joint venture of Star Hi-Tech, a holding company dominated by the Shenzhen municipal government, and TCL Group, a large Chinese consumer electronics producer. CSOT recruited LCD talent in Japan, Korea, and Taiwan and formed collaborations with a number of foreign suppliers.[146] By 2012, China Star was the second biggest LCD maker in China and the only Chinese firm operating an advanced 8.5 GEN fab.[147]

Kunshan Hisense Electronic Co., Ltd (Hisense)

Hisense was formed in 2007 by Chinese and U.S. nationals who had studied in U.S. universities or worked in U.S. and European high-technology companies.[148] Hisense develops and produces materials associated with printed electronics, including nanosilver conductive ink, conductive silver paste, and printed electronic inkjet printing systems. Managed and staffed primarily by "returned

---

[144] "Printed Battery Company in China," *Printed Electronics World,* July 22, 2013.

[145] <*http://www.ksfpe.com*>.

[146] Foreign affiliations included Corning (glass substrates); Toppan and Dai Nippon Printing (color filters); and Sumitomo Chemical, Toyota Gosei, and JSR (photoresist materials). Mei Chih Hu, "Technological Catching-up in East Asia," in *International Economic Development: Leading Issues and Challenges,* eds. Fu Lai Tony Yu, Wai Kee Yuen, and Dinna S. Kwan (New York and Abington: Rutledge, 2014), 86.

[147] "Rise of Chinese LCD Firms Threatens Local Makers," *Taipei Times,* November 12, 2012. In 2012, Taiwan's AUO sued two of its former executives in Taiwanese court alleging that they had leaked company secrets to China Star including AMOLED related technologies. "AUO Former Executives Are Suspected of Selling OLED Technology to Chinese Companies," *OLED-Info.com,* October 16, 2012.

[148] Hisense chairman Xu Haisheng is a Chinese professor at East China University who studied for a doctorate in the Material Research Institute and Department of Electrical Engineering at Penn State, and subsequently served as a senior scientist with Intel. General Manager Zhou Yang, an American, has an engineering doctorate from Dartmouth and held a series of engineering positions with U.S. companies. <*http://www.en.kshisense.com*>.

students," the company is located in the National Business Incubator for Returned Personnel Kunshan Development Zone, Jiangsu. In 2012, Hisense unveiled a new inkjet deposition and analysis system, IJDAS-300, which is capable of jetting a wide range of functional fluids using multiple deposition print heads interchangeably.[149]

Government Supported Research Institutes

China's manufacturing industries are commonly supported by government-backed industrial research institutes.[150] Several such institutes already exist to support China's flexible electronics industry.

Industrial Institute of Printed Electronics

In September 2013, the Industrial Institute of Printed Electronics was established in Changzhou, Jiangsu Province, between Shanghai and Nanjing, by the municipal government and high-technology industrial park of Changzhou.[151] The local government has committed to a first-phase investment of $16.4 million with "much more" in the second phase. The Institute will join with the Printed Electronics Fund Ltd. to form a consortium that will create companies to produce sensor and smart RFID tags, flexible TFTs, smart tags, flexible displays, and inkjet print heads. The fund company reportedly has registered capital of $49 million, which can be increased to $162 million "very soon." The Institute's function will be "to establish a platform for printed electronics tests, prototypes including a certain amount of production for sales, and to provide printed electronics companies with technical support. . . . The research focus will be centered on the integration of printed electronics applications and roll-to-roll mass production."[152] The Dean of

---

[149] "Kunshan Hisense Electronics Introduces New Cost Competitive and Versatile Materials Printer," *Printed Electronics Now,* September 10, 2012.

[150] For example, in one of China's most dynamic manufacturing sites, the city of Shenzhen in Guangdoing province, the Chinese Academy of Sciences, a state organization, has established the Shenzhen Institutes of Advanced Technology (SIAT), which consist of five research institutes and numerous ancillary laboratories and research facilities. SIAT seeks to enhance the innovative capability of local manufacturing and service industries and to contribute to human resource development. SIAT maintains extensive research cooperation ties with foreign research organizations. It has established partnerships with more than 500 local enterprises and incubated 60 high-technology companies. "Into the Time Shenzhen Institutes of Advanced Technology," *<http://english.siat>*.

[151] Changzhou National Hi-Tech District is a national-level high-technology industrial development zone. It is the site of activity by more than 10,000 companies, including 1,500 foreign-invested entities, including Komatsu, ThyssenKrupp, Ashland, Fujitsu, Nikon, Hitachi, Hyuadai, and Denso. Broken down by sector, its companies are engaged in chemicals and new materials (23.1 percent), equipment manufacturing (41.3 percent), electronics (18.6 percent), and photovoltaics (13 percent). *<http://www.invest-in-Cnd.cn/>*.

[152] "China Establishes Industrial Institute of Printed Electronics," *Flexible Substrate,* November 2013; "Industrial Institute of Printed Electronics," *Printed Electronics World,* October 29, 2013.

the new Institute is Dr. Zhang Xiachang, who was one of the 2002 co-founders of Enfucell, the first Finnish firm to be named a "technology pioneer" at the World Economic Forum in Davos.[153]

Hong Kong Nano and Advanced Materials Institute

In 2006, the government of Hong Kong (Innovation and Technology Commission) set up five R&D centers for applied research in order to bridge the gap between scientists' research and industrial application.[154] One of these was the Nano and Advanced Materials Institute Limited (NAMI), an entity funded by the Hong Kong government and owned by the Hong Kong University of Science and Technology (HKUST).[155] Key research themes have included nanotechnology-enabled/enhanced display and lighting devices, new high-performance light-emitting material, and nanoscale engineering in electronics.[156] In 2013, NAMI was reportedly working on development of transparent conductive films based on silver nanowire and a scalable printing technique.[157]

Universities

South China University of Technology

The South China University of Technology's School of Materials Science and Engineering operates a number of laboratories and centers pursuing research themes relevant to flexible electronics. Its State Key Laboratory of Luminescent Materials and Devices reportedly developed China's first flexible AMOLED display in 2013.[158] The Guangdong Key Laboratory of High Performance and

---

[153] Dr. Zhang was educated in China as a biochemical engineer. He moved to Finland to study the work of Dr. Aarne Halme, who was concentrating on research themes that later proved relevant to the development of flexible batteries. "Great Potential Seen in Soft Batteries," *Helsingin Sanomat,* March 20, 2007.

[154] Hong Kong Innovation and Technology Commission, "LegGo Members Visit Hong Kong Science Park," Press Release, July 18, 2011.

[155] "Printable Transparent Conductive Films and Its Applications in Organic Solar Cells," *Flexible Substrate,* January 2013. Total government funding for NAMI during its first 5 years of operating was HK $400 million ($52 million). Dr. Jiyun Feng, *Commercialization of Nanotechnology Products in Hong Kong: Strategy and Progress* (presentation at International Nanotechnology Business Summit, Tokyo, February 21, 2007).

[156] Feng, *Nanotechnology Products in Hong Kong.*

[157] "Flexible Transparent Conductive Films and Its Application in Organic Solar Cells," *Flexible Substrate,* January 2013.

[158] The device was reportedly 4.8 inches, rollable, flexible and bendable, and resistant to mechanical shock. It was developed jointly with Guanzhau New Horizons Optoelectronic Technology Co. Ltd. "China Successfully Developed its First Flexible Display," *ChinaAbout.net,* August 12, 2013.

Functional Polymer Materials conducts research on themes that include polymer photoelectric materials and devices.[159]

City University of Hong Kong

In 2011, it was disclosed that Arizona State University (ASU) would collaborate with the City University of Hong Kong in the field of flexible microelectronics. The two institutions collaborated on a proposal for the National Science Foundation (NSF) Engineering Research Center on large area sensing arrays, which would use bendable transistors in a plastic sheet to detect radiation and biomarkers. ASU proposed to provide pilot time manufacturing capability in flexible electronics, and the City University offered expertise in flexible nanowires.[160]

Zhejiang University

Zhejiang University in Hangzhou has reported results of research to develop materials for microencapsulation of electrophoretic displays.[161]

Shanghai Jiao Tong University

Researchers at Jiao Tong have reported use of vacuum filtration to deposit single-wall carbon nanotube films on paper substrates, enabling the creation of patterned nanotube films fabricated as flexible electrodes. The paper-based flexible electrodes were said to "show great potential in flexible display technology."[162]

TAIWAN

Taiwan's entire national developmental effort in flexible electronics is centered on a single organization, the government Industrial Technology Research Institute (ITRI), which is recognized as one of the premier institutes of applied research in the world. ITRI, which translates basic research into commercial products and industrial processes, is perhaps the principal reason Taiwan has secured a leading position in some flexible electronics product areas (e-paper) and relevant industrial processes (adapting conventional LCD production lines to the production of flexible displays). Taiwanese companies have proven adept at moving products to market quickly, driving down manufacturing costs, and

---

159 *<http://202.38.194.245/hgschool/materials>*.

160 "ASU and University of Hong Kong to Work Together on Advanced Flexible Electronics," *Flexible Substrate,* December 2011.

161 "Materials for the Microencapsulated Electrophoretic Display," *Flexible Substrate,* June 2012.

162 "Flexible Electrodes Using Single-Wall Carbon Nanotubes Patterned as Paper Substrate," *Flexible Substrate,* March 2012.

building the necessary supply chains. Japanese electronics firms, facing an existential crisis in the face of Korean competition, are increasingly seeking support and tie-up with ITRI and Taiwanese companies.

Notwithstanding its demonstrable strengths, Taiwanese leaders worry that their country lags in innovative thinking and that young people, in particular, are too risk averse with respect to startups. The Taiwanese IT manufacturing method of "accepting detailed computer performance benchmarks from American customers and then figuring out the least expensive way to meet those benchmarks . . . may also be an obstacle to innovation now. The least expensive approach often involves barely exceeding the benchmarks, and not coming up with solutions that may cost more but dazzle consumers."[163] Most of all, Taiwanese leaders wonder whether their domestic industries can successfully match the financial resources, IP portfolios, and bold investments of Korea's chaebol firms.

Taiwanese scientists have been engaged in basic research relevant to flexible electronics since the 1970s. Research with respect to potential industrial applications began around 2000.[164] In 2004, Chen Liang-gee, Director of the Electronics and Optoelectronics Research Laboratories (EOL) at ITRI, proposed that his organization launch a research effort in flexible electronics:

> Taiwan's high-tech sectors must think about what new area deserves attention if we want another wave of growth in 10 years. . . . [W]hat is the future of the electronics industry after the semiconductor and flat-panel display booms? We must utilize Taiwan's experience in the field and find a new direction for it.[165]

ITRI established a Flexible Electronics Technology Division in July 2005 and within a year it was staffed by 31 professionals. In August 2005, for the first time at the annual meeting of the Cabinet-level Science and Technology Advisory Group (STAG), flexible electronics was the focus of discussion. Following the meeting the government decided to begin earmarking funds to ITRI to pursue research in the field, with an emphasis on flexible displays.[166]

Government Entities Supporting R&D

Ministry of Economic Affairs (MOEA)

MOEA is Taiwan's central entity responsible for economic planning, industrial policy, and program implementation. It supervises and provides one-half of the financial support for the ITRI, Taiwan's largest research organization, which

[163] "In Taiwan, Lamenting a Lost Lead," *The New York Times,* May 12, 2013.

[164] Professor Chen Show-an, Department of Chemical Engineering, National Tsing Hua University, in "Bendable, Rollable, Profitable," *Taiwan Review,* July1, 2006.

[165] Chen-Liang-gee in "Bendable, Rollable, Profitable," *Taiwan Review,* July 1, 2006.

[166] Ibid.

TABLE 6-7 MOEA Funding of Flexible Electronics R&D (2010)

Allocation	Amount (Millions of Dollars)
To research institutes (ITRI)	30
To academia	3.5
To industry	6.1

SOURCE: Dr. Janglin (John) Chen, *Flexible Electronics Development in Taiwan,* September 24, 2010.

spearheaded the development of the country's semiconductor and electronics industries. MOEA is the principal source of funding for Taiwan's R&D in the field of flexible electronics. (See Table 6-7.) According to 2010 data from the leading international association for organic electronics, OE-A, between 2006 and 2013 Taiwanese government investments in printed electronics were about $200 million.[167]

Industrial Technology Research Institute (ITRI)

ITRI is a not-for-profit institute for applied industrial research and development. One-half of its operating budget of $510 million is provided by MOEA and half by the private sector. Most of its operations consist of R&D projects involving collaboration with and funding by private companies, to which research results are transferred for development into commercial products. Some ITRI research units have been spun off to form successful private companies, including the United Microelectronics Corporation (UMC), Taiwan Semiconductor Manufacturing Corporation (TSMC), Vanguard International Semiconductor Corporation, and the Taiwan Mask Corporation. ITRI also provides logistical support for startup enterprises at its Incubation Center and its Open Laboratory program. Greater than 60 percent of ITRI's 6,000 employees hold graduate degrees in the sciences. ITRI has amassed more than 10,000 patents and assisted in the creation of more than 165 spinoffs and startups. Seventy CEOs of high-technology companies are former ITRI staffers.[168]

ITRI's managers characterize their organization as a "technology intermediary" between the research community and Taiwanese industry. ITRI's primary function is not research—although it performs substantial research—but adaptation and transfer of research results and technology from Taiwanese and foreign laboratories (public and private) to domestic companies and industry. Technology is transferred to Taiwanese companies through licensing agreements, various co-development projects, spinoffs, patent auctions, migration of ITRI personnel to companies, and research consortia and alliances. (See Figure 6-3.)

[167] Andrew Hannah, CEO, Plextronics, "The Global View of Printed Electronics and What It Could Mean to the U.S," September 24, 2010.

[168] ITRI, "What is ITRI?" <*http://www.itri.org.tw*>.

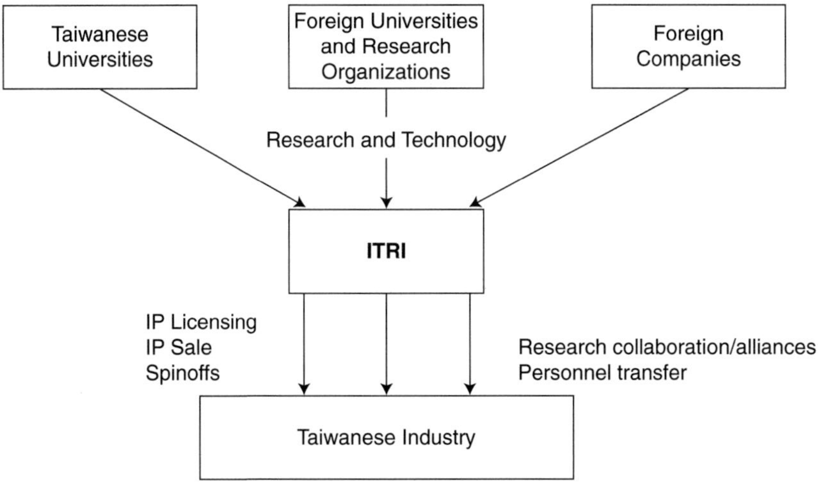

FIGURE 6-3 ITRI as a technology intermediary.

ITRI is "arguably the most capable institution of its kind in the world in scanning the global technological horizon for developments of interest in Taiwanese industry, and executing the steps required to import the technology—either under license or joint development—and then absorbing and adopting the technology for Taiwanese firms to use."[169] ITRI maintains an extensive network of technology collaborations and alliances with first-tier high-technology multinationals and research organizations.[170]

Taiwan's entry into the field of flexible electronics was made possible by friendly transfer of technology from Eastman Kodak Company to ITRI. Dr. John Chen, a Taiwanese scientist who served in various R&D roles at Kodak from 1982 through 2006, was able to facilitate transfer of technology for large area R2R fabrication of flexible displays from Kodak to ITRI after Kodak decided to abandon commercialization and sell technology "to someone who was competent."[171] Chen led an ITRI team to Kodak's laboratories in Rochester where a Kodak team worked with them to facilitate the technology transfer, which included transfer of know-how and equipment. ITRI's subsequent development of flexible displays has surpassed the highest technological levels achieved by Kodak but was "all

[169] John A. Mathews and Dong-Sung Cho, *Tiger Technology: The Creation of a Semiconductor Industry in East Asia* (Cambridge: Cambridge University Press, 2000).

[170] See National Research Council, *21st Century Manufacturing: The Role of the Manufacturing Extension Partnership Program* (Washington, DC: The National Academies Press, 2013).

[171] See National Research Council, *21st Century Manufacturing: Flexible Electronics for Security, Manufacturing, and Growth in the United Sates: Summary of a Symposium.* Washington, DC: The National Academy Press, 2013, p. 76.

based on Kodak technology." Chen observes that "this was the beginning of flexible displays in Taiwan."[172]

ITRI's principal research activities are conducted in six thematic "core" laboratories: and roughly a half dozen "technology centers" that focus on strategic topics and seek to integrate the multiple disciples of the various core laboratories. The core laboratories pursue "deeper and newer" ideas in areas such as materials, electronics, measurement, and machinery whereas the technology centers combine and integrate those competencies with an eye toward commercialization in sectors such as cloud computing, biomedicine, and flexible displays. In flexible electronics, the ITRI core laboratories pursuing relevant research themes are the Electronics and Optoelectronics Research Laboratories (EOL), the Material, Chemical and Nanotechnology Research Laboratory, and the Mechanical and Systems Laboratory.

ITRI's Electronics and Optoelectronics Research Laboratory has developed electro-optical substrate and "FlexLite" technologies that use "soft silicone packaging technology and circuit design to develop a flexible lighting technology with a thickness of 0.5 cm."[173] FlexLite enables designers to "sculpt any lighting forms they wish, and any soft light source applied to the design can be used."[174] Two core technologies are involved:

- The LED is a light source that emits heat, requiring a high thermal conductivity structure, so ITRI used graphene structure technology to the back of the light, achieving the requisite thermal conductivity.
- An optical absorption structure was established in the front of the lamp housing, allowing use of a flexible honeycomb structural design to achieve uniform effects without glare.

ITRI's Display Technology Center (DTC) is one of several internal "Technology Focus Centers" established to promote interdisciplinary collaborations involving ITRI's specialized research laboratories. DTC's main facility is a 3,124 square meter second generation laboratory pilot line (glass substrate size 370×470 mm^2) that has been used to produce 20-inch TFT-LCDs. Clean room class ranges from 10 to 10,000, covering array, cell, and module assembly processes.

DTC is capable of addressing the entire vertical chain of display technology, including design, fabrication, packaging, materials, component, process, and equipment to the system level. (See Figure 6-4.) DTC commonly engages in contract services, joint R&D, technology transfer, cross-licensing, and evaluation and verification of customers' flexible displays materials, equipment, and systems.

[172] Interview with John Chen, Director, ITRI Display Technology Center, Hsinchu, Taiwan, February 14, 2012.

[173] ITRI, "Flexible Electronics Packing a Punch," March 31, 2011.

[174] Ibid.

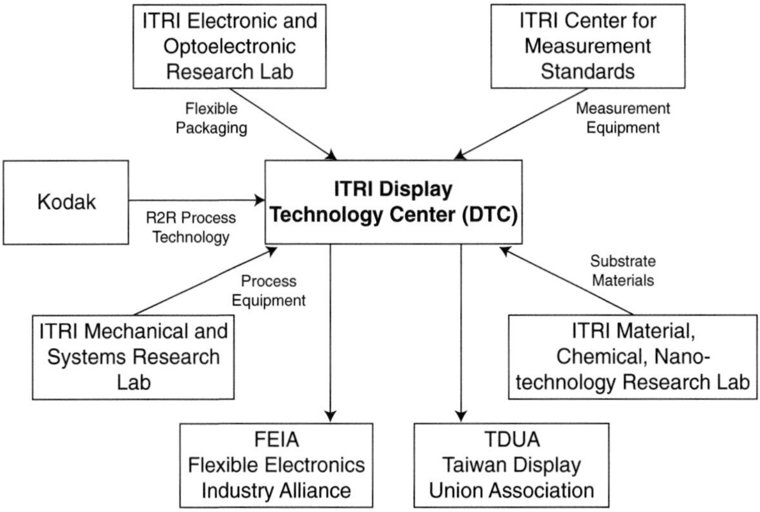

FIGURE 6-4 ITRI Flexible Display Program integration.
SOURCE: Dr. Janglia (John) Chen, ITRI Display Technology Center, "ITRI Display Programs," February 14, 2012.

DTC's objective is to position Taiwan as a leader in the field of flexible displays, which it forecasts will achieve global revenues of $4 billion by 2015. ITRI research groups are addressing a number of themes in flexible display technology:

- ***Flexible substrate technology.*** ITRI has developed flexible substrate process technology using amorphous silicon (a0Si). TFT and an organic TFT process have been developed to meet the requirements of TFT-LCD/AMOLED applications. ITRI is also developing process technology for gas barrier layers for AMOLEDs.[175] ITRI developed polymer-based ultrasonic paper (PUP) for use in physiotherapy and sensing, which is believed to have potential as a flexible substrate for various printed electronic components. Polymer-based ultrasonic paper was the first flexible technology in the field of capacitive micromachined ultrasonic transducers (CMUTs)[176] A flexible CMUT technology could permit

[175] In May 2011, ITRI received the silver award for "Display Component of the Year," given by the Society for Informath Display, the leading global display group, for its flexible substrate for displays. According to ITRI the substrate is the first and only technology that permits the mass production of flexible and transparent displays of all sizes. "ITRI Wins Prominent Display Technology Award from Industry Group," *Central News Agency,* May 19, 2011.

[176] CMUTs are devices that generate and receive sound waves at ultrasound frequencies. They are made of small and thin membranes suspended over a conductive silicon substrate by insulating posts. The membranes that comprise the CMUT are micromachined onto a silicon substrate. See generally

each ultrasonic transducer to remain within an array perpendicular to the relevant object's surface during ultrasonic transmission and reception.[177]

- ***Flexible display media technology.*** ITRI has developed a monochrome cholesteric liquid crystal display (Ch-LCD) flexible display built on a plastic substrate based on R2R manufacturing technology, suitable for large-size public displays. ITRI has developed a 10.4-inch color display featuring plastic substrates and a total thickness of less than 10 mm. ITRI has developed electrowetting display (EWD) technology that can create an optical shutter with switchable liquids or change the visual area of colored liquids allowing wide-angle display applications. In June 2011, ITRI won *R&D Magazine's* R&D 100 Award for development of rewritable e-paper, i2R e-paper, based on a new type of polarizer protector film named HyTAC. The technology utilizes organic and inorganic nanomaterials, a highly transparent optical film and low-toxicity manufacturing processes.[178]
- ***Flexible thin-film transistor array technology.*** ITRI has developed flexible silicon-based TFT array technology. It has integrated highly accurate photolithography processes on flexible substrates using organic thin-film transistor (OTFT) and OLED technology.
- ***Flexible display system technology.*** ITRI has developed "the electronic devices and circuits to integrate a complete system for image processing, power supply," microprocessor and memory control; driving systems for displays; and "a color image quality assessment instrument and processing technology to evaluate and monitor performance of flexible color displays."[179]

ITRI has taken a leadership role in organizing Taiwanese industry to commercialize emerging flexible electronics technology. In 2005, ITRI set up the Flexible Electronic Promotion Alliance in collaboration with the government, academia, and private manufacturers. ITRI hosted a flexible electronics technology forum, with more than 150 invitees from government, academia, and industry and "invited local makers, from upstream, to downstream, to set up the alliance, hoping to integrate related material, production processes, devices, finished products and equipment to form a platform for academy-industry interchanges."[180] At the time ITRI's EOL then-Director General, Chen Lian-gee, commented that

---

Arif S. Ergun et al., "Capacitative Micromachined Ultrasonic Transducers; Theory and Technology," *Journal of Aerospace Engineering,* April 2003.

[177] "Flexible Electronics Pilot Lab and New Electronic Paper in Taiwan," *Printed Electronics World,* June 8, 2007.

[178] "ITRI Wins Further Prestigious R&D Awards," *Central News Agency*, June 23, 2011.

[179] Ibid.

[180] "ERSO to Set UP Flexible Electronics Alliance in July," *Taiwan Economic News,* May 19, 2005.

TABLE 6-8 ITRI Assigns Technological Roles to Individual Companies

Topic	Companies
Materials	Chang-Chun
Equipment	Shuz-Tung Gallant Precision Machining
Panels	AUG CMO CHI Mei Optoelectronics
System	Elan Microelectronics Corp.

SOURCE: ITRI Display Technology Center Presentation (February 14, 2012).

> [i]t is important to build a supply chain if we want to create an industry. We hope that the alliance acts as a window through which enterprises likely to get involved in the emerging industry can have a look at the new technology and see opportunities for themselves. The alliance is also a forum in which companies from different parts of the chain can discuss a future division of labor.[181]

Janglin Chen, Vice President and General Director of DTC, concurrently serves as Chairman of the Taiwan Flat Panel Display Materials and Devices Association. Chen points out that ITRI assigns technological roles to individual companies and that DTC's research projects in flexible electronics represent "a complete manufacturing chain in its early stages."[182] ITRI thus works with not only displays companies such as AU Optronics, but also materials and systems suppliers. (See Table 6-8.)

In 2010, DTC announced the development of FlexUPD, a technology that enables manufacturers of display panels "to convert their existing production lines and panels for glass panels to flexible displays with minimal investment in new equipment."[183] FlexUPD was honored as *The Wall Street Journal*'s 2010 TIA Gold Winner and was one of *R&D Magazine*'s R&D 100 winners. Because FlexUPD can be introduced by adapting existing mainstream LCD production lines, by adding coating and debonding equipment, it is a technology that requires minimal new investments by the display industry in order to manufacture bend-

[181] "Bendable, Rollable, Profitable," *Taiwan Review,* July 1, 2006.

[182] Interview with DTC Director Janglin Chen, Hsinchu, Taiwan, February 14, 2012.

[183] ITRI, "Flexible Electronics Packing a Punch." FlexUPD is based on an ultra-thin, transparent, soft plastic substrate. Dr. Janglin Chen, General Director of the DTC, described the process steps in 2013: "Once the transistors are layered on the plastic substrate and enclosed, the substrate can be cut and peeled off the glass substrate to make a flexible display. The secret lies in the instant debonding/removal process. It is like a layer of non-stick material between a crepe and a pan. In addition to letting the crepe slip off the pan easily, the non-stick layer does not damage the filling in the crepe. This innovative technology is more advanced and cost-efficient than the metal film substrates and laser removal technology used by some leading display manufacturers." "Q&A: Dr. Janglin Chen of ITRI (Part I)," *The Emitter,* July 3, 2013.

able, foldable, and rollable electronic devices. Dr. Chen stated in a July 2013 interview that since ITRI began winning awards for FlexUPD,

> ITRI has received widespread interest from well-known multinational enterprises that range from the display to the aviation, automotive, and mobile device industries. Due to confidentiality considerations, I am unable to release any more information; all I can say is that flexible displays will bring the display industry to a new frontier, unleashing designers from factors that have restricted their imaginations since mankind entered the digital age. Due to ITRI's first obligation to the domestic industry, we are currently in negotiation with various enterprises in Taiwan on the possibility of either spinning in or off a team together with this technology to scale up this process for production.[184]

In 2007, ITRI established a Flexible Electronics Pilot Laboratory to perform R&D on flexible electronics and develop international cooperation in the field.[185] Its initial activities included development of production process R&D for flexible electronic circuits, flexible solar cells, flexible reactors, and flexible displays. The Pilot Laboratory prioritized development of machinery for continuous processes. ITRI indicated it would establish an industry alliance to carry out joint efforts to establish high-performance flexible electronic production lines. The Pilot Laboratory established a production process for 25 centimeter wide lithography and three processes for inkjet printing and screen printing.[186]

The Emerging Alliance with Japan

In 2013, the Taiwanese magazine *Business Today* published a cover story accusing Samsung of plotting to "kill Taiwan," targeting the island's technology-intensive industries. Although a number of Taiwanese CEOs expressed doubt that Samsung had any such intention, the article underscored the extent to which Taiwan is concerned about the competitive challenge posed by Korean firms in advanced technologies, particularly semiconductors, computers, displays, and smartphones.[187] Taiwanese electronics companies are increasingly responding to Korean competition by forming alliances with companies in Japan, in particular, and to a lesser degree, the United States and Europe, involving the thematic areas relevant to flexible electronics.

Taiwanese high-technology companies have been pursuing technology alliances with Japanese firms for many years, and such collaborations have become more feasible as Taiwan has caught up with Japan in the information

[184] "Q&A: Dr. Janglin Chen of ITRI (Part I)," *The Emitter,* July 3, 2013.

[185] Domestic and foreign companies, universities, and research organizations are welcome to conduct R&D at the Pilot Laboratory, which was tasked with forming international R&D alliances.

[186] "Flexible Electronics Pilot Lab and New Electronic Paper in Taiwan," *Printed Electronics World,* June 8, 2007.

[187] "Samsung Cannot 'Kill Taiwan' CEOs," *Taipei Times*, March 23, 2013; "Taiwan Tries to Shore Up its Defenses Against Taiwan," *The New York Times,* April 21, 2013.

TABLE 6-9 Comparison of Strengths and Weaknesses of Japanese and Taiwanese Electronics Companies

Japanese Companies	Taiwaneses Companies
Strengths	Weaknesses
• Global brands	• Obscure brands
• Large scale	• Often too small
• Deep technology portfolios and competencies	• Dependence on foreign technology
Weaknesses	Strengths
• Conservative business approach	• Pioneers new business models
• Sluggish decision-making process	• Nimble decision making
• Concern over "China Risk"	• Strong established business base in China

SOURCE: Based on Hiroko Shimpo, *Business Alliances Between Japanese and Taiwanese Companies* (Osaka Sang Yo University, November 2, 2011).

technologies.[188] When Taiwanese firms entered the LCD market in the late 1990s, in almost every case they formed strong affiliations with Japanese firms.[189] In contrast to relations between some former colonies and their previous rulers, relations between Taiwan and Japan have been close historically.[190] Taiwanese and Japanese electronics companies enjoy mutual complementarities that can potentially offset the weaknesses of the other. (See Table 6-9.)

Terry Gou, the CEO of Hon Hai Precision Industry, the Taiwanese firm commonly known as Foxconn, the world's biggest contract manufacturer of electronics, explained the logic of a Taiwan-Japan tie-up in 2011 as follows:

1. Both Japanese and Taiwanese companies take advantage of their own advantages. The advantage of a Japanese company is the technology cultivated for a long time and the seriousness, and the advantage of a Taiwanese company is the high flexibility.

[188] "Taiwan Enters Japan's Semiconductor Supply Chain," *Taiwan Insights,* April 4, 2012. In 2011, Taiwan's Presidential Office and Cabinet formed a Taiwan-Japan exchange promotion office with the goal of securing "long-term partnership between Taiwan and Japan, not just for a few years." "More Japan Firms Planning to Invest in Taiwan," *China Post Online,* March 3, 2011.

[189] Mei Chi Hu, "Technological Catching-up in East Asia," in *International Economic Development: Leading Issues and Challenges,* 84.

[190] Japanese rule (1845-1945) was characterized by significant investments in industry, utilities, agriculture, and educational infrastructure, providing a solid base for Taiwan's modernization in the post–World War II era. Japanese colonial administration was relatively enlightened and did not result in the kind of lasting antipathy toward Japan found elsewhere in East Asia as a legacy of the Greater East Asia Co-Prosperity Sphere. Thomas R. Howell, Alan Wm. Wolff, Brent L. Bartlett, and R. Michael Gadbaw, eds. *Conflict Among Nations: Trade Policies in the 1990s* (Westview Press, 1992), 312–317.

2. Both companies respect intellectual property rights. We never copy and we pay the appropriate royalty. Hon Hai also possesses much intellectual property.
3. Both cultures are similar in that each respects trust and honesty.
4. A Japanese company has its own brand but we do not have our own brand.[191]

For a variety of historical and practical reasons, potential technological alliances between Taiwanese or Japanese firms, on the one hand, and Korean firms, on the other hand, do not share the same complementarity and are therefore less likely to materialize.[192]

Collaborations between Taiwanese and Japanese electronics firms began with an alliance in semiconductors between Japan's Elpida Memory, Inc. and Taiwan's Powerchip Technology Corporation in the early 2000s. Powerchip subsequently established a joint venture with Japan's Sharp and Peneasas in LCD drivers. In 2010, Hon Hai bought a majority interest in Hitachi Displays Ltd. with the intention of establishing a factory in Japan to produce small LCD panels.[193] In the wake of the 2011 Fukushima earthquake and destructive floods in Thailand, Japanese electronics firms, confronting disrupted supply chains as well as the need to control costs, have outsourced key manufacturing operations to Taiwan, including production of 28 mm memory devices, and have increased their investments on the island.[194] These moves have been paralleled by the development of closer ties between ITRI and Japanese businesses, with ITRI "increasingly becoming the place from which Japanese medium-sized businesses seek advice," ITRI's venture capital arm, the Industrial Technology Investment Corporation, established a fund in 2011 with Japan's Mitsui Sumitomo Insurance Venture Capital that invests in Japanese and Taiwanese companies.[195] In August 2013 ITRI and Japan's Komari Machinery Co. disclosed joint development of a non-lithography

[191] Hiroko Shimpo, *Business Alliances Between Japanese and Taiwanese Companies* (Osaka Sang Yo University, November 2, 2011), citing Nikkei Business, *<http://www.nikkeibp.co.jp/article/NEWS/20110614/1925531>*.

[192] Some major Japan-Korea company alliances exist, such as the mutual-shareholding arrangement between steelmakers Nippon Steel Corporation and the Pohang Iron and Steel Corporation (POSCO). However, electronics companies in Japan and Taiwan commonly compete head to head with Korean firms and regard them as major competitors. Many Koreans harbor nationalistic suspicion of Japan, a fact that hinders both sides from building a long-term, stable and cooperative relationship." Shimpo, *Business Alliances,* 52.

[193] "Hon Hai to Invest $1.2 Billion in Hitachi LCD Unit," *Reuters,* December 27, 2010.

[194] "Taiwan Enters Japan's Semiconductor Supply Chain," *Taiwan Insights,* April 4, 2012. In 2011, Toray, a major Japanese maker of photovoltaic thin-film devices, announced plans to establish a manufacturing plant in Taiwan. "More Japan Firms Planning to Invest in Taiwan," *China Post Online,* May 3, 2011.

[195] Chang-Chien-Chung and Ann Chen, "Top Industrial Research Body Praised for Helping Japanese Firms," *Central News Agency,* February 5, 2013.

R2R printing technology that can be used to fabricate touch panels, cutting manufacturing time and cost.[196]

ITRI-Kaneka

In 2013, ITRI disclosed that in collaboration with Japanese polymer substrate firm Kaneka Corporation it had co-developed a new flexible oxide semiconductor TFT backplane for applications in mass-produced AMOLED displays. The two parties have reportedly overcome the technical challenge posed by the fact that stress occurring in the TFT fabrication easily destroys flexible TFT devices. The new substrate reportedly meets the requirements of high transparency and high temperature resistance.[197]

ITRI-Komori

In 2013, ITRI and Japan's Komori International disclosed development of a new fine-line printing technology that will enable replacement of seven different pieces of equipment with one direct-printing station. The new technology is expected to enable printing of fewer than 10 microns, fine metal lines and enhance materials utilization in R2R manufacturing operations.[198]

E Ink/Sony

Taiwan's E Ink Holdings developed the original concept of electronic paper and the technology used for e-ink screens on dedicated e-readers. In 2013 it introduced E Ink Mobius in conjunction with Sony, which created the device's electronics and which will use it in a new series of e-reader products it will commercialize in 2014. E Ink is reportedly working with Sony to develop large-screen display e-readers for educational use. The E Ink Mobius e-reader features a flexible display encapsulated in a rigid glass. Its creator, Giovanni Mancini, says that the critical issue is not whether the screen can be folded, but whether it is lighter and more durable and consumes less power than conventional backlit screens on tablets and phones.[199]

---

[196] Tsutomu Niitsuma, representative director of Komari, commented that "our great collaboration proves that printed electronics can lead to innovations for existing products." "ITRI Lifts Lid on Advanced Printing Tech for Lower Priced Touch Panels," *The China Post Online,* August 27, 2013.

[197] "ITRI and Kaneka Unveil New Flexible Display Technology," *ITRI Today,* April 3, 2013.

[198] "ITRI and Komori Announce New Fine-Line Printing Technology," *Flexible Substrate,* November 2013.

[199] "E Ink, Sony Debut New Flexible Screen Technology," *Publisher's Weekly*, May 23, 2013; "The Glass is Half Full …," *Flexible Substrate,* May 2013.

AUO-Idemitsu

In 2012, Taiwan's AU Optronics entered into an alliance with Japanese display maker Idemetsu Kosan Co. Ltd. to jointly develop next-generation OLED displays "in an attempt to catch up with rivals [e.g., Samsung] making the high-resolution panels for smartphones and TVs."[200] AU Optronics' OLED manufacturing technology was reportedly about 2 years behind that of Korea. Under the firms' agreement, Idemitsu delivers high-performance OLED materials to AU Optronics for applications in OLED panels for smartphones and tablet PCs.[201] The alliance is intended to "allow AUO to make up its shortcoming in materials and patent layout by shortening its development cycle."[202]

AUO-Sony/Panasonic

In 2011, Japan's Sony Corporation ended a 50-50 joint venture with Samsung for the production of FPDs, and Sony was reportedly looking for ways to diversify its supply sources of LCD panels.[203] In June 2012, Sony and Panasonic disclosed that they would collaborate on OLED television development in an effort to counter Korean competition.[204] In early 2012, it was reported that Taiwan's AU Optronics was collaborating with Sony on OLED production and that Sony would buy all of AU Optronic's AMOLED production over the next term.[205] In January 2013 AU Optronics announced that it had successfully co-developed with Sony a 56-inch OLED TV panel, then the world's largest, with 4K resolution.[206]

Toppan-Chi Lin

In 2011 Japan's Toppan Printing concluded a collaboration agreement with Taiwan's Chi Lin Technology Co., Ltd., a member of Taiwan's Chi Mei Group, to co-develop e-paper for applications in electronic price tags, logistics instruction labels, and retail point-of-sale displays.[207]

---

[200] Lisa Wang, "AUO Signs Accord with Idemitsu to Manufacture OLEDs," *Taipei Times Online,* February 3, 2012.

[201] "AUO Signs OLED Cooperation Agreement with Idemitsu Koscan," *The Emitter: Emerging Displays Technology Monthly Report,* February 2012.

[202] "Taiwan Enters Japan's Semiconductor Supply Chain," *Taiwan Insights,* April 4, 2012.

[203] "Sony More Open to TV Panel Sourcing," *Central News Agency,* December 28, 2011.

[204] "Panasonic and Sony Confirm Plans to Join Forces for OLED TV Development," *Whathifi.com,* June 25, 2012.

[205] "Sony to Buy All of AUO's OLED Capacity for Its Mobile Phones?," *OLED-Info.com,* April 6, 2012; "AUO to Raise Money for OLED TV Production, Collaborates with Sony?," *OLED-Info.com.*

[206] "AUO, Sony Co-develop World's 1st 4K OLED TV Panel," *Central News Agency,* January 9, 2013.

[207] Pursuant to the agreement, Toppan Printing acquired shares in a newly formed Chi Lin subsidiary, Pervasive Displays Inc., and Toppan committed to sell Pervasive Displays products in the

Other International Alliances

Taiwanese entities have also entered into significant collaborations with partners based in the United States and Europe. In 2010, Taiwan's AU Optronics Corporation, one of the world's leading makers of TFT-LCDs, announced that it had become an industry partner of Arizona State University's Flexible Display Center (FDC). The two entities will "collaborate on the development of mixed oxide thin film transistors to accelerate the commercial availability of active-matrix organic light-emitting diode (AMOLED) flexible displays."[208] Yong-Hong Lu, Vice President of AU Optronic's Technology Center, said that "the FDC has significant experience in adapting standard flat panel display manufacturing technologies for use with flexible substrates, which is a critical aspect of being able to bring flexible AMOLEDs to market. Working with the FDC offers us an opportunity for flexible substrates and [to] participate in the development of viable approaches to commercialization."[209]

ITRI and Corning have been collaborating since 2011 to develop R2R process technology for flexible glass. ITRI's competency in R2R processing of plastic substrates is complemented by Corning's glass handling expertise. The two parties have developed an R2R process for 100 µm flexible glass substrates and R2R machines that produce touch panel modules on Corning Willow Glass, a flexible-display-quality substrate. In addition to touch screens, ITRI indicates the new processing technology can be used for applications in photovoltaics and OLED lighting.[210]

Sollink is a Taiwan-based joint venture between Qualcomm MEMS Technology, a subsidiary of Qualcomm Communications specializing in displays for mobile devices, and Taiwan's Cheng Uei Precision Industry Co., Ltd., a contract electronics manufacturer. Sollink is working to combine Qualcomm's Mirasol display technology with flexible e-paper.[211]

In 2008, Germany's Bayer Group disclosed plans to open two R&D centers in Taiwan, one of which would focus on industrial applications of functional films that are widely used in flexible and printed electronics applications. The center was expected to expand cooperation with local universities in its developmental efforts.[212]

Japanese market. "Toppan Printing and Taiwan's Chi Lin Collaborates to Launch Electronic Paper Business Targeting Use in Industry," *Printed Electronics Now,* May 12, 2011.

[208] "The Flexible Display Center and AUO Enter Strategic Partnership to Accelerate Flexible AMOLED Development," *Nanowerk,* November 16, 2010.

[209] Ibid.

[210] "ITRI and Corning Collaborate on Roll-to-Roll Flexible Glass for Touch," *The Emitter: Emerging Display Technologies,* November 30, 2012.

[211] Mirasol technology features ultralow power consumption and good viewing quality even in sunlight. "E-Paper Makers Set Sight on Flexible Material Technology," *Taiwan Economic News,* December 7, 2010.

[212] "Bayer to Establish Two New R&D Centers in Taiwan," *Central News Agency,* October 9, 2008.

Key Companies

E Ink Holdings (Prime View International)

E Ink (formerly Prime View International) is the world's leading producer of ePaper, holding greater than 90 percent of the world's reader market, comprised of companies such as Amazon, Sony, and Barnes and Noble. E Ink also produces ePaper for applications in watches, signage, electronic shelf labels, credit cards, and battery and memory indicators. Based in Hsinchu Science Park, E Ink is a subsidiary of the Yuen Fuong Yu Group, Taiwan's largest paper producer. Founded under the name Prime View International in 1992, the company entered the market for ePaper display modules. In 2009, Prime View acquired Massachusetts-based E Ink Corp. for $215 million. E Ink produced e-paper displays for the Amazon Kindle and the Sony Reader, and following the acquisition, Prime View became the largest supplier of e-paper in the world, with a 90 percent market share in 2010.[213] In 2010, Prime View changed its name to E Ink Holdings, although it is still sometimes referred to as PVI or Prime View. In 2011, E Ink entered into a collaboration with Japan's Seiko Epson, the world's largest supplier of controller chips, to develop and produce e-paper displays for sale in the Japanese and Chinese markets.[214]

In 2012, E Ink concluded an agreement to acquire SiPix Technology Inc. and its wholly owned subsidiary, SiPix Imaging Inc., from AU Optronics. The acquisition gave E Ink "complete control of the electrophoretic display technology that dictates the ePaper field," and one observer commented that, with the acquisition, "if challenging E Ink's supremacy in the ePaper market was hard before, it just became Sisyphean."[215] E Ink currently holds more than 2,000 patents on ePaper, LCD technology, and other technologies. Underscoring the strength of its intellectual property portfolio, in 2013 E Ink filed a patent infringement lawsuit in Germany against Trekstor and declared its readiness to "protect and defend its intellectual and other property worldwide."[216]

---

[213] "AUO Aims for 10-fold Increase in E-Paper Shipments This Year," *Taiwan Economic News,* February 10, 2011.

[214] The two firms will jointly develop 300-dpi high-resolution displays for use in commercial and educational e-book readers with screen sizes above 11 inches in size. The technology is intended to achieve better visibility of Japanese and Chinese characters. Epson will manufacture and provide a high-speed display controller platform for e-paper displays, incorporating a display controller integrated circuit (IC), processor, power supply IC, and associated software. PVI will manufacture the e-paper display.

[215] "E Ink Acquires SiPix, May Dominate ePaper Universe," *Engadget,* August 4, 2012.

[216] "E Ink Files Patent Infringement Lawsuit Against Trekstor," *Flexible Substrate,* January 2013.

AU Optronics Corp.

AU Optronics Corp. is the largest TFT-LCD manufacturer in Taiwan.[217] In March, 2009, AU Optronics acquired a 31.58 percent equity stake in SiPix Imaging, based in Fremont, California.[218] SiPix is a major developer of e-paper technologies, including R2R production technologies and novel materials for e-displays. SiPix has "filed over 100 patent applications and has developed a core expertise in roll-to-roll based display solutions and integration."[219]

During the first half of 2010, AU Optronics was forced to shut down its production line for SiPix e-paper due to "technical problems" that were subsequently resolved. The company indicated in February 2011 that it foresaw an 8- to 10-fold increase in e-paper sales in 2011 over 2010 levels and that it expected to achieve a 20 percent market share in the global e-paper market in 2011.[220] The sale of SiPix to E Ink followed in 2012.

Chin Lin Technology

Chin Lin Technology, part of Taiwan's Chi-Mei Group, is a "leading design, engineering and manufacturing company specializing in backlighting technology, material sciences and advanced display systems."[221] Chin Lin is one of the world's most vertically integrated manufacturers of FPDs. In May 2011, Chin Lin announced the creation of a new subsidiary, Pervasive Displays Inc., for the design and manufacture of e-paper modules for use in commercial and industrial displays. "Pervasive Displays will develop low-power small-to-medium display components for applications in logistics, price tags, medical devices, automation, smart labels and energy control panels." In April 2011, Chin Lin and Japan's Toppan Printing Co., Ltd. entered into an agreement pursuant to which Toppan will sell Chin Lin manufactured e-displays into Japanese industrial markets.[222]

Delta Electronics, Inc.

Delta Electronics, based in Taiwan, is the world's largest provider of switching power supplies and DC brushless fans. It also produces electronic displays and components, renewable energy products, and industrial automation

[217] AU Optronics was formed in 2001 through the merger of Acer Display Technology, Inc. and Unipac Optoelectronics Corporation. In 2006, the combined entity acquired Quanta Displays, Inc.

[218] The SiPix shares, valued at $30 million, were acquired by two AU Optronics subsidiaries, Konly Venture and Ronly Venture.

[219] *<http://Sipix.com>*.

[220] "AUO Aims for 10-fold Increase in E-Paper Shipments This Year," *Taiwan Economic News,* February 10, 2011.

[221] German Trade Office, Taipei, "Bayer Material Science, Bayer Taiwan, and Chi-Lin Technology Launch a New Era in Taiwanese Opto-electronics Industry," March 11, 2011.

[222] Yahoo Finance, "Chi-Lin Announces Formation of Pervasive Displays Inc.," May 3, 2011.

technology. In January 2011, Delta announced that it had developed a range of color e-paper technologies in collaboration with Bridgestone Corporation based on the latter company's Quick Response Liquid Powder Display (QR-LPD) technology. "Delta's focus is on downstream development [of this technology], such as product design, software and hardware integration, and the development of applications for the consumer electronics and business-professional markets."[223] Bridgestone will market its e-paper modules under its AeroBee brand name,[224] and "Delta will use the modules to develop next generation products and applications."[225]

University R&D

National Tsing Hua University

National Tsing Hua University, located in Hsinchu, is one of the most highly regarded universities in Taiwan, with particular strengths in the sciences and engineering.[226] Tsing Hua's Department of Materials Science and Engineering is engaged in R&D in themes with application to flexible electronics. In 2010, researchers from this department reported that they had succeeded in developing a direct growth fabrication method for paper-based electronics. They grew vertically aligned, highly crystalline, and defect-free single crystal zinc oxide nanorods and nanoneedles on paper to form prototype hybrid junction diodes and UV photodetectors. The scientists found that bending and twisting devices fabricated in this way had a negligible effect on electrical/mechanical fatigue properties of the diodes. "Repetitive bending of the diode affected the performance only marginally."[227] In 2011, National Tsing Hua University researchers reported development of a technology to make key components of flexible e-books and displays from silk proteins.[228]

National Taiwan Normal University (NTNU)

National Taiwan Normal University, based in Taipei, is one of Taiwan's foremost universities. In January 2011, a research team led by Professor Kao

[223] "Delta Electronics Blazes a Trail in Color QR-LPD Electronic Paper Applications," *Central News Agency,* January 5, 2011.

[224] "New Technology Makes E-Paper Animation Displays Possible," *Central News Agency,* January 4, 2011.

[225] "Delta, Bridgestone to Develop New E-Paper Products," *Central News Agency,* November 9, 2010.

[226] One graduate Lee Yuan-tseh is the holder of a Nobel Prize in chemistry, the first Taiwanese to receive a Noble Prize.

[227] "National Tsing Hua University Develops Direct Growth Fabrication for Paper-Based Electronics," *Flexible Substrate,* November 2010.

[228] "Taiwan Research Team Turns Silk into E-Display Material," *Asia Pulse,* March 2, 2011.

Wen-chung at NTNU's Silicon-on-Chip (SOC) Lab at National Taiwan Normal University unveiled a new technology that will make animation displays possible on e-paper and in electronic books. The e-paper developed in this project can be bent and folded and will allow moving pictures to be visible on items such as electronic tickets, electronic menus, or other sites where information displays are needed. Mass production of the new e-paper is seen as 4 to 6 years away.[229] In 2012, a research team at the university reported development of an easy and inexpensive method to construct optical thin films.[230]

JAPAN

Japan's longstanding prowess in consumer electronics, microelectronics, and advanced materials should seemingly make it a leader in the emerging field of flexible electronics. However, Japan's large electronics groups are reeling from the competitive challenge from Korea. The country's relatively low level of inward foreign direct investment, although increasing, is a handicap in any technology-intensive industry characterized by global patterns of innovation.[231] In a field in which startups such as Plastic Electronics, Heliatek, and Cambrios are playing a major role in defining the competitive landscape, Japan's business culture continues to inhibit startups, notwithstanding government promotional measures.[232] In a widely quoted comment in 2013, Tetsuya Ohashi, Public Relations Manager of Tera Motors, a Japanese startup that makes the world's first smartphone connected e-scooter stated:

---

[229] "New Technology Makes E-Paper Animation Displays Possible," *Central News Agency,* January 4, 2011.

[230] "National Taiwan University Develops Stacked Nanoparticle Layers," *Flexible Substrate,* January 2013.

[231] Japan's inward FDI was less than 4 percent of its GDP at the end of 2011, compared with 48.8 percent for the UK. The OECD's index of regulatory restrictions on FDI, which includes limits on foreign equity holdings, screening, and approval procedures, and rules on repatriating capital and hiring foreigners, indicated that Japan was the OECD's most closed economy in 2012. "Little Sign That Abe Can Shake Up Japan's Inbound FDI," *Reuters,* May 20, 2013. Japan's leaders have been seeking to change this dynamic for many years. In a 2003 speech to the Diet, Prime Minister Junihiro Koizumi said, "Foreign direct investment in Japan will bring new technology and innovative management methods, and will also lead to greater employment opportunities. Rather than seeing foreign investment as a threat, we will take measures to present Japan as an attractive destination for foreign firms in the aim of doubling the cumulative amount of investment in five years." Prime Minister's office, "General Policy Speech by Prime Minister Junichiro Koizumi to the 156th Session of the Diet," January 31, 2003.

[232] Nato Kan, who in 2009 was serving as Japan's Deputy Prime Minister and Minister of State for Science and Technology Policy, commented in that year that "unfortunately we do not yet have an environment in Japan that is suitable for venture companies." "Entrepreneurs Lack Serious Support," *Japan Times,* November 27, 2009. See generally National Research Council, *S&T Strategies of Six Countries: Implications for the United States* (Washington, DC: The National Academies Press, 2010), 46–58.

> There are limitations for young people in Japan. Bosses don't take risks. Japanese workers can't challenge the boss. If you give opinions, they don't listen. Bosses don't give young people opportunities. Only old men get to do interesting work.[233]

Exceptions exist to the foregoing generalizations, some of which are noted in this study, and Japanese policy measures have improved the climate for innovation, but caution by some Japanese businesses inhibits the country's ability to assert leadership in this field.[234]

In general, Japanese firms' perspective on flexible electronics appears to be one of caution. In late 2013 the consultancy IDTechEx reported on "an intense study of printed electronics in Japan," including 45 recently completed visits to companies and institutions in the country. IDTechEx reported, "We have not found significant success with printed transistor, photovoltaics or other printed components or circuits although printed and conductive patterns are popular in Japan." No firm appeared to be selling printed transistor circuits. There was "little interest in printed electronics on paper in Japan and this is misguided." Company focus was "exclusively on high volume potential applications because the inevitably huge companies involved want nothing less. . . . Flexible OLEDs are now a particular focus." Many companies are moving into printed electronics but "they rightly perceive that the biggest profits will be in materials not the final devices when it comes to displays. . . ." The Japanese reportedly "do not share the enthusiasm for graphene seen in the West."[235]

In Japan, the challenges and opportunities presented by flexible electronics are overshadowed by the existential crisis of the country's once-vibrant electronics industry. An examination of the underlying causes of the startling collapse of the once-mighty Japanese electronics giants is beyond the scope of this study, but observers offer explanations that include the strong yen, conservative management, failure to grasp the implication of digital technology and the Internet, slow decision making, reluctance to withdraw from declining product areas that were once a major source of profits, and excessively broad product portfolios.[236] The Japanese government has responded to the deepening industrial decline by forming a large investment fund, the Innovation Network Corporation of Japan (INCJ), which has invested in troubled companies and facilitated the consolidation of money-losing business units of Japanese electronics firms. (See Figure 6-5.)

At the end of 2011, Sony, Toshiba, and Hitachi concluded an agreement to spin off and merge their LCD units in a new entity, Japan Display Corporation, which is 70 percent owned by the government Innovation Network Corporation of

233 "Hey Japan, What's With Your Startup Culture," *Fastcolabs,* September 6, 2013.

234 "Start-up Spirit Emerges in Japan," *The New York Times,* December 25, 2013.

235 IDTechEx, "Intense Study of Printed Electronics in Japan," September 16, 2013.

236 "The Taiwanese make 14 decisions in one phone call. 'Yes, we will deliver.' The Japanese just can't do that." Ta-Lin Hsu, founder of private equity firm H&Q Asia Pacific, in "Japan's Once Mighty Tech Industry Has Fallen Far Behind Silicon Valley," *San Jose Mercury News,* October 12, 2012.

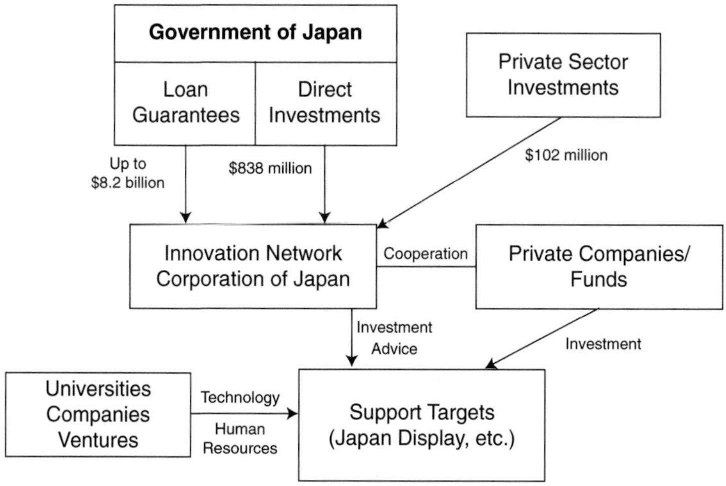

FIGURE 6-5 Innovation Network Corporation of Japan.
SOURCE: Japan Ministry of Finance.

Japan. Sony, Hitachi, and Toshiba will each hold 10 percent stakes in the new entity.[237] The new company "must tackle South Korea's Samsung Electronics Co., which has extensive resources to invest in new technology as well as other Asian rivals that offer less expensive, if generic, products." Hiroshi Hayase, an analyst at the research firm display Search, commented in 2012 that "if Japan Display fails, the country's whole electronics industry will be left with few alternatives."[238]

Japan Display's initial moves indicate that while its main business will remain conventional LCDs over the near term, it will be engaged in technology areas relevant to potential flexible electronics applications. In 2012, Japan Display indicated it had developed prototype video e-paper that it indicated would be ready soon for production.[239] In 2013, the company indicated it "had developed a high pixel density 5.2 inch OLED display with full high-definition display resolution of 1,080 horizontal pixels × 1,920 vertical pixels."[240] In 2012, Japan Display joined Osaka University's OLED project, the Center for Organic Photonics and Electronics Research.[241]

A collaboration between Sony and Panasonic initiated in 2012 to develop OLED TVs using printing-based product technologies fell apart at the end of

[237] "Sony, Toshiba, Hitachi, Unload LCD Units to Japan Government-Backed Fund," *Bloomberg,* August 31, 2011.

[238] "Japan Takes a Gamble on Displays," *The Wall Street Journal,* April 10, 2012.

[239] "Japan Display Develops Video-Rate E-Page," *Plastic Electronics,* November 6, 2012.

[240] "Development of 5.2-Inch full-HD OLED Display," *Printed Electronics World,* May 27, 2013.

[241] "Japan Display Joins Kyushu University's OLED Project," *OLED.si,* August 1, 2012.

TABLE 6-10 Japanese Supply Chain Firms—Foreign Alliances

Company	Foreign Partner	Form of Collaboration	Technology Area
Sumitomo Chemical	Cambridge Display (UK)	Acquisition	Flexible displays, OLEDs
Toppan Printing	Chi Lin (Taiwan)	Equity investment, collaboration agreement	e-paper
Showa Denko KK	Nova Centrix (US)	Collaboration/licensing agreement	Conductive inks
Teijin	DuPont (US)	DuPont Teijin Films (JV)	PET/PEN films
Konica Minolta	Konarka (US)	Strategic partnership	Organic thin-film PV
Toppan Printing	Plastic Logic (UK)	Collaboration agreement	Flexible large area signage

2013.[242] The companies "were unable to make the panels durable enough nor to cut production costs." A likely additional consideration cited by observers was the fact that LG and Samsung's recently introduced 55-inch OLED TVs with "initial price tags of $10,000 that kept sales low and business in the red."[243]

The travails of Japan's electronics majors tend to mask the solid and steady technological and commercial achievements of less well-known Japanese makers of materials, equipment, and printing technologies relevant to flexible and printed electronics. Although few if any of these firms can match the Sonys and Hitachis in sheer size and global reach, most of them are large entities with extensive industrial competencies and intellectual property portfolios, and some, such as Seiko-Epson in the area of piezo-crystal-based inkjet printers, are true technological pioneers. Moreover, Japan's materials, equipment, and printing companies are defying the national stereotype for insularity by forging technology and commercial alliances with major foreign firms. (See Table 6-10.)

Government Entities Supporting R&D

Ministry of the Economy Trade and Industry (METI)

METI was created in 2001 when its institutional predecessor, the storied Ministry of International Trade and Industry (MITI), was merged with the then Economic Planning Agency and other economic agencies. METI is responsible for industrial, energy, and trade policy. Its promotional activities in the field of flexible electronics are generally carried out through subordinate organizations,

[242] "Sony, Panasonic OLED Link is Official," *Advanced Television,* June 26, 2012.

[243] "Technical Difficulties Foil Sony-Panasonic OLED Effort," *Flexible Substrate,* January 2014.

TABLE 6-11 Distribution of NEDO Support for Flexible Electronics Research

Research Field	Theme(s)
Information and telecommunications	Organic semiconductor flexible display
Nanotech and materials	Electronic paper transparent conductor organic semiconductor
Manufacturing technology	Flexible device

the National Institute of Advanced Industrial Science and Technology (AIST) and the New Energy and Industrial Technology Development Organization (NEDO).

New Energy and Industrial Technology Development Organization (NEDO)

NEDO is an incorporated government agency under METI's supervision. Its mission is to develop new energy and energy conservation technologies, verify technical results, and promote the dissemination of new technologies with an eye to their commercialization. NEDO coordinates research activities in the academic, industrial, and governmental sectors and arranges and promotes R&D projects.

NEDO's support for research with flexible electronics applications has been scattered among a range of R&D fields.[244] (See Table 6-11.)

A British Parliamentary Committee observed in 2009,

> During our visit to Japan, the impact that strategic investment in the plastic electronics sector can have was apparent. The Japanese Government has acted to ensure strategic capability in the OLED industry of the future. For instance, the Ministry of Economy, Trade and Industry (METI), through the New Energy and Industrial Technology Development Organisation (NEDO), is providing ¥35 billion (£173 million) to fund a collaborative project between Sony, Toshiba, Panasonic, Sharp and other partners to develop 40-inch and larger OLED television panels to a pre-competitive stage.[245]

The reference was to a Japanese government effort launched for the 2008-2013 timeframe involving Japanese electronics firms in an effort to develop a 40-inch OLED display sometime after 2015 in an effort to "get the jump on South Korean TV heavyweights such as Samsung Electronics Co. and LG Display Co."[246]

[244] Kaoru Hunjo, NEDO Executive Director, "Government-Industry R&D Partnerships . . . Japanese Experiences . . . Introduction of NEDO." (January 2006).

[245] House of Commons, Innovation, Universities, Science and Skills Committee, *Engineering: Turning Ideas into Reality,* vol., March 18, 2009, 45.

[246] "Japan Backs OLED Display Research" *CBC News,* July 10, 2008.

National Institute of Advanced Industrial Science and Technology (AIST)

AIST, a government "independent administrative institution," is Japan's largest research organization.[247] It operates under METI's supervision and derives most of its funding from the government. It has more than 40 autonomous research units and roughly 2,400 resources, of which about 80 are foreign. A number of its research units are involved in R&D efforts involving flexible electronics:

- ***Flexible Electronics Research Center.*** This unit was established in April 2011 to pursue "green innovation with thin, light, flexible devices." Its Director is Toshohide Kamata, a professor at Tsukuba University with a research background in printed electronics and organic electronics and displays. He is currently supervising research programs to develop flexible printed devices for displays, tags, and sensors. The Center currently has five research teams pursuing flexible electronics themes including one team headed by Professor Kamata himself. (See Table 6-12.) In 2014 it was disclosed that the Flexible Electronics Research Center had developed software for rapid, easy, and precise simulation of ink droplets printed on substrates with surfaces patterned with wettability, a tool that is expected to substantially accelerate the development of printed electronics manufacturing.[248]
- ***Photonics Research Institute (PRI)*** PRI conducts R&D on advanced information technology and telecommunications themes, including flexible displays and printed organic devices.
- ***Nanotube Research Center*** In 2008, AIST established its Nanotube Research Center, a follow-on to the Nano Carbon Research Center (2001-2007). The Nanotube Research Center was headed by Sumio Ijima, who discovered carbon nanotubes. The Center is pursuing R&D into the synthesis of carbon nanotubes "with an emphasis on applications technology for the commercialization of CNT (carbon nanotubes)."[249]
- ***Carbon nanotube plant*** In February 2011, AIST disclosed that it had established the world's largest production plant for single-walled carbon nanotubes, conductive materials that are being used in fabricating

---

[247] Independent Administrative Institutions (IAI) are governed by Japan's Basic Law on Reforming Government Ministries (1998). The designation of IAIs was part of a broader scheme to separate government organizations into planning and operational units. Planning functions remained in government ministries while operating functions were transferred to IAIs, which tend to utilize private sector–type management methods and to operate in relative autonomy.

[248] "AIST Develops Droplet Simulation Technology for Printed Electronics," *Flexible Substrate,* March 2014.

[249] "AIST Nanotube Research Center Fabricates and Develops Ultrafine Processing Technology for CNT Wafer; One Step Closer to Mass Production and Practical Application of MEMS Devices," *Tokyo Semiconductor FDP World,* January 5, 2009.

TABLE 6-12 Research Teams at the AIST Flexible Electronics Research Center

AIST Team	Research Themes
Functionalizing process	1. Upgrading super inkjet 2. Fulfillment of material functions under low oxygen partial pressure limit 3. Upgrading noncontact printing technology 4. Development of new materials under low oxygen partial pressure limit
Functional display device	1. Developing low damage formation technology for functional thin film 2. Developing manufacturing technologies for high-efficiency nanoparticles and thin-film forming materials 3. Developing usability innovation for display devices
Advanced surface processing	1. Development of large area and high-definition printing 2. Development of standard evaluation method for printed electronics 3. Development of manufacturing and assembling technologies for flexible devices
Flexible organic semiconductor	1. Developing printable functional materials 2. Innovation of printing processes
Printed electronics device	1. Developing component technology for printing process 2. Developing technology to produce flexible information input devices 3. Developing fundamentals of evaluation technology for device/process/material

SOURCE: National Insitute for Advanced Industrial Science and Technology of Japan <*http://unit.aist.go.jp/flec/en/teams/index.html*>.

flexible and stretchable electronic devices.[250] In 2004, AIST developed a "super growth" technology for the synthesis of large quantities of single-walled carbon nanotubes at an ultrahigh purity of 99.9 percent, reducing production costs to one-thousandth of the previous cost. Output capacity of the new facility is 600 grams per day. The facility was created pursuant to a collaboration with Zeon Corporation, a Japanese maker of specialty chemicals.[251]

- ***Ultra Flexible Display Component Project*** Beginning in 2006, AIST and the Japan Chemical Innovation Institute (JCII) pursued developments in array technology based on printing methods together with inks for TFTs. This project was part of NEDO's Technological Development of Superflexible Display Components program (2006-2009). This effort

[250] "Japanese Build World's Biggest Carbon Nanotube Plant," *Jiji,* February 14, 2011.

[251] The research team succeeded in fabricating a 100 ppi organic thin-film transistor array on a 15 cm square plastic substrate using a printing method in which all materials were converted into inks and no vacuum processes were necessary. The team also successfully used microcontact printing to print gate electrodes, S-Delectrodes, and silver nanoparticles for wiring. The team achieved narrow line widths that sufficiently maintained conductivity at a thickness of 600 nm. AIST, "Development of an Inexpensive Method for Mass Synthesis of Single-Walled Carbon Nanotubes," February 7, 2007.

was led by Kiyoshi Yase, Deputy Director of AIST's Photonics Research Institute, and Hirobomi Ushijima, Leader of the Bio-Photonics Group of the Photonics Research Institute.

AIST characterized the results as "a major step toward commercially viable processes for producing flexible displays and organic devices by large area high-speed printing methods such as roll-to-roll printing."[252]

- ***Flexible solar submodules.*** AIST's Research Center for Photovoltaics has developed flexible solar submodules with the world's highest photovoltaic energy conversion efficiency among thin-film solar cells using CIGS thin film.[253] The research was conducted pursuant to a contract with NEDO as part of the latter's Research and Development of High Performance Technologies on CIGS Solar Cells project (FY2006-2009). The flexible solar cells can be applied to various types of substrate materials. AIST indicated that "we will drive forward the research and development toward their applicability of larger-area substrates as well as the realization of lower-cost and higher performance integrated type flexible solar cell modules and industrialization through collaboration with companies."[254]
- ***TFT OTFT backplanes.*** AIST exhibited at Nano Tech 2010 a number of flexible organic thin-film transistor (OTFT) backplanes that had been developed jointly with JCII, Toppan Printing Co. Ltd., DIC Corp., and the Konica Minolta Technology Centre Inc. One OTFT backplane had a resolution of 1,600 × 1,200 pixels and was believed to be the world's largest OTFT sheet manufactured using printing technology.[255]

Ministry of Education, Culture, Sports, Science and Technology (MEXT)

MEXT, also known as Monkashō, is a Ministry that arose out of the former Ministry of Education. It is one of two entities administering the Grants-in-Aid for Scientific Research program, which provides for grants to projects organized by individual researchers or research institutes engaged in basic research, particularly critical fields.

[252] Kiyoshi Yase, "Organic Transistor Frontline, Joint Research Between AIST and the Japan Chemical Innovation Institute, Technology for Fabrication of Organic TFT Arrays Using Full Printing Method, Step Closer to Realization of Printable Fabrication," *Tokyo Semiconductor FDP World,* December 2, 2008.

[253] "Flexible CIGS Solar Cell Submodule Achieves Record Energy Conversion of 15.9%," *Nanowerk,* April 5, 2010.

[254] CIGS is an acronym for copper indium gallium selenide, compound semiconductor material that is used as light absorber material in thin-film solar cells.

[255] "AIST Develops Large Area (A4) OTFT Backplane," *OSA Direct Newsletter,* February 22, 2010.

Japan Society for the Promotion of Science (JSPS)

JSPS, originally a nonprofit foundation for promoting scientific research, is now a government "independent administrative institution." Virtually all of its operating budget is derived from the Japanese government. Together with MEXT it is one of two government institutions administering the Grants-in-Aid for Scientific Research to support university R&D.[256]

Other Research Organizations

Japan Chemical Innovation Institute (JCII)

JCII is a nonprofit research organization promoting cooperative R&D in chemistry involving government, industry, and academia. Beginning in 2006, JCII collaborated with AIST in the development of technology permitting the printing of thin-film transistors on flexible polymeric substrates.

E-printing Standards Development Group

In November 2010, it was disclosed that about 100 companies in Japan would collaborate with the University of Tokyo and the University of Osaka to draw up an international evaluation standard for next-generation electronic printing technology, which prints electronic circuits as flexible materials. The purpose of the collaboration is to enable "Japan to take the lead in helping evaluate electronic printing technology in a fair manner." Current valuation criteria differ between countries and between groups of researchers, "making fair evaluations difficult." The collaboration was an outgrowth of a study group formed in May 2010 by Katsuaki Suganuma of Osaka University consisting of experts from industry, academia, and the government.[257]

University Research

University of Tokyo R&D projects

The University of Tokyo is highly regarded for its development research in the area of printed electronics. In 2008, University of Tokyo scientists revealed that they had devised a reliable method to inkjet print dots of 1 micron width onto flexible film, an achievement that was described in the *Proceedings of the National Academy of Sciences.* The new method will conserve the very expensive inks used in printed electronics and will improve the electronic properties

---

[256] "Droplet Simulation Technology for Printed Electronics," *Printed Electronics World,* January 21, 2014.

[257] "Japan Universities, Firms to Draw Up E-Printing Standard," *Jiji,* November 8, 2010.

of printed electronic devices by reducing key dimensions such as the channel length of transistors.[258]

Single-Walled Carbon Nanotubes

Since 2005, a research team at the University of Tokyo under the direction of Associate Professor Tokao Someya has been developing highly elastic conductive rubber using single-walled carbon nanotubes emphasizing creation of large area stretchable and bendable integrated circuits, large area sensors, and large area activators. The team is using inkjet printers at room temperature to deposit organic semiconductors using carbon nanotubes as conductive dopants on a rubber substrate. When the rubber is stretched, the nanotube-based conductive network is deformed, but remains conductive. The group is working to develop carbon nanotube–based stretchable wires and fully printed stretchable integrated circuits.[259] The research is being supported by funding from the Grants-in-Aid for Scientific Research administered by the Ministry of Education, Culture, Sports, Science and Technology.[260] A potential application of nanotube-based stretchable electronics includes the development of "electronic skin" for humanoid robots (sensitive to touch, temperature, sound and light). Someya indicates that "e-skin" could enable the creation of robots that could help senior citizens and disabled people and play with babies.[261]

Collaborative Institute for Nano Quantum Information (Nano Quine)

In 2006, the University of Tokyo established the Collaborative Institute for Nano Quantum Information (Nano Quine), an institution to foster industry-academic cooperation in nanoelectronics, quantum-encrypted communications, and quantum computers. Nano Quine has formed partnerships with other academic institutes in Japan and overseas and with Japanese companies.[262] (See Table 6-13.)

Nano Quine promotes "T-style partnerships between academics and industry, with the base of the T representing the research themes that are jointly set by groups from universities and companies." One of the research themes selected through this process is the application of organic transistors to advanced flexible electronics and optoelectronics. Nano Quine reports that its

[258] "Much Finer Detail Possible with Inkjet in Japan," *Printed Electronics World,* March 28, 2008.

[259] "Horizontally Aligned Growth of Carbon Nanotubes Holds New Possibilities for Integration; Fusion with Silicon LSI Anticipated," *Tokyo Semiconductor FPD World,* January 5, 2009; Interview with Tokao Someya in *Discovery News,* February 5, 2010.

[260] "Methods to Horizontally Align Single-Walled Carbon Nanotubes on Amorphous Substrate," *Journal of Novel Carbon Resource Sciences,* September 2010.

[261] Interview with Tokao Someya, *Discovery News,* February 5, 2010.

[262] "University of Tokyo Forms Industry-Academic Cooperation Group," *Tokyo Semiconductor FDP World,* July 31, 2007.

TABLE 6-13 Nano Quine Academic and Corporate Partners

Academic Partners	Cooperating Companies
Institute of Industrial Science (U. of Tokyo)	Sharp Corporation
Research Center of Advanced Science and Technology (U. of Tokyo)	NEC Corporation
School of Engineering (U. of Tokyo)	Hitachi Ltd.
Graduate School of Information Science and Technology (U. of Tokyo)	Fujitsu Ltd.
School of Science (U. of Tokyo)	
Kyoto University	
Keio University	
Hokkaido University	
Stanford University	
Cambridge University	
Munich University of Technology	
Delft University	
University of Salenio	

research group "Next Generation Flexible Electronic Devices" has successfully developed an extensible active matrix integrating carbon nanotube extensible conductors and organic transistors. The same group has also demonstrated low-voltage operating organic CMOS circuits on flexible substrates.[263] Nano Quine's R&D activities are supported by the Special Coordination Funds for Science and Technology administered by the Ministry of Education, Culture, Sports, Science and Technology.

Kumamoto University

Kumamoto University in Kyushu is conducting research with Fuji Electric Systems to develop technology to fabricate solar cells on flexible substrates. Existing solar cell technology is based on crystallized silicon cells placed between rigid glass substrates. Fuji Electric is reportedly manufacturing an amorphous (noncrystalline) solar cell, FWAVE, a thin, flexible, black sheet covered with small holes encased in a transparent plastic film. It bends easily and can be fitted to domes and other curved or irregular surfaces. The Kumamoto prefectural government is reportedly establishing a $9.8 million center to support the development and production of thin-film solar cells.[264]

[263] Nano Quine, *Nano Quine: Quantum Information Electronics,* *<http://www.nanoquine.iis.u-tokyo.as.jp/NQ/brochure.pdf>*.

[264] "National Kyushu Taking the Lead in Thin Film Solar Cells," *Asahi Shimbun,* November 13, 2010.

Nagoya University

In 2013, in a new study researchers at Nagoya University reported development of integrated circuits composed entirely of carbon-based materials that can be molded into various shapes like plastic. The discovery may enable the easy integration of electronic circuits to a wide variety of plastic products.[265]

Kyushu University

Researchers at Kyushu University report discovery of a new molecule that increases the efficiency of florescent OLEDs without use of heavy metals.[266] In 2013, Kyushu researchers announced development of a new technique for making germanium crystals at low temperatures, making more feasible the use of germanium in thin-film transistors required for flexible electronics.[267]

Companies

In addition to the major Japanese producers of displays, Japan has a broad array of companies developing and producing materials, equipment, and process technology with flexible electronics applications.

Semiconductor Energy Laboratory Co. Ltd.

Semiconductor Energy Laboratory Co. Ltd. (SEL) is a research company founded and majority owned by Dr. Shunpei Yamazaki, who is widely acknowledged as one of the most prolific inventors in history. SEL develops patentable technologies without engaging in manufacturing or sales.[268] SEL has an extensive technology portfolio in the area of thin-film transistor devices and manufacturing processes, much of which it licenses to Japanese makers of displays, and it has collaborated with DuPont in the development of technology integrating TFT and OLED.[269] In 2013 SEL disclosed development of a flexible lithium-ion rechargeable battery that could be bent to a curvature radius of 40 mm 10,000 times without deterioration in its properties.[270] SEL has reportedly developed a

---

[265] "Nagoya University Researchers Develop Plastic Products That Are Moldable All-Carbon Circuits," *Flexible Substrate,* November 2013.

[266] "Efficient OLED from Kyushu University Gets Rid of Heavy Metals," *Flexible Substrate,* February 2013.

[267] "Kyushu University Grows Thin Films of Germanium," *Flexible Substrate,* October 2013.

[268] "Message from the President" SEL—Creating Technology to Server Society," *<http://www.sel.co.jp/en/01_AboutSEL/1-2_FromthePresident.html>*; "Shunpei Yamazaki Holds Almost 2000 Patents," *Wired,* October 26, 2007.

[269] IDTech Ex, *Printed, Organic & Flexible Electronics* (2011) op. cit., 272.

[270] "SEL Shows a Flexible Rechargeable Li-Ion Battery Prototype," *Flexible Substrate,* November 2013.

range of flexible OLED displays using c-Axis Aligned Cristal (CAAC) oxide semiconductors, a material that reportedly enables creation of larger and more reliable flexible displays.[271]

Fuji Electric Systems

Fuji Electric Systems, a subsidiary of Japan's Fuji Electric Group, is a producer of power semiconductors, solar cells, and electronic systems with applications in automobiles, industrial systems, and consumer electronics. The company has been pursuing R&D into amorphous silicon solar cells using plastic substrates since 1993, with the objective of developing a large surface area solar cell that is lightweight and bendable, in contrast to glass substrate solar cells. Fuji Electric began mass production of thin-film amorphous solar cells in 2007 and stated its intention to invest $320 million by 2011 to expand its production capacity for light, flexible solar cells by 12-fold.[272]

Toppan Printing

Toppan Printing is a "huge force in printing,"[273] with 2011 sales exceeding $18 billion, and also produces semiconductor photomasks, smart cards, RFIDs, and materials for displays. Toppan has been developing large area electrophoretic display signage since 2004. In 2012 it entered into a collaboration with Plastic Logic of the UK to develop flexible, reflective digital signage more than 40 inches in size, a prototype of which was exhibited in 2013.[274]

Teijin Limited

Teijin is a Japanese pharmaceutical and chemical company that produces high-performance aramid and carbon fibers and polyester films and plastics. In February 2009, Teijin formed a technology development agreement with Silicon Valley-based Nano Gram Corporation to conduct R&D on the optimization of silicon nanoparticles and inks and to develop process technology to sinter silicon nanoparticle film at the relatively low temperature of 200°C. As a result of this research Teijin was able to sinter silicon nanoparticle film on a polycarbonate resin, and the technology is expected to have applications for flexible displays and

---

[271] "SEL Develops Flexible OLED Displays with New Materials," *Flexible Substrate,* March 2014.

[272] "Fuji Electric Systems Fully Utilizing Lightweight, Flexible Characteristics of Film Type Solar Cells to Expand Into Various Applications," *Tokyo Semiconductor FDP World,* June 26, 2007; "Fuji Elect Systems to Lift Capacity for Bendable Solar Cells," *Asia Pulse,* October 3, 2007.

[273] "Toppan Printing, Plastic Logic Collaborate on OTFT-Based Large Area Flexible Display," *Printed Electronics Now,* March 2013.

[274] "Ibid.

printed semiconductors.[275] In 2010, Teijin wholly acquired Nano Gram. Teijin has set a target of capturing 80 percent of the global market for silicon inks with semiconductor properties by 2017.[276]

In 2010, Teijin began manufacturing and selling two types of transparent conductive films for use in new versions of electronic paper:

- SS120 uses polycarbonate (PC) film as a substrate, is suited for electronic paper that uses liquid crystals, and is expected to have applications with respect to OLEDs.
- HP125 uses polyethylene teraphthalate (PET) film as a substrate and is suited for electronic paper that uses electrically charged particles.[277]

Showa Denko

Showa Denko is a major Japanese chemical engineering firm, with more than 180 subsidiaries. In 2013, it disclosed that it had developed printable silver nanowire ink with which highly stable, transparent conductive patterns on flexible films. The new ink is expected to replace indium tin oxide (ITO) conductive film for touch screen applications. The new ink is a hybrid made by combining a small amount of silver nanoparticles with copper nanoparticles, creating a hybrid ink that can be used as a low-cost alternative to conventional silver ink and paste. The new ink was jointly developed with Osaka University.[278]

DIC Corp

DIC Corp., until 2008 known as Dainippon Ink and Chemicals, is a manufacturer of printing inks, synthetic resins, and other chemicals including materials with electronics applications. DIC is the world's largest manufacturer of ink. In June 2011, DIC disclosed that it had developed a weatherproof coating agent that can be used instead of glass in solar cells, permitting the fabrication of flexible solar cells that weigh half as much as glass versions.[279]

Taiyo Ink

Taiyo Ink is the world's leading supplier of resist and conductive inks for printed circuit boards. The company has five R&D centers around the world and a U.S. subsidiary, Taiyo America, based in Nevada. Taiyo has developed conductive

---

[275] "Teijin's World First Silicon-on-Plastic Integration Technology," *Printed Electronics World,* November 28, 2009.

[276] "Teijin Acquires Nano Gram Corporation," *Printed Electronics World,* August 11, 2010.

[277] "Teijin Chemicals Enters Electronic Paper Market," *Printed Electronics World,* July 6, 2010.

[278] "Showa Denko Develops Printable Silver Nanowire Ink," *Flexible Substrate,* January 2013.

[279] "Breakthrough for Flexible Solar Cells," *Smarthouse,* June 23, 2011.

inks for printed electronics and solar applications on both flexible and rigid substrates. The company's conductive inks consist of individual metals or metal blends (such as silver with copper) or with ceramic additives.[280] Taiyo America is reportedly developing flexible dielectric materials that utilize either UV or low temperature curing for use on flexible substrates made of PET or PEN.[281]

Dai Nippon Printing (DNP)

DNP is a major Japanese printing company that makes components and materials for electronic displays. DNP has been conducting research on OTFTs for application in printed e-paper displays.[282] DNP has also developed a high-function waterproof flexible film to protect components of organic electro-luminescence devices and solar panels, a potentially attractive alternative to bulky and rigid glass panels. The advent of a thin, flexible, and highly effective film is expected to enable the introduction of smaller and lighter components for use in a new generation of ultrathin solar panels.[283]

In 2009, DNP disclosed that it had developed a transparent conductive film designed to replace ITO films in e-paper, touch panels, and OLED panels. DNP used silver for conductive particles formed into patterns through a printing process on flexible film.[284]

Soken Chemical and Engineering Co

Soken Chemical is a Japanese developer and producer of pressure-sensitive adhesives, fine particles (including toner, photographic materials, and LCD materials), specialty-coated products, as well as chemical production plants. In 2008, the company disclosed that in a research collaboration with the University of Tokyo it had developed a new type of display material for e-paper. The material consists of resin spheres made with two colors mixed with silicone rubber and shaped into a sheet that is injected with silicone oil to enable the spheres to rotate freely. "Patterns are generated by selectively applying voltage to make the spheres turn so that either one color or the other faces forward. When power is cut, the spheres remain in place and the pattern is retained. According to Soken

---

[280] "Printed Electronics at Taiyo," *Printed Electronics World,* October 14, 2009.

[281] "Taiyo America Adopts Knowledge from PCB Market to Printed Electronics," *Printed Electronics Now,* January 2014.

[282] "Printed Electronics Asia: Visits to Local Center of Excellence, Part 2," *Printed Electronics World,* October 15, 2008.

[283] DNP's new protective film produces a completely dust-proof, hermetically sealed environment with moisture barrier properties 100,000 times superior to that of conventional films used in food packaging, making it one of the most waterproof membranes in the world. "Waterproof Film for Solar Panels," *Japan Market Information,* July 25, 2009.

[284] "DNP Produces Printed Transparent Conductive Film to Replace ITO," *Flexible Substrate,* June 2009.

the spheres are cheaper to make than capsules typically used for e-paper, and permit manufacture of large sheets of e-paper because panel production involves simply stretching it over a substrate."[285]

Konica Minolta Holdings Inc.

Konica Minolta is a Japanese manufacturer of imaging equipment, measuring instruments, optical devices, and printers. It announced in 2010 that it would undertake mass production of OLED lighting products applying its photographic film technology. The same year it entered into a strategic partnership with U.S.-based Konarka Technologies Inc. to develop and manufacture organic thin-film photovoltaic cells and panels.[286] In 2013 it reported successful demonstration of printed electronics technologies to achieve commercially viable OLEDs for general lighting applications.[287]

Seiko Epson Corporation

Japan's Seiko Epson is one of the world's largest manufacturers of computer printers and imaging equipment, and it also produces integrated circuits and LCD components. The company pioneered inkjet printing utilizing piezo crystals, a technique it has carried forward into printed electronics, and was one of the first to build industrial-scale, high-precision machines capable of printing electronic components.[288] In 2009, Seiko Epson announced development of a "breakthrough technology that uses inkjet technology to uniformly deposit organic material for the production of OLED TVs in large screen sizes.[289]

Asahi Kasei Finechem

Asahi Kasei Finechem is a subsidiary of Japan's Asahi Kasei Chemicals Corporation, a producer of petrochemicals, polymers, and performance chemicals. Asahi Kasei Finechem is developing organic materials for electronic and optical materials. In 2009, the company disclosed that it had developed polyvinyl sulphonic acid (PVS) for use as a dopant for conductive polymers achieving substantially higher conductivity, more than 100 times greater than an alternative

---

[285] "Japan's Soken Chemical Develops New Display Material for E-Paper," *Asia Pulse,* October 22, 2008.

[286] "Konica Minolta and Konarka Join Forces to Develop Organic Thin Film Photovoltaics," *Nanowerk,* March 4, 2010.

[287] "Application of Printed Electronics Technologies to OLED Lighting," *Flexible Substrate,* May 2013; "Recent Progress on High-Performance OLED Technologies for Lighting Applications," *Flexible Substrate,* January 2014.

[288] "Industry Leaders Discuss Growth and Future of PE," *Printed Electronics Now,* June, 2010.

[289] "Seiko Epson Announces OLED TV Breakthrough," *Flexible Substrate,* June 2009.

method using polystyrene sulphonate (PSS) as dopant. The company plans to apply polymers produced by the PVS method on OLED devices and solar cells.[290]

International Collaboration

Nitto Denko/Singapore

Nitto Denko is a Japanese producer of LCDs, tapes, flexible circuits, semiconductors, insulation, vinyl, and reverse osmosis membranes for desalinization of water. In 2008, Nitto Denko established the Nitto Denko Asia Technical Centre (NAT) in Singapore's Fusionopolis, a large multidisciplinary R&D hub. NAT is a collaboration between Nitto Denko and Singapore's A*STAR Data Storage Institute (DSI), the A*STAR Institute of Materials Research and Engineering, and Nanyung Technologica University.[291] NAT was established to foster the development of organic electronics technologies for use in sensor products that are flexible and less expensive to produce than are inorganic conductors such as silicon or copper. The biosensors are expected to have applications such as early detection of diseases, electronic paper, and lightweight plastic solar cells.[292]

Kisco/Singapore

Kisco Ltd. is a Japanese producer of basic materials for the electronics, automobile, housing, and pharmaceutical industries. In 2011, Tera-Barrier Films, a portfolio company of Exploit Technologies Pte. Ltd., the commercialization arm of Singapore's Agency for Science, Technology and Research, invested an undisclosed amount in Kisco and became Kisco's exclusive distributor for Asia Pacific.[293] The funds provided to Kisco by Tera-Barrier will enable the company to continue its efforts to commercialize its proprietary moisture-resistant films for applications in flexible displays and organic solar cells. Tera-Barrier has also developed technology for flexible optoelectronics products, flexible solar cells, and disposable or wraparound displays.[294]

[290] "Asahi Entrances Polymer Conductivity by 100 Times," *Flexible Substrate.*

[291] A*STAR is a government agency promoting scientific research and education.

[292] Singapore's Economic Development Board (EDB) has identified organic or plastic electronics as key elements in the country's strategy to promote new activities in high technology. NAT will be able to draw on biomedical expertise at the A*STAR research institutes in the nearby Biopolis, as well as using nearby hospitals to field test its innovations. "Nitto Denko Invests $10 Million to Pioneer Organic Electronic Device Research in Singapore," *JCN Network,* November 30, 2008.

[293] Applied Ventures LLC, the venture capital arm of Applied Materials, Inc., also has significant holdings in Tera-Barrier. Tera-Barrier was originally a spinoff of A*STAR, Singapore's agency promoting scientific research.

[294] "Tera-Barrier Investment & Exclusive Distribution Agreement with Kisco," *Printed Electronics World,* January 26, 2011.

SDK/NovaCentrix

In April 2011, Japan's Showa Denko KK (SDK) and U.S.-based NovaCentrix agreed to collaborate in the field of printed electronics. Pursuant to a licensing agreement SDK will manufacture and sell conductive inks developed by NovaCentrix, and SDK and NovaCentrix will jointly develop conductive inks to be used with NovaCentrix's Photonic Curing process technology, a proprietary manufacturing process involving high-speed sintering with visible-light flash lamps, which restricts the rise in temperature enabling the use of plastic substrates that can be produced in an R2R process. SDK will contribute its metal, inorganic, and organic materials technologies to the joint development of conductive inks. SDK's consolidated subsidiary, Shoko Co., Ltd., will serve as sales agent for NovaCentrix in Japan.[295]

[295] "SDK and NovaCentrix to Cooperate in Printed Electronics," *JCN Network,* April 11, 2011.

7

United States

The United States enjoys numerous advantages in the emerging field of flexible electronics. It has the largest and best system of research universities in the world, many of which are currently engaged in research projects with themes relevant to flexible electronics. In a field in which materials science is unusually important, North America has world-class companies with competencies, equipment, process technologies, and intellectual property relevant to flexible and printed electronics, and a number of them are engaged in significant precompetitive research and development (R&D). The U.S. defense establishment, which has fostered many successful high-technology industries, has demonstrated that it is interested in flexible electronics technologies for military use and that it will support development of the necessary research infrastructure. A number of U.S. states that are at the forefront of U.S. innovation have established research centers for flexible and printed electronics.

However, in a field in which government support appears to be a necessary driver in early-stage industrial developments, the United States appears to be being outspent by governments in Europe and East Asia. Moreover, as in Europe, no major company has emerged in North America as a large-scale device aggregator producing consumer products incorporating flexible electronics technologies. Although several innovative small firms are producing consumer products such as e-writing displays and medical and athletic patches domestically, most current onshore manufacturing consists of materials/equipment supply chain production supporting manufacturing operations outside the United States.

FEDERAL SUPPORT FOR FLEXIBLE ELECTRONICS

The U.S. federal government is promoting the development of U.S. capabilities in flexible electronics through numerous institutional channels in the defense, civilian, and dual-use spheres. Much of this effort reflects mission orientations of individual federal agencies rather than an effort to foster industrial development, per se.[1] Government procurement, which has played a major role in the early-stage development of a number of U.S. high-technology industries, represents a potentially significant source of initial demand pull.[2]

The U.S. Armed Services

The U.S. Armed Services have been in the forefront of support for development of flexible electronics in the United States. The Army Research Laboratory and other U.S. Army organizations, the Office of Naval Research (ONR), the Air Force Research Laboratory (AFRL), and the U.S. Special Operations Command (SOCOM) currently support numerous research efforts in the field with an emphasis on defense applications, particularly displays. The Armed Services' research efforts in flexible electronics arose out of earlier programs in the field of displays, an area in which the services led U.S. research through the end of the 1980s.[3] Following a 13-year interval during which the Defense Advanced

[1] Most federal organizations, including the national laboratories, have focused missions and are not necessarily sensitive or responsive to particular industry strategies and imperatives. NIST is exceptional because its own mission is actually to support industry, and its laboratories' focus is on so-called "infra-technologies" that serve to develop industry standards, comprising pre-competitive public goods. Apart from NIST, the mission orientation of the principal federal R&D agencies limits the extent to which they can support industry innovation. See generally Congressional Budget Office, *Federal Support for Research and Development,* 2007; Rand Corporation, *Federal Investment in R&D* (Science and Technology Policy Institute, study prepared for the President's Council of Advisors on Science and Technology, 2002).

[2] See Stewart W. Leslie, "The Biggest 'Angel' of Them All: The Military and the Making of Silicon Valley," in *Understanding Silicon Valley: The Anatomy of an Entrepreneurial Region*, ed. Martin Kenney (Stanford: Stanford University Press, 2002). A number of studies have concluded that over the longer term, government procurement "triggered greater innovation impulses in more areas than did R&D subsidies." Jakob Edler and Luke Georghiou, "Public Procurement and Innovation—Resurrecting the Demand Side," *Research Policy* 36 (2007): 949. See also Michael Borrus, James E. Millstein, and John Zysman, "Trade and Development in the Semiconductor Industry: Japanese Challenge and American Response," in *American Industry in International Competition: Government Policies and Corporate Strategies,* eds. John Zysman and Laura Tyson (Ithaca and London: Cornell University Press, 1983), 154–56; James Utterback and Albert Murray, *The Influence of Defense Procurement and Sponsorship of Research and Development on the Development of Civilian electronics Industry* (Center for Policy Alternatives, MIT, 1977).

[3] The active matrix liquid crystal display (AMLCD) was created pursuant to the Air Force–supported projects to improve cockpit displays. The Army supported research to develop inorganic electroluminescent (IEL) display technology for land combat systems. Between 1989 and 2001, defense-related R&D in displays was led by DARPA. Darrel G. Hopper, "Defense Display Strategy and Roadmaps," in *Cockpit Displays IX: Displays for Defense Applications* (Proceedings of SPIE,

Research Projects Agency (DARPA) spearheaded U.S. defense-related display research, research leadership by the services was resumed.[4] At that point the Army, seeking to improve and modernize its capabilities pursuant to the Future Combat Systems (FCS) and other similar programs, began to undertake significant developmental investments in flexible displays, most notably the establishment of the Flexible Display Center at Arizona State University.[5] Dr. David Morton, program manager for flexible displays at the Army Research Laboratory (ARL), summarized the potential significance of flexible displays for U.S. soldiers:

> The soldier is going to have a display that is essentially embedded on his or her uniform that will provide information when it is needed. The system will determine what information is needed so as not to overload the soldier with additional information. If a soldier needs friend-or-foe information or instructions on what to do, it will be provided instantly. The display that's on the soldier will not break. It will use very low power, and it's not going to wear out. More important, from a systems standpoint, it's made in a commercial environment. It didn't cost too much to insert, which means we can give it to all soldiers.[6]

Dr. Eric W. Forsythe, staff physicist for the ARL, indicated in 2013 that the Army's development strategy is to leverage existing intellectual property with a focus on technology that can be manufactured at reasonable cost. "Once the manufacturing processes are in place, the goal is to transition them to industry so the DoD [Department of Defense] can purchase the technology back at costs comparable to Apple iPads."[7]

The AFRL, a unit of the Air Force Material Command, and the Air Force Office of Scientific Research (AFOSR), performs technological research and contracts with U.S. entities for research in a variety of areas related to printed and flexible electronics. A current Air Force priority is human monitoring, including pilots and support personnel, and in 2013 the AFRL announced it would

---

vol. 4712, 2002, 1–7); Darrel G. Hopper and Daniel D. Desjardins, "Analysis of the Defense Display Market," *Information Display* 18, nos. 4 and 5 (2002): 40–44.

[4] Darrel G. Hopper, "Display Science and Technology for Defense and Security," in *Organic Light-Emitting Materials and Devices VII* (Proceedings of SPIE, 2004).

[5] David Morton and Eric Forsythe, "Flexible-Display Development for Army Applications," *Information Display,* October 2007.

[6] Dr. Eric W. Forsythe, "Flexible Communications," in *Army AL&T Magazine,* July-September 2012.

[7] "Army Targets Flexible Electronics, Displays," *Micromanufacturing,* November-December 2013. Historically, it has not always been possible to devise technologies that are suitable for defense applications that can concurrently be manufactured for civilian use on a mass basis. Some essential military technologies exist for which there is simply no consumer demand, such as radiation-hardened, high-power microwave and millimeter-wave integrated circuits. Many military applications require custom-designed features for which limited civilian demand exists. Commercial manufacturers whose competitiveness rests on high-volume production "are not amenable to custom product production, putting them at odds with DoD's leading-edge technology, custom-design, small volume product needs." Defense Science Board *Task Force on High Performance Microchip Supply,* February 2005, pp. 23–24, 41.

provide $2.2 million in funding to co-fund an industry manufacturing consortium that will "operate at the junction of nanotechnology, biotechnology, additive manufacturing and flexible electronics" to develop technologies permitting the remote, wireless monitoring of physical functions in real time.[8] In 2010, the AFRL cited as one of its significant achievements in nanotechnology the development of new processing techniques in the formation, doping, and transfer of silicon nano-membranes, which were used to produce what were then the world's fastest flexible electronic devices, an effort conducted in collaboration with the University of Wisconsin and the AFOSR.[9] The Air Force has provided funding to a number of U.S. companies to develop process technologies and tools to enable the manufacturing of flexible electronic devices.[10]

Defense Advanced Research Projects Agency (DARPA)

DARPA is a Department of Defense agency that develops new technologies with defense and dual use applications. DARPA's High Definition Systems (HDS) program, which operated between the late 1980s and 2001, invested more than $650 million in the development of display technologies for military use. DARPA's effort was undertaken, in substantial part, because little U.S. capability existed with respect to manufacturing displays, and Japanese producers of liquid crystal displays (LCDs), who dominated world markets in the 1980s, refused to work with U.S. defense research organizations for "cultural" reasons. South Korean and Taiwanese producers of LCDs, who displaced Japanese companies as market leaders, were willing to work with DARPA and in some cases were funded by DARPA directly. By 2001, DARPA observed that "the market for LCDs is highly competitive, presently a robust marketplace in which DoD suppliers have

[8] "Air Force Award to FlexTech Alliance Will Accelerate Development of New Nano-Bio Devices," *Red Orbit,* January 30, 2013.

[9] AFRO, *AFRL Nanoscience Technologies: Applications, Transitions and Innovations,* 2010, 15. This project featured the work of University of Wisconsin Professors Zhenqiang Ma and Max Lagally. Subsequent financial support from AFOSR enabled Professor Ma to develop curved night vision goggles using flexible germanium nanomembrane semiconductors. "Military Projects Push Boundaries of Flexible Electronics in Imaging Technologies," *University of Wisconsin, Madison News,* December 28, 2012; "Scientist Demonstrates Record Speed for Bendable Electronics," Advanced Materials, Manufacturing, Testing Information Analysis Center, January 30, 2008.

[10] AFORS funded a 2008-2012 project involving Cornell University and DuPont to develop a process for spraying a film composed of carbon nanotubes onto a substrate to create thin, flexible electronic devices. "Flexible Circuits Unfold," *AFCEA International,* June 2009. In 2013 the AFRL awarded a contract to Optemec, a spinoff from Sandia National Laboratory, to enhance its "Aerosol Jet" technology to enable large area printing of high-performance, carbon nanotube–based, thin-film transistors. "Air Force Research Lab Awards Optemec Contract to Advance Fully Printed Transistor Technology," *Printed Electronics Now,* December 12, 2013; "Optemec wins Contract to Advance Fully Printed Transistor Technology," *Flexible Substrate,* December 2013.

ready access to the most advanced technologies."[11] DARPA's Flexible Emissive Display program was launched in fiscal 1999 to develop and demonstrate large area, high-resolution emissive, rugged displays for defense applications, including use in aircraft, ships, vehicles, and infantry uniforms and equipment.[12]

National Institute of Standards and Technology (NIST)

NIST, an arm of the Department of Commerce, has a mandate to promote U.S. industrial competitiveness through the application of standards, measurement, science, and technology. NIST funds R&D by universities and companies and operates its own laboratories featuring cutting-edge equipment that are available to industrial users.

NIST's Technology Innovation Program (TIP) provided grants usually ranging from $3 million to $5 million to support research and innovation in areas of critical national need. The program supported innovation in the flexible electronics sector, but was effectively terminated in 2011.[13] NIST has developed proprietary flexible electronic technologies in its own laboratories, such as a flexible memory device similar to a memristor fabricated out of inexpensive common materials in 2009.[14] NIST's measuring equipment has been deployed to assist technology developers pursuing themes on organic photovoltaics, including printable and flexible thin-film organic and hybrid solar cells.[15]

National Science Foundation (NSF)

NSF is a government agency that funds basic research and education in non-medical areas of engineering and science. With an annual budget of $7.5 billion, it is purely a funding agency for external research and does not operate its own

[11] Testimony of D. Jane A. Alexander, Acting Director, DARPA, Before the Subcommittee on Emerging Threats and Capabilities, Committee on Armed Services, United States Senate, June 5, 2001, 24.

[12] Ibid.

[13] NIST extended a $3 million matching TIP grant to Ohio-based Kent Displays in 2011 to enable the company to optimize its manufacturing capabilities for flexible LCDs. "Northeast Ohio's Flex Matters Cluster Attracts $14 Million in 2010," *Printed Electronics World,* February 11, 2011.

[14] Nadine Gergel-Hackett et al., "A Flexible Solution-Processed Memristor, *IEEE Electron Device Letters,* July 2009.

[15] Advanced organic photovoltaics rely on complex nanostructured shapes with multiple components including combinations of materials types with nanoscale structures that are not fully known. Performance of devices varies from line to line without sufficient control or understanding. NIST seeks to address such basic, critical unknowns using sophisticated measuring tools that include x-rays, synchrotrons, scanning tunneling microscopes, and acoustic surface measurement. The objective is to understand the materials science underlying the behavior of materials in a manufacturing context. Eric K. Lin, NIST, "Advancing Technology Through Measurement Science at NIST," National Research Council, *The Future of Photovoltaics Manufacturing in the United States: Summary of Two Symposia* (Washington, DC: The National Academies Press, 2011), 111–112.

laboratories. NSF primarily funds flexible electronics through its Division of Electrical, Communications and Cyber Systems. Depending on the precise definition of "flexible electronics," NSF supports about 200 research projects in the field, usually small, single-investigator efforts, which include thematic areas such as transistors, zinc oxide, organic light-emitting diodes (OLEDs), and printed electronics.[16] Each of these projects is funded at a level of about $300,000 over a 3-year period. Many of those projects are undertaken in collaboration with industry.[17]

Advanced Research Projects Agency-Energy (ARPA-E)

The Advanced Research Projects Agency-Energy is an organization which was formed within the Department of Energy (DOE) in 2007 and was modeled on DARPA to fund high-risk, high-reward R&D in the energy field that might not otherwise be undertaken in the private sector because of the technological and financial risks. Funded at $275 million in 2012, it extended financial support to nearly 300 research projects in the energy field between 2009 and mid-2013.

ARPA-E is funding a number of research projects with flexible electronics themes. It provided a $6 million grant to ITN Energy Systems, Inc. (ITN) to support development of flexible electrochromatic coating materials, which use a small electrical charge to change the tint of windows to control light and heat. The new technology, which will be produced in a roll-to-roll process, will be formed to fit a variety of surface shapes and will represent a lower-cost alternative to materials that must be applied to flat glass during the manufacturing process.[18] In 2013 the Palo Alto Research Center (PARC) launched a project with ARPA-E and Lawrence Berkeley National Laboratory, the "Printed Integral Battery Project," using PARC co-extrusion technology to demonstrate a manufacturing process that deposits an entire functional lithium-ion battery in a single pass.[19]

Small Business Administration (SBA)

The SBA is an independent agency of the federal government established to advise, assist, and protect the interests of U.S. small businesses. The SBA provides financial assistance to regional innovation clusters for business training,

---

[16] Presentation of Dr. Pradeep Fulay, NSF, National Research Council, *Flexible Electronics for Security, Manufacturing and Growth in the United States: Report of a Symposium* (Washington, DC: The National Academies Press, 2013). NSF's Emerging Frontiers in Research and Innovation (EFRI) is funding the BioFlex initiative, which is developing biocompatible flexible electronic systems with healthcare applications, including patient-friendly monitoring, detection, and treatment functions. "NSF Funds Flexible Electronics and Self-Folding Structure Projects," *Flexible Substrate,* November 2012.

[17] Presentation of Pradeep P. Fulay.

[18] ARPAE, "Electrochromatic Film for More Efficient Windows," *<http://aspa-e.energy.gov/?q=arpa-e-projects/electrochromatic-film-more-efficient-windows>*.

[19] "PARC Launches Printed Battery Project," *Flexible Substrate,* August 2013.

commercialization and transfer services, and other services for small companies. In 2010, the SBA awarded 1 of 10 Regional Innovation Cluster contracts to support FlexMatters, an emerging flexible electronics cluster in northeast Ohio. The 1-year award of $500,000 was provided to NorTech, a regional nonprofit development organization to help large organizations source products and services from small flexible electronics firms in the region. In 2011, an additional $500,000 was awarded to NorTech to continue the work. In 2012, the SBA awarded one of seven Regional Innovation Cluster contracts to NorTech to help small firms in the FlexMatters cluster engage market-leading companies to become anchor customers.[20] The SBA awarded NorTech $385,000 in 2012 and $370,000 in 2013. An additional 3 years of funding is optional.

Economic Development Administration (EDA)

As the only federal government agency focused exclusively on economic development, the U.S. Department of Commerce's Economic Development Administration plays a critical role in fostering regional economic development efforts in communities across the nation. Through strategic investments that foster job creation and attract private investment, the EDA supports development in economically distressed areas of the United States. In 2011, the EDA led a multiagency effort "to support the advancement of 20 high-growth, regional industry clusters. Investments from three federal agencies and technical assistance from 13 additional agencies supported development in areas such as advanced manufacturing, information technology, aerospace and clean technology, in rural and urban regions in 21 states."[21] A NorTech-led consortium of organizations received funding of $2 million to create the Northeast Ohio Speed to Market Accelerator (STMA). The mission of the STMA is to provide manufacturing, market development, and workforce development services to support the advanced energy and flexible electronics (FlexMatters) clusters in northeast Ohio.

Small Business Innovation Research Program

The Small Business Innovation Research (SBIR) Program is a federal requirement that government agencies with large research budgets use 2.5 percent of their extramural R&D budgets for research contacts or research grants to small businesses. SBIR Phase I awards of up to $150,000 may be followed, in appropriate cases, with Phase II awards of up to $1 million. Commercialization (Phase III) must be funded by nonfederal entities. Federal SBIR awards are frequently augmented by state funding, and a number of states administer programs

---

[20] "NorTech Wins U.S. Small Business Administration Contract to Grow Flexible Electronics Cluster," *Flexible Substrate,* February 2013.

[21] Department of Commerce, "New Obama Administration Initiates to Spin Economic Growth in 20 Regions across the Country," September 22, 2011.

to assist local small businesses to apply for federal SBIR funding.[22] A number of SBIR awards have supported research projects with flexible electronics themes.[23] (See Table 7-1.)

U.S. National Laboratories

The U.S. national laboratories, administered by the Department of Energy, support research related to national defense, energy security, and public health. The national labs offer the university and industrial community access to large, cutting-edge research facilities, such as high-performance computers, light sources, simulation tools, and specialized synthesizing facilities, to address major technological challenges. A number of the laboratories are conducting research directly applicable to the field of flexible electronics.

National Renewable Energy Laboratory (NREL). NREL is the principal U.S. laboratory for R&D with respect to renewable energy and energy efficiency. In 2010, NREL entered into a Cooperative Research and Development Agreement (CRADA) with Solarmer Energy, Inc., to develop ways to extend the useful life of plastic solar cells.[24] In 2013, NREL disclosed that it was using flexible glass developed by Corning, Willow Glass, to develop thin-film cadmium telluride photovoltaic cells, which are sufficiently thin and durable to permit direct installation on rooftops as solar shingles.[25]

Lawrence Berkeley National Laboratory. In 2013, Lawrence Berkeley National Laboratory (Berkeley Lab) disclosed that it had developed a microscale activator that flexes like a finger and may have applications in artificial muscles, microfluidics, and drug delivery in humans.[26] Berkeley Lab's work with respect to materials that respond physically to light could lead to applications such as "smart curtains" that bend, straighten, or open and close in response to light, and light-driven motors and robots that move toward or away from light.[27] Berkeley Lab,

---

[22] National Research Council, *Best Practices in State and Regional Innovation Initiatives* (Washington, DC: The National Academies Press, 2013), 93–95; National Research Council, *SBIR: An Assessment of the DOD Fast Track Initiative* (Washington, DC: National Academy Press, 2000).

[23] In 2013, Plextronics, a Pennsylvania-based maker of polymers and inks for organic electronic applications, disclosed that it had been awarded a $221,000 SBIR grant from the Department of Energy for the design and development of low-cost processors of printed electrodes for OLED lighting. This project will be undertaken in partnership with Electroninks, a maker of printable electronic inks that was spun off from the University of Illinois. "Plextronics Awarded SBIR Grant for Reducing OLED Lighting Costs," *Flexible Substrate,* August 2013.

[24] "NREL and Solarmer Energy to Extend Lifetime of Plastic Solar Cells," *Printed Electronics World,* July 5, 2010.

[25] "NREL Uses Corning's Flexible Willow Glass to Develop Cheap, Efficient Solar Cells," *Flexible Substrate,* October 2013.

[26] "Lawrence Berkeley National Laboratory Develops Micro-Activator That Flexes Under Laser Light," *Flexible Substrate,* January 2013.

[27] "Lawrence Berkeley Engineers Use Nanotubes to Create Light-Activated 'Curtains,'" *Flexible Substrate,* January 2014.

TABLE 7-1 U.S. SBIR Awards—Flexible Electronics Themes

Year	Sponsor	Recipient	Amount (Thousands of Dollars)	Theme
2013	DoE	Universal Display	250	Energy saving phosphorescent OLED lighting
2013	DOE	Plextronics	221	Printed electrodes for OLEDs
2013	NIH	Applied Nanotech	175	Sensors for detecting bedsores
2012	USSOCOM	eMargin	1,120	Optimize OLED microdisplay for mass production
2012	DOE	Universal Display	150	Outcoupling solution for OLED lighting
2012	Air Force	Ascent Solar	750	Demonstrate PV product using CIGS technology
2012	Army	Applied Nanotech	730	Thermal management for portable energy systems
2011	DOE	Universal Display	1,000	Thermal management—phosphorescent OLEDs
2010	DOE	Universal Display	100	Enhance performance—white PHOLED devices
2010	DOE	Universal Display	500	Protective barrier for flexible OLED displays, lighting
2010	DOE	Applied Nanotech	1,600	Pilot line for conductive inks
2010	Army	Applied Nanotech	100	Develop anode for lithium battery with novel material
2009	Army	Universal Display	334	Prototype flexible displays on metal foil
2009	DOE	Universal Display	200	Demonstrate white PHOLED technology
2009	Army	Nano Dynamics	733	Fabrication of nanomaterials for infrared obscurants
2009	Air Force	Universal Display	750	Flexible OLED displays

SOURCE: *Printed Electronics NOW*, May 2, July 4, 2013; February 6, November 5, December 5, 2012; October 13, 2011; February 18, April 10, June 23, September 16, 2010; January 15, June 30, October 7 and 22, 2009.

a leading source of x-ray and ultraviolet light beams for research, recently discovered that impure domains in polymer PV cells can improve the performance of the cells if reduced sufficiently in size, with an efficiency gain of 42 percent.[28]

[28] "New Path to More Efficient Organic Solar Cells Uncovered at Berkeley Lab's Advanced Light Source," *Flexible Substrate,* February 2013.

Oak Ridge National Laboratory. Oak Ridge National Laboratory has advanced capabilities in manufacturing and materials research and operates a manufacturing demonstration facility. It is conducting research in roll-to-roll manufacturing using pulse thermal processing and other technologies to develop efficient manufacturing techniques for flexible electronics, photovoltaics, and battery systems.[29] Oak Ridge has worked with Texas-based Novacentrix in the development of Pulse Forge tools, which dry, cure, sinter, or anneal high-temperature materials on low-temperature substrates, with applications in the manufacture of flexible electronics, an innovation that received a regional Federal Laboratory Consortium Award in 2013.[30] In 2011, researchers at Oak Ridge National Laboratory developed a way to make conductive coatings comprised of carbon nanotubes more transparent, reducing the problem associated with their tendency to absorb too much light in the visible region of the light spectrum. The discovery enhances the extent to which nanotube-based coatings can compete with the transparent conducting material indium tin oxide, which requires indium, a material that is expensive and sometimes difficult to obtain.[31]

Ames Laboratory. The Ames Laboratory is a national laboratory located on the campus of Iowa State University in Ames, Iowa. In 2012, its scientists disclosed that they had discovered a new technique for using an established polymer, PEDOT:PSS, in OLEDs in a manner that could eliminate the need for use of indium tin oxide.[32]

FLEXTECH ALLIANCE

The FlexTech Alliance is the only organization in North America devoted wholly to promoting the growth of the electronic displays and printed electronics industry chain.[33] The Alliance originated in 1992 as the U.S. Display Consortium (USDC) an industry-led nonprofit public–private partnership modeled on Sematech to develop tools, materials, and components for what was hoped would be an emerging U.S. flat-panel display (FPD) industry. Originally USDC

---

[29] *<http://web.ornl.gov/sci/manufacturing/mdf.shtml>*.

[30] "ORNL Wins Tech Awards," *Knoxville News-Sentinel,* March 21, 2013; "Novacentrix Wins 2010 R&D Award for Low Cost Metal on ICI Copper conductive Ink," *Printed Electronics Now,* July 20, 2010. In 2008, Oak Ridge granted Novacentrix an exclusive license for a patent for pulsed thermal processing of functional materials to enable the company to "commercialize this important intellectual property for solar and other printed electronics markets including RFID, smart packaging and displays." Dr. Ron Off, ORNL, in "Oak Ridge National Laboratory Announces Granting of Exclusive Solar and Printed Electronics License to Novacentrix," Novacentrix Press Release, November 13, 2008.

[31] "ORNL Develops Carbon Nanotube Conductive Coatings for Flexible Electronics," *Flexible Substrate,* September 2011.

[32] Min Cai et al., "Extremely Efficient Indium-Tin-Oxide Free Green Phosphorescent Organic Light-Emitting Diodes," *Advanced Materials* 24, no. 31 (2012): 9; "Ames Laboratory Scientists Develop Iridium-Free Organic Light-Emitting Diodes," *Materials Views,* December 4, 2012.

[33] FlexTech Alliance website, "About the Flex Tech Alliance," *<http://www.flextech.org/au-overview.asp>*.

was funded by DARPA, which was seeking to create a supply capability in the U.S. for FPDs with military applications.[34] Although LCD technology had been invented in the United States, over time U.S. companies abandoned their research efforts and licensed their technologies to Japanese companies, who commercialized a broad array of mass market consumer products incorporating FPDs, with the result that by the early 1990s, they controlled more than 95 percent of the world market for the most prevalent FPD devices, that is, LCDs.[35] Japan's market position has eroded in the face of competition from South Korea and Taiwan, but virtually all FPD manufacturing remained in the Far East. DARPA ended its support for USDC in 2001. At that point the ARL agreed to partner with USDC and to take over the administration of the contract with USDC being managed by DARPA.[36]

Currently the FlexTech Alliance seeks to identify technology gaps, partner with industry to establish pilot manufacturing sites, and develop the associated supply chain. It receives funding support from the Army and Air Force research laboratories and, as of 2013, from the U.S. Special Operations Command.[37] The FlexTech Alliance has sponsored more than 150 technical projects in flexible electronics. One of the most noteworthy of these efforts was the award of a grant to Corning to develop commercially viable techniques for continuous printed electronics manufacturing on glass substrates, an initiative that led to the creation of Corning Willow Glass, a flexible glass substrate with electronics applications.[38] Another FlexTech grant assisted Polyera Corporation in the development of high-performance solution-processable organic semiconductors for printed thin-film transistor devices.[39] In 2013 a FlexTech-sponsored project involving Physical Optics Corporation and E Ink developed a medical triage bandage

---

[34] Jay Stowsky, "Secrets to Shield or Share? New Dilemmas for Dual Use Technology Development and the Quest for Military and Commercial Advantage in the Digital Age," BRIE Working Paper 151 (2003).

[35] In 1993, the U.S. FPD industry was limited to "several small companies making screens for niche markets and for the military." The biggest U.S. producer, Wisconsin-based Standish Industries, had annual revenues of approximately $20 million. "Opening the Screen Door: New Consortium Hopes to Take United States into Crucial Flat-Panel Display Territory—Before it's too Late," *Austin American-Statesman* (June 28, 1993). By this point the important military applications of FPDs were becoming evident through their use in systems such as cockpit displays for fighter pilots and portable navigation devices for soldiers in the field. Stowsky, "Secrets to Shield or Share?" 15.

[36] Ibid.

[37] "FlexTech Alliance Announces 2013 RFP for Funding Opportunities," *Flexible Substrate,* January 2013. FlexTech R&D awards are based on open solicitations and feature 50/50 cost share between FlexTech and its research partner(s).

[38] The research was conducted in collaboration with Binghamton University's Center for Advanced Microelectronics Manufacturing (CAMM) and Western Michigan University's Center for the Advancement of Printed Electronics (CAPE). "Flexible Displays, Lighting and Solar Cells: Developing Technologies and Building the Supply Chain," *Flexible Substrate,* September, 2012.

[39] Ibid.

that detects ECG impulses, skin temperature, and respiration rates, which could increase the ability of first responders to treat patients in the field effectively.[40]

In 2013, the AFRL announced an award to the FlexTech Alliance of $2.2 million to help launch a new manufacturing consortium operating at the junction of nanotechnology, biotechnology, additive manufacturing, and flexible electronics. Twenty consortium partners that include GE, DuPont Teijin Films, and Lockheed Martin will contribute an additional $3.3 million. "Nano-bio manufacturing" refers to the integration of multiple functions, such as power, sensing, and communications as a flexible electronics platform. Potential applications include embedded sensors on an aircraft's surfaces to measure stress and to conduct real-time analysis of structural integrity, as wells as medical patches monitoring human performance via wireless technology.[41]

At the end of 2013, the FlexTech Alliance announced two R&D awards to Soligie, a Minnesota-based firm that provides design and manufacturing services on a foundry-type basis to flexible electronics companies.

- A Soligie-led team comprised of Boeing, American Semiconductor, and Imprint Energy will develop and demonstrate a sensor platform that incorporates a power source, microcontroller, display, and wireless communications functions utilizing printed components and silicon-on-polymer technology. This project is supported by the Army Research Lab.
- Soligie has also received a grant supported by the U.S. Special Operations Command to design, develop, and fabricate scatterable media cards based on printed and flexible electronics technologies capable of delivering a 30-second audio message that could replicate the paper flyers currently mass-distributed by SOCOM forces.[42]

The FlexTech Alliance's solicitation for 2014 R&D projects is backed by $3.5 million in new funds from the ARL for development of cutting-edge tools, materials, and processes associated with specific types of flexible and printed electronics.[43] There are four themes:

- Develop a tool to process electronic components and hybrid integrated circuits (ICs) on 3-D surfaces utilizing 3D-additive manufacturing processes.

---

[40] "FlexTech Alliance Announces Completion of Project 154 for a Flexible Medical Triage Bandage," *Flexible Substrate,* October 2013.

[41] "Air Force Award to FlexTech Alliance Will Accelerate Development of New Nano-Bio Devices," *Flexible Substrate,* February 2013.

[42] "Soligie Receives Two R&D Awards from FlexTech Alliance," *Flexible Substrate,* December 2013. FlexTech Alliance, *2014 Targeted and Open Solicitation Request for Proposals (RFP),* issued December 10, 2013.

[43] FlexTech Alliance, *2014 Targeted and Open Solicitation Request for Proposals (RFP),* issued December 10, 2013.

- Develop flexible integrated systems for energy storage, energy harvesting, and wireless transmission for diverse military applications.
- Develop new solutions for integrating hybrid silicon CMOS technologies, including new manufacturing systems and system design for interconnect technology, with applications in medical monitoring, on-body sensors, logistics tagging, and structural integrity monitoring.
- Develop product demonstrators for printed and flexible electronics applications in power generation, energy storage, sensors, communications, and lighting.

The FlexTech Alliance is currently forming "user groups" emphasizing flexible electronics applications, which are intended to facilitate roadmapping, networking with supply chain companies, and precompetitive R&D in thematic areas such as wearable and disposable electronics.[44]

ASU FLEXIBLE ELECTRONICS AND DISPLAY CENTER

The Flexible Display Center (FDC) is a research and innovation center focusing on the commercialization of flexible displays formed through a collaboration between the Army and Arizona State University (ASU). In 2004, the Army awarded ASU $43.7 million in a 5-year cooperative agreement to establish the DTC, with an optional additional $50 million for a follow-on 5-year period.[45] The FDC has evolved into a major collaboration involving the Army, 18 U.S. foreign companies, and 7 additional universities. (See Table 7-2.)

The FDC has eight academic partners that include seven universities and RTI International, a nonprofit R&D organization based in North Carolina's Research Triangle Park. Most of these institutions are being funded by the Army, and their work at the center is coordinated by their Army managers. The universities do not contribute funds to the FDC or vice versa.[46]

The participating companies represent a number of stages in the supply chain. (See Table 7-3.)

The FDC is equipped with a Generation (GEN) II pilot production line that is unique in the Western Hemisphere. It can be employed to produce OLEDs as well as any other organic electronic device involving precision processing of high-purity organic thin-layer-dependent surfaces. The line enables the FDC to prove and demonstrate that the technologies it develops can be manufactured, as well as

---

[44] "FlexTech Alliance Introduces Wearable and Disposable Electronics User Groups," *Flexible Substrate,* January 2014.

[45] "Army Awards ASU $43.7 Million for Flexible Display Development," ASU Press Release, February 10, 2004.

[46] National Research Council, *Best Practices in National Innovation Programs.*

TABLE 7-2 ASU Flexible Display Center Participants

Government	Industry	Academia
The Army Research Laboratory	AKA America	Oregon State University
The Natick Soldier RD&E Center	Corning	Lehigh University
The U.S. Army Manufacturing Technology Center	Dpix	University of Texas at Dallas
The Office of the Assistant Secretary of the Army for Acquisition, Logistics and Technology	DuPont Teijin Films	Princeton
The US Army Research, Development & Engineering Command	E Ink	SUNY—Binghamton
	Etched in Time, Inc.	North Carolina A&T
	EV Group	Penn State University
	FlexTech Alliance	
	Henkel	
	HP	
	Honeywell	
	ITO America	
	L-3 Communications Display Systems	
	Plastic Optics Corporation	
	Plextronics	
	Raytheon	
	Universal Display Corporation	
	RTI International	

SOURCE: <*http://flexdisplay.asu.edu.partners*>.

TABLE 7-3 Participating Companies in the Supply Chain

Materials	Manufacturers	System Integrators	Equipment
Cytec Industries	Dpix	BAE Systems	EVG
DuPont Teijin	E Ink	Boeing	Etched in Time, Inc.
Honeywell Electronic Materials	HP	General Dynamics	ULVAC
PETEC	LG Display	Honeywell	ART America
	QD Vision	L-3	FlexTech Alliance
	Sharp	Lockheed Martin	Macon
	UDC	Raytheon	Ito America
			Particle Measuring Systems

to manufacture limited batches of displays if needed.[47] The FDC ultimately seeks to develop a transferable manufacturing process for flexible displays.

The FDC has begun to record technological achievements. In 2012, it disclosed that it had manufactured the world's largest flexible color display, an OLED prototype measuring 7.4 inches diagonally that met DOD's request for a full-color, full-motion video flexible display unit.[48] In 2013, the FDC and PARC

[47] Ibid.

[48] "Arizona State Researchers, Army, Build World's Largest Flexible Display," *Phoenix Examiner,* June 12, 2012.

disclosed that they had "successfully manufactured the world's largest flexible x-ray detector prototype using advanced thin film transistors, measuring 7.9 diagonal inches."[49] In July 2013, ASU disclosed that it had manufactured an even larger flexible color OLED display, measuring 14.7 inches.[50]

The FDC includes foreign participants on the same terms as U.S.-based members, which is not viewed as a security concern. The FDC's goal is to enable the development of displays that will be produced commercially and competitively priced. Security concerns would only arise when the displays are incorporated in military systems, which is outside the scope of the FDC's activities. Nicholas Colaneri, Director of the Center, comments that

> [w]e're trying to design components that Apple Computer, and Samsung, and anybody else will be putting into their products. Then the Army can buy them for the same $12 that those guys do rather than paying $20,000 a display, which is the old paradigm.[51]

State-Led Initiatives—The Northeast Ohio Flexible Electronics Cluster

Significant initiatives to support development of flexible electronics are under way at the state as well as the federal level. One noteworthy example is the activity of the Northeast Ohio Technology Coalition (NorTech), a technology-based economic development organization, which is an initiative to accelerate the flexible electronics cluster in northeast Ohio. The FlexMatters cluster was one of the first regional innovation clusters recognized and funded by the federal government and today is linked to a number of federal manufacturing initiatives such as the Jobs and Innovation Accelerator Challenge Clusters (JIAC) and the National Additive Manufacturing Innovation Institute (NAMII). In addition to federal grants and contracts, NorTech and FlexMatters receive significant financial support from regional philanthropic foundations and Ohio Third Frontier.[52] As of September 2013, the cluster consisted of 70 member organizations including large companies such as Avery Dennison and American Greetings, regional research institutions, and small- to medium-sized technology businesses representing various segments of the flexible electronics value chain.

---

[49] "ASU Center, PARC Produce World's Largest Flexible X-Ray Detector," *Targeted News Service*, March 19, 2013.

[50] "ASU's Flexible Display Center Hits Milestone with New Technology," *Phoenix Business Journal*, July 3, 2013.

[51] National Research Council, *Best Practices in National Innovation Programs*.

[52] Presentation by Byron Clayton, NorTech, National Research Council. *Flexible Electronics* (2010) op. cit. The Third Frontier program, created in 2002, is a state fund supporting early-stage research and development in areas that the private sector may not otherwise invest. During the period 2002-2015 its budget was $2.3 billion. As of 2010, it had invested approximately $60 million in flexible electronics in Ohio. See National Research Council, *Best Practices in State and Regional Innovation Initiatives* (Washington, DC: The National Academies Press, 2013), 115–116, 121, 124.

FlexMatters Roadmap. A strategic roadmap of northeast Ohio's flexible electronics cluster was completed in early 2013.[53] It identified a $42 billion dollar market by 2019 for FlexMatters' core and emerging competencies. (See Table 7-5.) Core competencies consist of the following:

Liquid crystal films: Liquid crystal films are electroactive liquid crystals encapsulated between two sheets of polymer. These functional film subsystems are integrated with electronic driver circuits and are packaged to create finished devices, particularly bistable display products such as e-readers, tinting eyewear, status indicators, or a host of others that do not require animated pixels and benefit from very low power consumption.

Complex flexible circuits: Complex flexible circuits employ the mounting and assembling of electronic components and circuits on flexible plastic substrates. The term "complex" indicates the incorporation of more than one electronics capability onto the circuit, resulting in the possible miniaturization of the electronic assembly or the introduction of robust features and functionality. Complex flexible circuits are used fairly widely over a range of electronics applications where the following attributes are desired: tightly assembled electronic packages, three-axis electrical connections, flexing of an assembly during normal use, replacement of heavier and bulkier wire harnesses, or where space of geometry constraints are driving factors.

Roll-to-roll (R2R) manufacturing: Driven in part by the need of regional liquid crystal film manufacturers to produce high-quality and cost-efficient devices, northeast Ohio has developed competencies in the commercial development and application of R2R processing manufacturing of flexible electronic components.

The NorTech roadmap also identified advanced and emerging competencies of FlexMatters member organizations, including the following:

Roll-to-roll manufacturing of liquid crystal devices: Several northeast Ohio companies with core competencies both in liquid crystal films and R2R manufacturing are developing and commercializing liquid crystal devices for the consumer, defense, and commercial markets.

Roll-to-roll manufacturing of functional films: Northeast Ohio is home to a network of companies and researchers that specialize in functional films and additives, as well as the National Polymer Innovation Center (NPIC) at the University of Akron, a state-of-the-art polymer synthesis and characterization facility that houses three R2R manufacturing systems.

Automated manufacturing of complex flex circuits: Northeast Ohio has substantial technical capabilities in the design and production of complex flex circuits, currently making products incorporated in sensing devices for major aircraft platforms, sophisticated medical devices, and heavy-duty industrial applications.

[53] NorTech FlexMatters Roadmap Final Report (2013). The report can be accessed at: <*http://www.nortech.org/flexmatters/reports-and-presentations/reports/flexmatters-roadmap-final-report*>.

Automated manufacturing of high-value flexible devices: The 2019 goal of the FlexMatters cluster is to achieve a global leadership position in R2R manufacturing of liquid crystal devices, R2R manufacturing of functional films, and automated manufacturing of complex flexible circuits. This will establish a complete regional value chain for manufacturing entire flexible devices. From 2019 forward, FlexMatters plans to focus on leveraging the entire regional value chain to achieve global leadership in producing high-value flexible devices.

Ohio Third Frontier program: The Ohio Third Frontier program was created in 2002 and has been extended through 2015 to support early-stage R&D efforts that the private sector might not otherwise support given that the payoff might be too far in the future. Its budget of $2.2 billion makes it the largest economic development program ever implemented in Ohio. It has made major investments in flexible electronics since 2008. (See Table 7-4.)

University engagement. The FlexMatters cluster is being supported by local research universities, and a number of the flexible electronics companies currently operating in the cluster trace their origins to one or more of these academic institutions. The combined polymers programs of the University of Akron and Case Western Reserve University are among the largest in the world, and those institutions' formal relationships with the Austen Bioinnovation Institute and the Cleveland Clinic, respectively, are likely to facilitate development of biomedical applications for flexible electronics. Kent State University, which pioneered the LCD, remains one of the foremost centers of LCD knowledge in the world.[54]

The University of Akron has established the National Polymer Innovation Center (NPIC) in partnership with the University of Dayton and The Ohio State University and 85 companies. Supported by the Ohio Third Frontier program, NPIC is a cooperative research center that works with regional companies to develop flexible electronics technologies. Its activities include research on polymer substrates, demonstration of R2R manufacturing processes on pilot manufacturing lines, and development of displays or electronic devices with double curvature or spherical devices that cover a face. NPIC's Dr. Miko Cakmak comments on the R2R facility:

> We are literally providing this capability, which is called an electromagnetic processing line, to the industry and to anyone who would like to use our facilities. In this center science and engineering research is being carried out to enable these technologies through partnerships with regional institutions including NASA Glen and local industry.[55]

---

[54] Miko Cakmak, University of Akron, "The Role of Regional Academic Institutions in Flexible Electronics Development," in National Research Council, *Building the Ohio Innovation Economy: Summary of a Symposium* (Washington, DC: The National Academies Press, 2013), 120–122; National Research Council, *Best Practices in State and Regional Innovation Initiatives,* 131–133.

[55] Miko Cakmak, "Role of Regional Academic Institutions," in National Research Council, *Building the Ohio Innovation Economy*, 120.

TABLE 7-4 Ohio Third Frontier Program Investments—Flexible Display and Electronics Cluster (2003-2010)

Program	Year	Amount (Dollars)
Action Fund	2003	375,725
Wright Projects	2003	1,640,000
Research Commercialization Grant Program	2005	174,657
Wright Projects	2005	100,000
Research Commercialization Grant Program	2006	700,000
Research and Commercialization Program	2007	23,937,840
Research Scholars Program	2008	15,292,382
Research and Commercialization Program	2008	5,000,000
Research and Commercialization Program	2009	4,900,000
Advanced Energy Program	2010	965,000
Advanced Materials Program	2010	2,918,000
TOTAL	2003-2010	56,003,604

SOURCE: NorTech, "A State's Initiative: Advancing Flexible Electronics in Northeast Ohio," 2010.

Industry engagement. NorTech periodically convenes cluster members to share resources, knowledge, and opportunities and provides services directly to cluster members for product development and scale-up, project and growth financing, and access to target markets. Dr. Bahman Taheri, representing Alpha Micron, a producer of curved, liquid crystal–based eyewear, comments that "our employees all came from Kent State University, and if they do leave, which they don't, they would all go to Al Green's company [Kent Displays in Kent, Ohio] and anyone who leaves Al's company comes to us."[56]

Kent Displays is a spinoff from Kent State University that makes flexible LCDs for nontraditional consumer applications. All of its products are manufactured on site in Kent, Ohio, using R2R processes. Dr. Albert Green of Kent Displays observes that "R2R manufacturing of displays is the holy grail of the display industry, and we're happy to be a pioneer in that space, along with Bahman Taheri and Alpha Micron."[57] He underscores the fact that whereas a conventional LCD manufacturing plant costs $1 billion or more, the Kent Displays facility was built with a capital investment of several million dollars.

[56] Bahman Taheri, "Manufacturing of Curved Liquid Crystal Devices," in National Research Council, *Building the Ohio Innovation Economy,* 125.

[57] Albert Green, "Roll-to-Roll Manufacturing of Flexible Displays," in National Research Council, *Building the Ohio Innovation Economy: Summary of a Symposium* (Washington, DC: The National Academies Press. 2013.)

TABLE 7-5 Core Competencies in the FlexMatters Cluster

Core Competency	Description
R&D pipeline	• Electro-optic materials • Semiconductor materials • Materials science and design • Device physics • Novel processes
Technology and innovation	• Particle processing • Coatings processing • R2R and wide web equipment/tools • Polymer film processing
Talent and intellectual capital	• Highly skilled faculty • Masters and PhDs • Engineers and scientists
Commercialized and near-commercialized products and components	• Electro-optic materials and films for curved product designs • Silver conductive inks and pastes for electrodes • Robust, low sheet resistance ITO conductive films • Nanotechnology-based coatings for screen protection • Solar devices/components • Lighting components • Thermal heat dissipation components • E-writers • Flexible electronic medical implants and devices • Flexible electronic tinted eyewear

SOURCE: NorTech, *FlexMatters Strategic Roadmap*, 2010, 26; Interviews in Kent, Cleveland and Akron, Ohio, June 4-5, 2013.

Kent Displays products include the Boogie Board, an e-writing tablet utilizing a flexible display with the feel of paper that allows the user to make notes using a stylus, finger, or any nonsharp object, with the writing and other images fully erasable. The company has also developed Reflex LCD Electronic Skins, which are ultra-thin, durable, single-pixel displays that can be cut to custom shape and conformed to a personal electronic device, with no power required from the host device to retain a displayed color image virtually indefinitely.[58] Kent Displays acknowledges the important role played in its growth by financial support from the Ohio Third Frontier program.[59]

Alpha Micron was founded by faculty members from Kent State University's Liquid Crystal Display Institute. The company received early-stage financing

[58] Interview with Kent Displays, Kent, Ohio, June 4, 2013; Albert Green, Kent Displays, "Roll-to-Roll Manufacturing of Flexible Displays," in National Research Council, *Building the Ohio Innovation Economy,* 122–124.

[59] "Colorful Skin: Kent Displays Nears Full Production of Cholestoric LCD Products," *Printed Electronics Now,* June 2009.

through federal SBIR grants and 6.1 and 6.2 military funding, followed by support from the Ohio Third Frontier program. The company has developed proprietary technology for transmissive LCD systems that control light as it passes through them: "Nobody in the world can put liquid crystals on curved surfaces by thermoforming without going through our patents."[60] The company is developing eyewear for U.S. special forces to enable them to go in and out of buildings without changing glasses and for Navy jet pilots who fly in and out of clouds and need to adjust the light-filtering qualities of their helmet eyewear. The company has also developed consumer products such as eyewear for skiers, auto-dimming mirrors for cars, and windows.[61]

Valtronic, based in Solon, Ohio, is a medium-sized company with branches in Europe that develops and manufactures flexible and semi-flexible electronic sensors and medical devices. Valtronic has identified applications in which the flexible form of a device adds real value. Its "caretaker" product is a band aid–like strip that can be worn on a finger that can detect not only blood pressure, heart rate, and respiration, but also the existence of internal hemorrhaging—making it valuable to first responders and the military. Valtronic has 21 new devices in clinical trials including a glucose measuring device that would obviate the need for finger prick blood tests for diabetics; "smart implants" that indicate months or years in advance whether the human subject will develop a medical condition; a wireless pacemaker that does not need battery replacement; and flexible retinal implants for the blind.[62]

Crystal Diagnostics in Kent, Ohio, was formed by professors from Kent State University and Northeast Ohio Medical University. The company uses liquid crystal technology to detect pathogens through a mechanical process that is much faster and lower cost than traditional methods based on biological sampling. It received assistance in acquiring equipment from the Ohio Third Frontier program. The company's technology is enabling development of small and relatively inexpensive detectors that can be used in applications such as medicine, bio-defense, food safety, and water quality.[63]

Akron Polymer Systems is a spinoff from the University of Akron that specializes in the synthesis of polymers for demanding applications. It provides custom polymer development for many uses including high-performance aerospace applications, flexible displays, and organic photovoltaics.

Leadership: NorTech collaborated with a core group of universities and companies to organize FlexMatters into a recognizable, place-based, emerging industry that attracts organizational members, public and private investors, and

[60] Interview with Alpha Micron, Kent, Ohio, June 4, 2013; Bahman Taheri, "Manufacturing of Curved Liquid Crystal Devices," in National Research Council, *Building the Ohio Innovation Economy,* 124–125.

[61] Bahman Taheri, op. cit.

[62] Interview with Valtronic, Solon, Ohio, June 5, 2013.

[63] Interview with Crystal Diagnostics, Kent, Ohio, June 4, 2013.

specialized talent. The 70 organizational members (and growing) are represented by a 15-member advisory committee consisting of academic and industry participants from the cluster. The committee meets on a quarterly basis to provide feedback and guidance to NorTech on how to provide the highest value to the FlexMatters cluster with the objectives of accelerating cluster growth and contributing to the revitalization of northeast Ohio's economy. The committee also served as the working group for the FlexMatters Strategic Roadmap.

UNIVERSITY RESEARCH

Many U.S. universities are engaged in basic and applied research in thematic areas relevant to flexible electronics and are generating a steady stream of new discoveries. U.S. universities have been a prolific source of spinoffs commercializing academic discoveries in flexible, printed, and organic electronics. (See Tables 7-6 and 7-7.)

Although most of these research achievements are attributable to individual universities, a number of multi-university collaborations have reported discoveries in thermatic areas relevant to flexible electronics.[64]

- ***Harvard*** is pioneering the 3-D printing of rechargeable lithium ion microbatteries as small as a single grain of sand, 1,000 times smaller than the smallest commercially available microbatteries, with potential applications in biomedical devices, micro-UAVs (drones), and "smart dust" (distributed sensor arrays). In 2013 Harvard's achievements in this area won the annual "Academic R&D Award" from the consultancy IDTechEx.[65]
- ***MIT*** engineers report creation of a new polymer film that can generate electricity from water vapor, changing its shape by curling up or down after absorbing miniscule amounts of evaporated water. Potential applications include harnessing this motion to power micro- and

[64] Stanford and the University of Nebraska-Lincoln recently jointly reported development of thin, transparent semiconductors that could provide the basis for inexpensive displays utilizing flexible plastic substrates. The University of Illinois at Urbana-Champaign and the University of Central Florida in Orlando jointly reported the discovery of a technique to create large sheets of nanotextured silicon microcell arrays that could be employed to create lightweight, bendable, efficient solar cells. "Universities of Illinois and Central Florida Develop Way to Make Solar Cells Thin, Efficient and Flexible," "Stanford and UNL Engineers Make World's Fastest Organic Transistor," *Flexible Substrate,* January 2014.

[65] "IDTech Ex Names Top Printed Electronics Developments of 2013," *Flexible Substrate,* December 2013; "Printing Batteries," *MIT Technology Review,* November 25, 2013. A team led by Harvard materials scientist Jennifer Lewis has developed an array of "functional inks" that can solidify into batteries, electrodes, wires, and antennae, as well as nozzles and extrudes that squeeze out batteries and other components from a 3-D printer. This technology works at room temperature, and the materials can be printed on plastic. "Harvard Uses 3D Printing on Lithium-ion Technology," *Flexible Substrate,* December 2013.

TABLE 7-6 Flexible Electronics—Spinoffs from U.S. Universities

University	Company	Technology
MIT Media Lab	E Ink	Electronic paper
Carnegie-Mellon	Plextronics	Conductive polymers/inks
U. Cal Berkeley	Imprint Energy	Flexible batteries
U. Illinois	MC10	Stretchable electronics
UCLA	Aneeve Nanotechnologies	Printed electronics using carbon nanotubes
MIT	Cambrios	Silver nanowires for touch screens
Kent State	Kent Displays	Flexible displays for consumer applications
Kent State	Alpha Micron	Transmissive LCD systems
U. Akron	Akron Polymer Systems	Polymers for flexible displays and organic photovoltaics
NYU	Tectonic Technologies	Large area multitouch sensors

nanoelectronic devices and controlling small robotic limbs. On a larger scale the technology could be embedded in clothing, harnessing evaporated perspiration to provide power for devices such as physiological monitoring sensors. Placed above a lake or river, the technology could be used for large-scale electricity generation.[66]

- ***Georgia Tech's*** Center for Organic Photonics and Electronics (COPE) won the 2012 Academic R&D Award from the consultancy IDTechEx for discovery of a universal technique for reducing the work function of organic electronics conductors, using a polymer modifier containing simple alphatic amine function groups. The modifiers are effective for a broad range of conductors, including graphene and conducting polymers, and are inexpensive, environmentally friendly, and compatible with R2R mass production techniques. Applications exist in OLEDs, organic solar cells, and organic thin-film transistors.[67]
- ***Stanford*** engineers reportedly have combined layers of flexible electronics and pressure sensors to create a wearable heart monitor that is thinner than a dollar bill. The monitors have potential to enable doctors to detect stiff arteries and other cardiovascular problems and to monitor safely key vital signs for newborn babies and high-risk surgery patients. Stanford researchers are working to make these devices completely

[66] "Acrobatic Polymer Film Developed at MIT Harvests Energy from Water Vapor," *Flexible Substrate,* February 2013.

[67] "IDTechEx Printed Electronics USA 2012 Award Winners," <*http://www.idtechex.com/research/articles/idtechex-printed-electronics-usa-2012-award-winners-00004993.asp*>.

TABLE 7-7 U.S. University Research Achievements, Flexible and Printed Electronics— 2013

Institution	Results Reported
MIT	New photovoltaic cell based on graphene sheet; one-molecule-thick material for ultrathin flexible solar cells and LEDs
Rice	Seamless graphene/hybrid electrode interface; nearly transparent films of conductive carbon nanotubes; inorganic flexible thin-film solar cell fabricated by solution processes
Princeton	Triple the efficiency of organic solar cells
U. Delaware	Stretchable power source for stretchable devices
Stanford	First all-carbon solar cell; carbon nanotube circuits
Penn State	Optical fiber for curved or twisted solar fabrics
North Carolina State	Elastic wires that reconnect when severed, silver nanowires for wearable sensors
Western Michigan U.	Etchant material for patterning indium tin oxide
UCLA	Stretchable polymer OLEDs, stretchable, foldable transparent electronic displays
Georgia Tech	Interfaces in organic solar cells; graphene structures suitable for room-temperature electronics
U. Buffalo	PV cells for application in liquid form
Harvard	New photonic fiber that changes color when stretched
UC Berkeley	Deposition over large areas for flexible medical applications, user-interactive sensor network on flexible plastic ("e-skin"), printed MEMs using metal inks; printing process for wall-sized displays, printed transistors on paper substrates
U. Illinois	Compound semiconductor nanowires grown on graphene sheet; stick-on electronic patches for health monitoring
Ohio State	1-atom thick germanium sheets for electronics
U. Michigan	Stretchable conductors utilizing gold nanoparticles; organic vapor jet printing enabling precise patterning of organic electronic devices
Purdue	Hybrid silver-graphene electrode for applications in flexible displays
Northwestern	Graphene-based conductive ink for flexible electronic applications
Stanford	Wearable, skin-like flexible heart monitor
U. of Pennsylvania	Computer model for designing flexible touch screens
UC Santa Barbara	Seamless ICs etched on graphene
New Jersey Inst. Tech.	Flexible battery made with carbon nanotubes
U. of Houston	Gold nanomesh stretchable, transparent conductor
U. of Texas Austin	2-D grapheme analogues for flexible solid-state thin-film supercapacitators
U. Pittsburgh	Polymers that move in response to light
NJIT	Flexible battery made of carbon nanotubes
Carnegie-Mellon	Energy-harvesting from user interaction with paper-like materials

SOURCE: *Flexible Substrate* (January, February, May, August, October, November, December 2013; January, March, April 2014).

wireless, with the expectation that doctors will be able to receive a patient's minute-by-minute heart status via a cell phone.[68]

- ***University of California (UC) Berkeley*** reports development of "electronic skin," a user-interactive sensor network on flexible plastic that responds to touch by lighting up, with the degree of emitted light increasing with the intensification of pressure. Potential applications include robots with increased touch-sensitivity, wallpaper that doubles as a touch screen display, and dashboard laminates that enable drivers to adjust electronic controls by waving hands.[69]
- ***Queens University*** in Canada has collaborated with Intel and Plastic Logic to develop a flexible "PaperTab" display, which developers believe could replace paper altogether. Intel has also expressed the view that this technology could replace conventional displays entirely. Devices incorporating PaperTab could reportedly be virtually unbreakable and as thin as a piece of paper. PaperTab supposedly can file and display thousands of paper documents obviating the need for a computer monitor and stacks of paper or printouts. Users would have 10 or more interactive displays ("PaperTabs"), each representing a different application.[70]

Research centers. Several universities have full-scale research centers devoted to flexible electronics–related topics such as R2R and printed electronics manufacturing.

Georgia Tech ***Center for Organic Photonics and Electronics (COPE)***, founded in 2003, is an R&D and educational center developing flexible organic photonic and electronic materials and devices with applications in IT, telecommunications, energy, and defense. COPE receives financial support from DOD (ARL, DARPA and ONR), DOE, and NSF.[71] Its industrial affiliates include a number of leading European firms including Novaled, Plastic Electronics, Solvay, and Beneq, as well as U.S. firms such as Boeing, Plextronics, and NextInput.[72] Dr. Bernard Kippelen, COPE's Director, has focused on the interdisciplinary character of COPE's research and has emphasized develop-

---

[68] "Stanford Engineers Monitor Heart Health Using Paper-Thin Flexible Skin," *Flexible Substrate,* August 2013.

[69] "UC Berkeley E-Skin Responds to Touch with Promise for Sensory Robotics and Interactive Environments," *Flexible Substrate,* August 2013.

[70] "Intel, Plastic Logic and Queen's University Reveal Bendable 'PaperTab' Display," *Flexible Substrate,* February 2013.

[71] *<http://www.cope.gatech.edu/partnerships/researchsponsors.php>*.

[72] COPE and Solvay began joint development of OLED technology in 2006, based on Georgia Tech's development of a unique material platform for OLEDs that could be deposited over large areas using inkjet printing and standard photolithography. The university's researchers discovered that exposing the material to ultraviolet light produces hardened materials that are insoluble and maintain stability at high temperatures. "Georgia Tech and Solvay Announce $3M Deal for OLED Research," *Eurekalert,* April 26, 2006.

ment of low-cost organic photovoltaics with progressively improving efficiency levels.[73] In 2012 COPE announced a "game-changing" innovation, the creation of the world's first plastic solar cell utilizing "inexpensive, environmentally friendly, easy-to-access materials compatible with existing R2R manufacturing processes.[74]

The ***Center for Advanced Microelectronics Manufacturing (CAMM)*** is an R&D center established at Binghamton University with the university's partners, Cornell, the FlexTech Alliance, and Endicott International Technologies. CAMM is part of the New York State Center of Excellence in Small Scale Systems Integration and Packaging. It received $12 million worth of equipment from the U.S. Display Consortium when it started operations. CAMM is engaged in demonstrating the feasibility of roll-to-roll electronics manufacturing by acquiring prototype tools and establishing production processes.

The ***Center for the Advancement of Printed Electronics (CAPE)***, located at Western Michigan University (WMU), is a collaboration that includes the university, Corning, Amway, Daetwyler R&D, and Neenah Paper Inc.[75] CAPE is a facility for R&D in materials used in the fabrication of flexible electronics devices through printing processes. CAPE draws upon cross-departmental faculty competencies and the university's pilot plant facilities to advance its research agenda.[76] Reflecting WMU's historic strengths in paper and printing innovation, CAPE is seeking to leverage its extensive embedded base of printing systems (including rotogravure, flexo, and inkjet) to develop multiple, inexpensive techniques for printing electronic devices.[77]

Sonoco Institute of Packaging Design & Graphics. The Sonoco Packaging Institute was established at Clemson University in 2009 to develop packaging design technology, a mission that is attending to the development of printed electronics packaging with applications such as smart packaging and the interaction

[73] "A Look at Printed Electronics: Printed Electronics Now Interview with Dr. Bernard Kippelen," *Printed Electronics Now,* July, 2011. Reporting on development of a new polymer with promise for encapsulation of metal conductors, Dr. Kippelen attributed the achievement in substantial part to input from "colleagues from Georgia Tech's Chemistry and Physics Department, who see much potential." "COPE Breakthrough May Simplify PE Manufacturing," *Printed Electronics Now,* April 2012.

[74] "New Technique Creates First Plastic Solar Cell," *Forbes,* April 25, 2012.

[75] Daetwyler R&D Corporation, renamed Ohio Gravure Technologies Inc. in 2011, is a partner in Western Michigan University's recently established Center for Advancement of Printed Electronics. The company is an engineering and software enterprise that has provided sophisticated technology to the gravure printing industry for 30 years. Ohio Gravure's engineering specialties are submicron positional cutting tools, pre-press layout software, and upgrading older printing equipment. Ohio Gravure designed and built the Star MicroEngraving system for optics and printed electronics applications. "WMU, Daetwyler R&D Announce Partnership in PE," *Printed Electronics Now,* November 25, 2009; "Daetwyler R&D Corp is Now Ohio Gravure Technologies Inc.," *Printed Electronics Now,* October 12, 2011.

[76] "WMU's CAPE Plays Key Role in Integrating PE, Printing," *Printed Electronics Now,* December 22, 2009.

[77] *<http://www.wmich.edu/engineer/cape>*.

of packaging with retail environments. Clemson University has a long background in developing and teaching printing and packaging technology and deep relationships in the traditional printing industry and supply chains as well as with end users.[78] It has been applying its expertise in flexographing printing to the printing of conductive inks suitable for RFID and other applications.[79] In 2013, the Sonoco Packaging Institute and PARC won an award from the FlexTech Alliance to develop technology to scale up and print functional devices using a commercial printing press.[80]

KEY COMPANIES

Many blue-chip U.S. companies are engaged in R&D related to flexible electronics themes, and U.S.-based firms are among the most important suppliers of process technology and specialized materials to the emerging global industries. With respect to applications, a number of startup firms have begun to commercialize niche products, but in general major U.S. companies have not yet invested in the commercialization of high-volume, mainstream consumer products based on flexible electronics technology.

Hewlett-Packard (HP)

HP is the world's largest PC maker and produces a broad range of information technology equipment, software, and services. Most of HP's information products incorporate displays, and for a number of years the company's central R&D division, the Information Surfaces Lab, has been pursuing device and process technologies to facilitate the replacement of glass-based displays with plastic displays, including bendable variants that can be produced with R2R processes.[81] HP has collaborated with the ASU Flexible Display Center in the development of e-paper and plastic displays, although in 2010 the company indicated that

[78] In 2008, *Printed Electronics Now* reported that "at Clemson University, in Clemson, SC, USA, researchers from several different departments have been at work developing conductive polymer ink systems, work that has resulted in the filing of US patents. According to Jay Sperry, of the Department of Graphic Communications, the university is in a position to collaborate with advanced materials and engineering technologies to bring package printing and display to a level that involves many projects including organic light emitting displays. This work has attracted allied packaging industries and some large consumer product companies." "A New Industry Shapes the Future of Printing," *Printed Electronics Now,* December 2008.

[79] "Clemson and Industry Backers Focus on Printed Electronics," *Printed Electronics Now,* January 2009; "Clemson's Sonoco Institute Offers Opportunities for PE," *Printed Electronics Now,* September 2010; "FlexTech Alliance Awards Clemson University with Contract to Benchmark Inks and Processes for PE Components," *Printed Electronics Now,* July 22, 2010.

[80] "Clemson University, PARC Receive Award from FlexTech Alliance to Transfer Functional Printing from Laboratory to Commercial Scale," *Printed Electronics Now,* January 7, 2013.

[81] Presentation of Dr. Carl Taussig, Director, HP Information Surfaces Lab, "Plastic Display Research at HP," 2011; "Inexpensive, Unbreakable Displays," *MIT Technology Review,* June 22, 2010.

its principal objective was to make displays that are thinner and lighter, rather than rollable or bendable, because the emerging technologies could not survive more than a few bends.[82] HP pioneered the development of self-aligned imprint lithography (SAIL) with PowerFilm Solar, a process for forming thin-film electronics on flexible substrates in an R2R production line.[83] It is currently working to upgrade SAIL to enable production of large, flexible OLED backplanes.[84]

Universal Display Corporation

Founded in 1994, Universal Display Corporation (UDC) is a world leader in the development of OLED technologies and materials. UDC currently owns or holds exclusive licenses and sublicenses for more than 3,000 patents issued or pending globally. UDC is the world's leading supplier of phosphorescent emitter materials to OLED product manufacturers. Most manufacturers of displays and lighting products that actually or potentially source UDC technologies and materials are based outside the United States, particularly in East Asia. UDC has research partnerships with USC, Princeton, and the University of Michigan, and its R&D efforts are supported by funding from the U.S. Army and the DOE.[85] UDC's OLED screens are used in Samsung's Galaxy series smartphones, and in 2012 UDC derived 68 percent of its consolidated revenues from sales to Samsung Display Co., Ltd.[86] All of UDC's proprietary OLED materials are manufactured by Pittsburgh-based PPG Industries, a major producer of coatings, chemicals, glass, and other specialty materials.[87] In 2013, UDC and PPG opened a new world-class OLED materials production facility in Ohio that will concentrate on UDC's phosphorescent OLED materials ("UniversalPHOLED").[88]

---

[82] "HP Flexible Display Unfurled on Video," *Engadget,* March 20, 2010. In 2010 HP's Chief Technology Officer, Phil McKinney, demonstrated a flexible display printed on mylar film that could be rolled up. McKinney noted that such displays could be printed with no size limit and joked of rooms wall-papered with digital displays. However, McKinney also pointed out kinks in the mylar that damaged the display, so that "first-generation versions of these flexible displays won't be as mobile as fruit-by-the-foot rolls that fit in your pocket." "Phil McKinney on Device Evolution and Flexible Displays," *San Francisco Examiner,* July 17, 2010.

[83] "Volume Production Necessary for Flexible Electronics," *Solid State Technology,* February 26, 2007.

[84] "Upgrading Self-Aligned Imprint Lithography (SAIL) in Preparation for Roll-to-Roll Manufacturing," *Flexible Substrate,* November 2013.

[85] UDC Form 10-K, February 27, 2013, 3–6.

[86] Ibid. 6.

[87] Ibid. 22.

[88] "PPG Expands OLED Production to Support Demand for OLED Display Products," *Flexible Substrate,* December 2013.

DuPont

E.I. du Pont de Nemours and Company was perhaps the preeminent corporate pioneer of systematic R&D for the purpose of generating a continuous stream of new products and processes.[89] Once heavily concentrated on the development of petroleum-based chemical products, it is making a strategic shift into green technologies based on plants and other renewable resources.[90] Various DuPont business units are developing materials and process technologies with applications in flexible electronics.

- ***DuPont Teijin Films***, a joint venture between DuPont and Japan's Teijin Limited, develops PET films for applications in plastic electronics and Teonex, a PEN film for applications in flexible displays involving potential exposure to extreme heat and/or harsh chemicals.[91] DuPont Teijin is participating in the EU's Clean4Yield R&D project headed by the Netherlands' Holst Centre and is reportedly developing a new "Clean-on-Demand" PET polyester film that could improve yields and reduce costs for R2R manufacturing of flexible electronics.[92]
- ***DuPont Microcircuit Materials***, which develops materials for microelectronic, photovoltaic, automotive, and consumer electronic applications, is developing and demonstrating R2R processes such as photonic curing, a thermal processing technique that enables printing of circuits on flexible substrates such as paper and plastic film that are normally vulnerable to high temperatures.[93] In 2011 DuPont Microcircuit Materials announced a collaboration with the Holst Centre in the Netherlands to develop technology for printed metallic structures on flexible substrates for applications in displays, lighting, RFID, biomedicine, and photovoltaics.[94]
- ***DuPont Displays***, which develops materials and processes for displays, has developed solution printing process technology for high-performance OLED displays that seeks to eliminate capital-intensive processing and

[89] Alfred D. Chandler, Jr., *Scale and Scope: The Dynamics of Industrial Capitalism* (Cambridge and London: Harvard University Press, 1990), 181–193.

[90] "DuPont Sees Green in Cleaning Up Its Act," *The Hamilton Spectator,* October 11, 2006; "DuPont Expands Renewable Polymer Portfolio," *Chemical Week*, October 31, 2007; "DuPont Focus on Plant Genetics," *Zecks.com*, March 20, 2010; "Developing Technology to Meet Market Needs is DuPont's Priority," *Supply Chains,* January 21, 2008.

[91] "DuPont Displays Eye-Catching Material Innovations of SID," *Display Central,* June 8, 2012.

[92] "DuPont Teijin Films Develops 'Clean-on-Demand' Film for Roll-to-Roll Flexible Electronics," *Flexible Substrate,* December 2013.

[93] "Advances in Conductive Inks," *Flexible Substrate,* August 2013.

[94] "DuPont Microcircuit Materials Expands Printed Electronics Research with Holst Centre Collaboration," Holst Centre Press Release, February 16, 2011.

reduce operating costs. DuPont has licensed this technology to at least one leading Asian OLED display manufacturer.[95]

At a 2009 National Academies symposium on the future of photovoltaic manufacturing in the United States, Dr. Stephen C. Freilich of DuPont outlined some of the R&D challenges facing DuPont in the area of thin-film photovoltaics on flexible substrates. Noting that his laboratory was developing flexible, durable, waterproof, polymer front-sheets for thin films that, "from a polymer perspective [was] essentially unheard of," he emphasized the vital importance of collaboration with university and/or national laboratory partners:

> We recognize that while we have a strong and vital research facility ourselves, we cannot possibly have all of the best people in the field. So we have to reach out to our industrial partners as well as the national laboratories and universities.[96]

Corning Incorporated

Corning is a maker of glass, ceramics, and related materials, primarily for scientific and specialized industrial applications. It has developed Corning Willow Glass, a thin, flexible glass that has multiple applications in flexible electronics, including touch screens, flexible solar cells, smartphones, tablets, and other types of displays. Using a research grant from the FlexTech Alliance, Corning recently collaborated with other organizations to demonstrate the compatibility of flexible glass with R2R production techniques and the printing of organic photovoltaic devices.[97] The U.S. government's National Renewable Energy Laboratory has reportedly built flexible solar cells out of Willow Glass that are sufficiently durable to eventually replace roofing shingles.[98] In addition to the development of glass with flexible electronics applications, Corning is pursuing other relevant research themes, such as the use of high-performance graphene field effect transistors on flexible substrates.[99]

---

[95] "Role of Solution Processing in the Future of OLED TVs," *Flexible Substrate,* February 2013.

[96] Dr. Stephen C. Freilich, "DuPont Reflections on Photovoltaics," in National Research Council, *The Future of Photovoltaic Manufacturing in the United States* (Washington, DC: The National Academies Press, 2011), 67–68.

[97] "Top Flexible Electronics Developments Win 2013 FLEXI Awards," *Flexible Substrate,* February 2013.

[98] "Corning Willow Glass Used to Make Flexible Solar Power Roofing Shingles, Could Lower the Cost of Solar Power Significantly," *ExtremeTech,* July 3, 2013.

[99] "Printed Graphene Transistors Promise High-Speed Wireless Communications," *OSA Direct,* August 21, 2013. This research is being undertaken in collaboration with 3M and the University of Texas (ibid.).

Palo Alto Research Center (PARC)

PARC, formed in 1970 as a research arm of Xerox Corporation, has a long history of pioneering innovations in the computer field, and PARC first suggested the concept of a flexible display. At the 2010 symposium convened for this project, PARC's Ross Bringans noted the destabilizing, revolutionary potential of flexible electronics but observed that "applications will drive the technology" and that the development of applications was "the big missing piece in the U.S."[100] In 2013 PARC joined with DoE's ARPA-E agency and the Berkeley Lab in a collaboration to develop manufacturing processes for printed lithium ion batteries.[101] PARC has been vocal in articulating the scale of the competitive challenge facing the United States in flexible electronics from abroad. In a 2009 National Academies symposium, PARC's Dr. Bob Street made a presentation on the R&D aspects of flexible electronics, noting in particular the importance of materials science and of U.S. leadership in that field. He observed that in Asia, displays are not only a major industry, but are supported by

> a whole ecosystem of equipment manufacturers, materials suppliers, and a stream of new technology that is being created in universities and research centers around the world. Because the industry in Asia is so big, it draws in new technology . . . from the Palo Alto center, from universities, from start-ups. And they have the manufacturing power. . . . I think this country needs to take this funnel of research and technology that is presently directed toward Asia and move it back into the United States and ensure that we have an industry here that can be the manufacturing focus for the new technology.[102]

Polyera Corporation

Polyera is an Illinois-based maker of functional inks for printed electronics applications. Polyera was spun off from Northwestern University in 2005 to commercialize nanomaterials developed by Professor Tobin Marks, a recipient of the U.S. Medal of Science in 2006, and by Dr. Antonio Facchetti, a Research Professor at Northwestern University, later CEO of Polyera.[103] Polyera engages industrial partners in co-development projects in which it takes the lead in materials design and ink formation for devices with applications in organic transistors, photovoltaics, and circuitry. In 2011 Polyera entered a partnership with Thin Film Electronics ASA of Norway to co-develop gravure-based inks for use in

[100] Ross Bringans, PARC, "Challenges and Opportunities for the Flexible Electronics Industry," September 24, 2010.

[101] "PARC Launches Printed Battery Project," *Flexible Substrate,* August 2013.

[102] Dr. Bob Street, "Flexible Electronics," in National Research Council, *Future of Photovoltaic Manufacturing,* 113–114.

[103] "Honey I Shrunk the Technology … NU Making a Big Difference in the World of Nanotech," *Chicago Sun-Times,* August 13, 2007.

high-throughput printing.[104] Polyera collaborated with the European chemical group Solvay and Belgium's IMEC to achieve a world-record efficiency level for organic solar cells in 2011.[105] Polyera develops new materials at R&D facilities in Illinois, while a subsidiary in Taiwan performs "formulation fine-tuning for large area deposition, device prototype fabrication, and customer support."[106]

Xenon Corporation

Xenon is a 50-year-old company based in Massachusetts that makes pulsed ultraviolet (UV) light equipment for a broad range of industrial uses ranging from curing to decontamination. Xenon flash lamps represent a low-temperature method for sintering functional inks in the manufacture of printed circuits at high speeds on paper, plastic, and other thin-film substrates. Xenon President and CEO Lou Paniro said in a 2013 interview that Xenon's first sales of production equipment for printed electronics had been to buyers in the United States for the manufacture of RFIDs and certain circuits and that

> based on these initial success stories, we feel both North American and European manufacturers can compete with the typically more aggressive Asian suppliers. Asian markets may work towards the high volume production but Western nations have invested heavily in the development of PE and there is bound to be significant growth even if it is on a high cost, low volume, highly custom applications.[107]

Xenon was the driving force behind the formation of the Printed Electronics Test Center Network, a global consortium of universities, manufacturers, and integrators who make their research facilities available to researchers and product developers to test ideas and processes in printed electronics.[108]

Cambrios Technologies Corporation

Cambrios was formed to commercialize technology for transparent conductors developed at MIT by Drs. Angela Belcher and Evelyn Hu. Cambrios has developed ClearOhm, a silver nanowire-based technology with touchscreen

---

[104] Thin Film Electronics' CEO commented that "Polyera's groundbreaking work on n-type organic transistors has paved the way for printed CMOS circuits—more energy efficient logic circuitry with simpler design." "Plastic Memory Firm Partners with Organic Ink Startup," *EETimes Europe,* December 1, 2011.

[105] "IMEC, Polyera and Solvay Set 8.3% Efficiency Record for Organic Solar," *EETimes Europe,* December 15, 2011. Solvay undertook a "strategic minority investment" in Polyera. "Corporate Spotlight Interview: Jordi Lopez Launes, Investment Manager, Solvay Corporate Venturing, Clean Technologies and Sustainable Industries Organization, December 12, 2012.

[106] "Startups in Materials: An Interview with Antonio Facchetti," *Materials Views,* February 5, 2013.

[107] "Interview with Lou Paniro and Saad Ahmed from Zenon Corporation," *Flexible Substrate,* November 2013.

[108] "Xenon Launches Worldwide PE Test Center Network to Help Drive Commercialization of Printed Electronics Industry," *Printed Electronics Now,* May 1, 2012.

applications. A number of companies, including 3M, have adopted ClearOhm because it enables, among other things, the efficient manufacturing of touch-enabled displays with curved or wrapped bezels.[109] ClearOhm won the 2012 Best Product Development Award at the annual meeting convened by the consultancy IDTechEx, Printed Electronics USA, and it was noted that ClearOhm is beginning to replace indium tin oxide (ITO) in touch screens, high-performance OLEDs, and photovoltaics devices.[110]

American Semiconductor

American Semiconductor is a Boise-based provider of semiconductor foundry services. It offers onshore fabrication services for flexible integrated circuits and flexible hybrid systems. It has received funding from the DoD to produce semiconductor devices to DoD specifications.[111] In 2013 American Semiconductor announced the FleX-MCU product line, featuring flexible microcontrollers fabricated through a proprietary silicon-on-polymer process. The company's roadmap envisions eventual development of flexible analog-to-digital converters, radio frequency wireless communications, and nonvolatile memory devices.[112]

MC10

MC10 was established in 2008 to commercialize the research of Professor John Rogers of the University of Illinois at Urbana-Champaign in stretchable electronics. The company is developing flexible electronic medical, consumer, industrial, and defense products designed to conform to the human body. The company has secured several rounds of equity funding and retained former Motorola executive Sanjay Gupta as Vice President for Product Development. Together with Reebok's Advanced Products Group, MC10 developed its first commercial product, a thin mesh skullcap that fits under sports helmets and signals impacts to the head. Other products reportedly in the pipeline include sensors for monitoring heart rate, brain activity, muscle function, body temperature, and hydration.[113]

---

[109] "Cambrios Becomes Established Brand for Touch Sensor Technology," *Aznano.com,* January 7, 2014.

[110] "How an Abalone Shell Turned MIT's Angela Belcher into One of the World's Leading Scientists," *Bostinno,* November 6, 2013. "Printed Electronics USA 2012 Awards Recognize New Developments," *Flexible Substrate,* January 2013; "Cambrios Partners with Novaled to Produce OLED Lighting Tile with New Highly Transparent Electrodes," *Flexible Substrate,* May 2013.

[111] "United States: Defense Money Goes to Idaho Projects," *TendersInfo,* December 18, 2009.

[112] "American Semiconductor Releases Industry's First Physically Flexible Microcontroller," *Flexible Substrate,* August 2013.

[113] "MC10: Reshaping Electronics," *Flexible Substrate,* September 2012; "Stretchable Electronics Maker Raises $19.8M," *Boston Business Journal,* December 27, 2013; "Stretchable Electronics Enable Minimally-Invasive Cardiac Electrophysiological Sensing and Actuations," *Flexible Substrate,* January 2014.

MC10 is reportedly working with Korea's Seoul National University to develop a flexible electronic skin patch with starch gauges to monitor tremors and heating elements to release drugs held inside nanoparticles.[114]

Microlink Devices Inc.

Microlink Devices, based in Illinois, specializes in metalorganic vapor deposition of semiconductor structures for wireless communications applications and the fabrication of solar cells for applications in space and unmanned aerial vehicles, as well as terrestrial uses. Microlink Devices produces solar sheets for UAVs, which are flexible and conform to curved surfaces.[115] These sheets are being used to generate power for drones that substantially enhances their performance and may eventually facilitate flights that last for days or even weeks.[116]

Soligie

Soligie is a Minnesota-based provider of design and manufacturing services for printed electronics, serving the industry in a manner analogous to that of a semiconductor foundry in microelectronics. Soligie was established in 2005 as a wholly owned subsidiary of Taylor Corp., a holding company with about 100 subsidiaries operating in niche markets in printing and media.[117] The company serves clients developing flexible electronics products with medical, security and logistics, and military applications.[118] In 2010 Soligie entered into an agreement with PARC to co-develop printed electronics products for the RFID, smart packaging, medical, and flexible interconnect markets.[119] In 2011 it introduced a sheetfed, flatbed screen printing line to accommodate clients with low-to-medium manufacturing volume needs.[120] In 2013 Soligie received two awards from the FlexTech Alliance to advance printed electronics manufacturing R&D and to establish project demonstrators in 2014.[121]

[114] "Seoul National University Develops Bandage that Senses Tremors and Delivers Drugs," *Flexible Substrate,* April 2014.

[115] "Flexible Materials and Devices for Advanced Power," *Flexible Substrate,* May 2013.

[116] "On a Bright New Wing," *The Economist,* September 7, 2013.

[117] "Soligie (Taylor Corp.) Progresses Co-deposition of Printed Electronics Components," *PIworld,* September 2008.

[118] *<http://soligie.com/content/28/markets-served>.*

[119] "PARC's Partnerships with Thinfilm, Soligie Hold Much Promise for PE," *Printed Electronics Now,* December 2010.

[120] "Soligie Enhances Printed Electronics Manufacturing Capabilities with Sheetfed, Flatbed Screen Printing Line," *Printed Electronics Now,* April 7, 2011.

[121] "Soligie Receives Two R&D Awards from FlexTech Alliance," *Flexible Substrate,* December 2013.

Imprint Energy

Imprint Energy is a startup founded by two Ph.D. students in 2010 to commercialize research developed at the University of California at Berkeley. Imprint manufactures paper-thin, bendable batteries based on zinc rather than lithium through a printing process, with applications in wearable consumer products. Imprint has received seed funding from Dow Chemical and the CIA's venture fund, In-Q-Tel.[122]

3M Company

3M is a Minnesota-based multinational that develops and produces a broad range of industrial and consumer products including electronic materials and circuits and optical films. The company has numerous areas of interest related to flexible and printed electronics technologies, including touchscreens, RFID, and displays. 3M has invested in a number of companies developing flexible electronics technologies, including Germany's Printechnologies GmbH (data memory and battery systems produced on paper), txtr GmbH (e-readers), and motionID technologies AG (RFID).[123] In 2012 3M introduced its FTB3-50 and -125 films—flexible, transparent films that protect electronics from water vapor and oxidation.[124] In 2013, 3M and Cambrios jointly introduced the proprietary 3M Patterned Silver Nanowire Film, comprised of silver nanowire conductive ink micropatterned on a polyester film substrate, for applications in touch sensors.[125]

Eastman Kodak Company

Through most of the 20th century Kodak was one of the leading imaging companies in the world and pioneered many of the technologies that are providing the foundation for the emergence of flexible electronics. Kodak developed and patented the original fluorescent OLED technology in 1987, and substantially all of its considerable OLED intellectual property was sold in 2009 to LG of Korea, which is now challenging Samsung for market leadership in consumer devices incorporating OLED displays.[126] Kodak's R2R printed electronics technology was transferred to Taiwan and has provided the underpinning for Taiwan's emerging flexible electronics industry.[127] Kodak filed for Chapter 11 bankruptcy protection in 2012 and, following divestiture of most of its business operations

[122] "A New Battery That Could Revolutionize Wearables," *Gigaom,* January 8, 2013.

[123] "Investments by 3M, Rusnano Show Interest in PE is Growing," *Printed Electronics Now,* June 2011.

[124] "3M Announces Commercial Availability of FTB3 Barrier Film," *Flexible Substrate,* June 2012.

[125] "Collaboration Produces Flexible Silver Nanowire Film for Touch Screens," *Printed Electronics World,* December 20, 2013.

[126] UDC 10-K, 15, 18.

[127] See Chapter 6, infra.

and intellectual property, emerged from bankruptcy in 2013 as a much smaller technology company specializing in imaging for business.[128] Kodak is currently collaborating with Kingsbury Corporation to produce touchscreen sensors using an R2R process. In 2014 Kodak and Xymox Technologies disclosed that as a result of a joint development effort, Kodak would commercialize a new highly conductive film (HCF) product, KODAK HCF-385 film, for use in capacitative touch sensors found in product packaging, signage, automotive displays, home appliances, machinery, and other applications.[129]

Kateeva

Kateeva, founded in 2008, is a Silicon Valley–based company developing inkjet printer technologies and equipment for printing flexible and large-scale OLED devices. Kateeva's Chief Technology Officer, Steven Van Slyke, was a co-inventor of the original OLED at Kodak in 1987.[130] In 2013, Kateeva introduced YIELDjet, an inkjet printing technology capable of producing flexible and large area OLED displays in high volume.[131] Kateeva began collaborating with South Korea's OLED Plus Co. Ltd., an OLED design and distribution company in 2011, and in January 2014 Kateeva acquired OLED Plus' assets, launching "Kateeva Korea" as a wholly owned subsidiary.[132] Kateeva, which has raised $75 million from venture capital firms and equipment makers Applied Materials and Veeco, believes that it has developed technology that will enable the mass production of much less expensive OLED displays than is currently feasible.[133]

IBM

IBM is a U.S.-based multinational with a long history of pioneering innovation in the computer and microelectronics industries. Although IBM is increasingly concentrating on the provision of information services and systems to large companies, it continues to perform R&D in microelectronics and related fields. In 2012 IBM demonstrated high-performance, state-of-the-art CMOS integrated circuits, including SRAM and ring oscillator devices, on flexible plastic substrates. IBM constructed the devices on silicon and utilized a low-cost, room-temperature

---

128 "New Kodak Comes into Focus After Bankruptcy," *Orlando Sentinel,* September 4, 2013.

129 "Kodak Adds to Transparent Conductive Films Portfolio," *Printed Electronics World,* April 29, 2014.

130 Van Slyke collaborated with Dr. Ching Tang, a professor of chemical engineering at the University of Rochester, to invent the OLED. "Organic Electroluminescent Diodes," *Applied Physics Letters,* 1987.

131 "Kateeva Introduces YIELDjet," *Printed Electronics Now,* November 27, 2013; "Kateeva Launches New Inkjet Print Tool to Bring Flexible OLEDs to Market," *Flexible Substrate,* December 2003.

132 "Kateeva Expands Operations in Korea," *Printed Electronics Now,* January 20, 2014.

133 "Kateeva Proposes Printing to Make Displays," *The Wall Street Journal,* November 20, 2013.

process ("spalling") to flake off the silicon substrate, while the devices were transferred to flexible plastic tape.[134] IBM researchers recently demonstrated the use of a single layer of graphene as an OLED transparent electrode for use on flexible substrates with potential applications in lighting and displays.[135]

General Electric

General Electric is a U.S.-based multinational corporation with widely diversified technology development and manufacturing operations, including fields in which flexible electronics will have applications, such as lighting, photovoltaics, and medical equipment. GE's Electronics Materials Systems (EMS) Advanced Technology Program is pursuing research themes in flexible electronics, including the development of medical sensors, OLEDs, and the R2R manufacture of OLED-based lighting devices.[136] In 2009, GE Global Research concluded an agreement with the firm Power Paper to jointly develop self-powered OLED lighting devices utilizing flexible thin-film batteries.[137]

Plextronics

Plextronics is a Pittsburgh-based developer and manufacturer of conductive polymers and inks for applications in organic electronics. It was spun out of Carnegie Mellon University in 2002 to commercialize technology developed by Dr. Richard McCullough. Its investors include Innovation Works, a state-sponsored development organization that invests in startups in southwestern Pennsylvania; Universal Display; Solvay; Applied Ventures (a subsidiary of Applied Materials); and private venture capital firms.[138] Plextronics won a 3-year contract from the Army in 2007 to develop electronic maps and other devices for soldiers.[139] Between 2002 and 2007 it raised $37 million in equity capital.[140] Plextronics' initial plan was to develop solar ink cells with photovoltaic applications, but as the price of competing silicon-based technologies fell, Plextronics transferred much of its activity to developing inks for OLED displays including flexible displays. But "Plextronics may have been too cutting edge for its own

---

[134] "IBM Demos High-Performance CMOS on Flexible Plastic Substrates," *Flexible Substrate,* November 2012.

[135] "High Performance OLEDs on Graphene Electrode and Thin C-Si TFT for Flexible Display and Lighting," *Flexible Substrate,* January 2014.

[136] *<http://ge.globalresearch.com/technologies/advanced-technologies/electronic-materials/>*.

[137] "Infinity Group Portfolio Company Power Paper and GE Collaborate," *Printed Electronics World,* December 11, 2009.

[138] *<http://www.plextronics.com/about>*; "A New Role for Plastic," *St. Louis Post-Dispatch,* July 12, 2004.

[139] "Plextronics Lands Deal with Army Research Lab," *Pittsburgh Post-Gazette,* June 26, 2007.

[140] "Plextronics Lands $21 Million to Fund Expansion, Marketing," *Pittsburgh Post-Gazette,* August 31, 2007.

good," with the result that the market did not develop at a pace sufficient to enable widespread adoption of Plextronics' OLED technologies. The company filed for Chapter 11 bankruptcy protection in early 2014.[141]

Novacentrix

Novacentrix is an Austin, Texas-based maker of specialty tools and materials for printed electronics applications. The company's proprietary Pulse Forge tools permit the application of thin films and functional inks at high temperatures on substrates without heating the latter, which are often sensitive to temperature extremes.[142] Novacentrix also produces silver and aluminum nanopowders and conductive inks and provides contract functional print manufacturing services.[143] In 2011 Novacentrix entered into a collaboration with DuPont Microcircuit Materials pursuant to which DuPont would use Pulse Forge tools to develop materials and processing technology for printed electronics.[144] In 2013, Novacentrix established a long-term collaboration with the German Muhlbauer group to commercialize new RFID antenna manufacturing technology.[145]

[141] "Local Tech Darling Files for Chapter 11," *Pittsburgh Post-Gazette,* January 26, 2014.

[142] The process "photonic curing" uses pulsed light from a flashlamp to heat films and inks in milliseconds, enabling their imprint on a substrate. K.A. Schroder, S.C. McCool, and W.F. Furlan, "Broadcast Photonic Curing of Metallic Nanoparticle Films," NSTI-Nanotech 2006, ISBN 0-9767985-8-1 Vol. 3, 2006.

[143] *<http://www.novacentrix.com/products/overview>*.

[144] "DuPont MCM Advances PE Development Efforts by Employing Novacentrix Pulse Forge Tools," *Printed Electronics Now,* August 15, 2011.

[145] "Mulhbauer, Novacentrix Enter Long-Term Collaboration for Developing a Flexible and Cost Effective RFID Antenna Printing Technology," *Printed Electronics Now,* April 26, 2013.

8

Findings and Recommendations

FINDINGS

A. ***Flexible electronics describes circuits that can bend and stretch, enabling significant versatility in applications and the prospect of low-cost manufacturing processes. They represent an important technological advance, in terms of their performance characteristics and potential range of applications, ranging from medical care, packaging, lighting and signage, consumer electronics, and alternative energy (especially solar energy).[1] What these technologies have in common is a dependence on efficient manufacturing that currently requires improved technology, processes, tooling, and materials, as well as ongoing research.[2]***

1. **Performance characteristics:** Flexible electronic devices will have performance characteristics that cannot be obtained from conventional rigid technologies. In addition, printed roll-to-roll processes, with their lower unit cost and large scale of production, promise major reductions in the cost of sophisticated electronic devices as well as significant environmental benefits.
2. **Applications:** The applications flowing from flexible and printed electronics are vast and diverse, from industrial to commercial, medical to military. Flexible electronics has the potential to generate new forms of display, sensing, and imaging and may be able to transform production

[1] See Chapter 2 for a review of the advantages and potential applications of flexible electronics.

[2] See Stephen Forrest, "The Path to Ubiquitous and Low-Cost Organic Electronic Appliances on Plastic," *Nature* 428 (April 2004): 911–918.

processes across the electronics industry.[3] Flexible electronics are expected to become a General Purpose Technology.

B. ***Flexible electronics markets are expected to grow rapidly and represent an important competitive opportunity for U.S. firms.***[4]

   1. **Displacing conventional electronics:** Although the global recession and technological hurdles have slowed growth of new industries based on flexible electronics during the past 5 years, over the long term these industries will compete with and partially displace conventional electronics technologies. This dynamic is already being manifested in RFID tags and OLED displays.[5]
   2. **Rapid growth:** Some industry experts predict that the market for global flexible electronics will experience a double-digit growth rate, reaching $250 billion by 2025.[6] Whatever the actual rate, substantial growth seems probable and is reflected in global investments.

C. ***The applications of flexible electronics technology in the military sphere are expected to significantly enhance the mobility, care, and capabilities of combat troops and equipment.***[7] ***Many of these applications also have significant potential in large-scale commercial markets*** (e.g., health care, power generation, environmental protection, and transportation). Conversely, commercial products increasingly lead military applications (e.g. smartphones) and can be adopted and/or adapted for military use, a trend which is likely to be observable in flexible electronics as the industry matures.

   1. **Lightweight and low energy:** Flexible, durable displays and sensors embedded in uniforms and mounted in vehicles and aircraft, hold the potential to convey critical information and communications with significant reductions to weight and with minimal energy consumption. Flexible batteries and conformable photovoltaic devices will improve the mobility of U.S. forces.
   2. **Ambulatory monitoring:** Flexible devices are also expected to provide ambulatory health monitoring of troops in combat and allow for the delivery of medication in emergencies.

---

[3] See also the summary of presentations by Ross Bringans of PARC, Julie Brown of Universal Display Corporation, and Carl Taussig of Hewlett-Packard Company in National Research Council, *Flexible Electronics for Security, Manufacturing, and Growth in the United States, Summary of a Symposium*, rapporteur S. Shivakumar (Washington, DC: The National Academies Press, 2013).

[4] See Chapter 2 for a review of the market growth of flexible electronics.

[5] Brown, "Impact of a Flexible Form Factor for Displays and Lighting," in National Research Council, *Flexible Electronics for Security, Manufacturing, and Growth*.

[6] TMR, Flexible Electronics Market—Global Industry Size, Share, Trends, Analysis and Forecasts 2012–2018 (2013).

[7] See the summaries of presentations by representatives from DARPA and the Army Research Laboratory in National Research Council, *Flexible Electronics for Security, Manufacturing, and Growth*. See also Chapter 4 of this volume for a summary description of military applications.

D. ***The U.S. flexible electronics industry has the potential to build on existing U.S. strengths and public and private investments in research and development.***

1. **U.S. strengths:** U.S. strengths include deep industrial competency in equipment development, microelectronics, printing, advanced materials and chemistry, and nanotechnology; an outstanding system of research universities with relevant curricula; government organizations with long experience working with industry to foster innovation; and arguably the world's best infrastructure and entrepreneurial culture for fostering innovative startups.[8]
2. **Public support:** The U.S. federal government promotes the development of capabilities in flexible electronics through partnerships with NIST and funding from the NSF, DARPA, and DOE.[9] A number of U.S. states have also established research centers for flexible and printed electronics. Support for the development of regional innovation clusters in flexible electronics is provided by the Small Business Administration and the Economic Development Administration of the Department of Commerce. Several federal laboratories also conduct research directly applicable to the field of flexible electronics.[10]
3. **Industry partnerships:** With support from the Army and Air Force Research Laboratories, the U.S. Special Operations Command, and its members, organizations such as the FlexTech Alliance promote collaboration among industry, academia, and research organizations to advance displays and flexible, printed electronics from R&D to commercialization and seek to foster the development of a domestic supply chain for flexible, printed electronics and displays.[11]

E. ***Seeking to capture the global market opportunity in flexible electronics, major U.S. competitors in Europe and East Asia have launched targeted, large-scale programs, with significant government funding to develop these new technologies, refine them, and ultimately manufacture them within their national borders. National and regional investment undertaken by our foreign competitors are significantly larger than comparable***

---

[8] For a review of U.S. competitive strengths and the changing global competitive challenge, see National Research Council, *Rising to the Challenge, U.S. Innovation Policies for a Global Economy*, eds. A. Wolff and C. Wessner (Washington, DC: The National Academies Press, 2012).

[9] For a review of federal and state initiatives, including the role of NIST, NSF, DARPA, and the Army, see National Research Council, *Flexible Electronics for Security, Manufacturing, and Growth in the United States*, Summary of a Symposium, 2013.

[10] See Chapter 7 of this report for a review of the sources of federal support.

[11] See Chapter 7 for a summary description of leading corporate research efforts and the role of the FlexTech Alliance. FlexTech members include startup companies, large companies (such as DuPont and Lockheed Martin), supplier companies (such as E Ink), universities (including Arizona State, Clemson, Georgia Tech, and Kent State), federal research laboratories, and regional nonprofit development organizations.

U.S. investment and more weighted toward later-stage applied research and development.

1. **East Asia:** East Asian firms dominate the manufacture of conventional displays and are using their installed manufacturing base in that field to leverage their entry into flexible displays with consumer applications.[12]
   i. **Technological competency:** Large Asian industrial groups enjoy not only ample financial resources but also deep industrial and technological competencies in relevant fields, such as microelectronics, optoelectronics, materials science, and printing.
   ii. **Government support:** The efforts of these firms are also backed by government programs, which emphasize applied research in industry and government research institutes that collaborate closely and effectively with industry.[13]
2. **European programs at the national and European Union levels**: These programs focus on enhancing collaborations between academia and industry and provide mechanisms to diffuse intellectual property.[14]
   i. **Infrastructure for applied research:** Europe enjoys not only a strong fundamental research base but also a formidable infrastructure for applied research in relevant technology areas, which includes inter alia Germany's Fraunhofer institutes, a new group of research centers in the United Kingdom, and world-class institutes such as IMEC in Belgium, the Holst Centre in the Netherlands, and Finland's VTT.
   ii. **Government support:** The European developmental effort is broad in both a geographic and technological sense. It is supported by successive layers of government at the national, regional, and local levels, and is engaging companies with a long tradition of collaboration to achieve technological objectives.
3. **U.S. investment in flexible electronics, in comparison to our national competitors:** The United States does not have a nationwide infrastructure supporting transitional innovation by small and large companies. Foreign national and supranational programs to support flexible electronics dwarf current U.S. efforts.[15] A point of comparison is the nearly $720 million in funding commitments by the European Union and various European

[12] See Chapters 4 and 6 for a review of review of Korean, Taiwanese, and Japanese initiatives in flexible electronics.

[13] See, for example, Taiwan's Industrial Technology Research Institute. For a detailed description of ITRI, see Appendix A3, "Taiwan's Industrial Technology Research Institute: A Cradle of Future Industries," in National Research Council, *21st Century Manufacturing, The Role of the MEP Program* (Washington DC: The National Academies Press, 2013).

[14] See Chapters 4 and 5 for a review of European initiatives in flexible electronics.

[15] See Chapter 7 for a review of leading U.S. efforts in flexible electronics. For a comparison of known government funding efforts in flexible electronics, see Table 3-1.

national governments for the period 2001 to 2013 versus the U.S. government commitment of $327 million over the same period.

F. ***A robust U.S.-based flexible electronics industry is in the national interest.***[16] ***A vibrant domestic flexible electronics industry could contribute to more on-shore manufacturing and employment.***[17]

1. **Capturing the benefits:** A key challenge is to capitalize on investments in research and development in flexible electronics through scaled-up production of applications in the United States.[18]
2. **Virtuous cycle:** Retaining such production onshore contributes to a virtuous cycle of manufacturing expertise, connected research, and supply chain development. Initiatives such as Ohio's FlexMatters and the Flexible Display Center at Arizona State University and many other smaller but impactful research groups with tight industrial partnerships demonstrate that research originating in U.S. universities can be translated into domestic manufacturing operations and jobs by companies that compete on a global basis.[19]

G. ***Significant U.S. expansion in the market for flexible electronics technologies is not likely to occur in the absence of mechanisms to address investment risks, the sharing of intellectual property, and the diverse technology requirements associated with developing and manufacturing flexible electronics technologies.***[20] ***Linking industry, university, and***

---

[16] Jonathan Epstein, "U.S. Interest, Security, Manufacturing and Growth," in National Research Council, *Flexible Electronics for Security, Manufacturing, and Growth.*

[17] See Chapter 1 of this report for a summary of the manufacturing challenge. See also Sridhar Kota, "The Flexible Electronics Opportunity and Industry Challenges: Perspectives from Industry," in National Research Council, *Flexible Electronics for Security, Manufacturing, and Growth.*

[18] See Richard M. Locke and Rachel L. Wellhausen, eds., *Production in the Innovation Economy* (Cambridge: MIT Press, 2014) and Suzanne Berger, *Making in America* (Cambridge: MIT Press, 2013). These MIT reports argue that the United States needs to actively move research ideas to market if it is to benefit from public and private investments in research and development. They note, moreover, that retaining manufacturing capabilities is important given the high level of innovation and learning involved in manufacturing, which feeds back into research and development. These MIT reports cite an influential 2012 report by the National Research Council that finds that given the high priority and significant resources provided in leading nations on translational research, applications, and manufacturing of high-technology products, U.S. investments in research and development no longer automatically translate into production in the United States and the concomitant benefits in high-value employment and economic growth. See National Research Council, *Rising to the Challenge.* For a review of the experience of the display industry, which though developed in the United States, migrated to East Asia, see Chapter 7 of this report.

[19] For a review of the FlexMatters initiative, see Chapter 7 of this volume. See also presentations in Panel VIII on the FlexMatters initiative summarized in National Research Council, *Building the Ohio Innovation Economy, Summary of a Symposium*, rapporteur C. Wessner (Washington, DC: The National Academies Press, 2013).

[20] Research consortia are one mechanism to address these needs. See the discussion in Chapter 1 on the role and effectiveness of industry consortia. For a review of the potential of industry consortia in flexible electronics, see the summary of a presentation by Malcolm J. Thompson, "A Consortium

government is a proven means to galvanize industry and promote co-operation in applied research and development.

1. **Fostering collaboration:** The development of materials, equipment, and processes cuts across many research areas, which are normally beyond the reach of any single company. Well-designed consortia are one of several mechanisms that can foster the collaboration needed to share costs, equipment, and pool precompetitive research, thereby addressing challenges that are common to all sectors and companies.
2. **Reducing development costs:** Consortia and other collaborative mechanisms can cut development costs and accelerate the development and commercialization of new technologies.[21] These mechanisms can also play an important role in workforce training and developing a domestic supply chain and relevant standards, while providing opportunities for benchmarking equipment from manufacturers and contributing to a robust research base.
3. **Advancing national programs:** Recognizing these advantages, foreign governments have made consortia a significant feature of national innovation programs to support flexible electronics. In Europe, for example, there are at present, more than 40 consortia of various sizes and focus that are advancing a variety of applications in flexible electronics.[22]

H. ***Collaboration among industry, universities, and government offers the best prospect for achieving the critical levels of investment and the acceleration of new technology development that is required to develop a vibrant flexible electronics industry. Consortia can reduce individual company risk, spread costs, facilitate development of industry road-maps and standards, provide a focal point for collaboration with public research organizations, enhance technology diffusion, and provide an early-stage basis for the development of industry supply chains. Best practices in innovation programs for flexible electronics include the following:***[23]

1. Consortia with industry, government, federal laboratory, and university participation.
2. Development of roadmaps for manufacturing technology and pre-commercial and near-to-market applications.

---

in Flexible Electronics for Security, Manufacturing and Growth in the United States," in National Research Council, *Flexible Electronics for Security, Manufacturing, and Growth.*

[21] Collaborative agreements can assume a variety of forms, ranging from informal networks and alliances to joint delivery of projects. For a summary of the academic literature on the effectiveness of industry consortia, see Table 1-1 in this volume. See also the summary of a presentation by Thompson, "Consortium in Flexible Electronics," in National Research Council, *Flexible Electronics for Security, Manufacturing, and Growth.*

[22] For a detailed review of Asian and European programs, see Chapters 4, 5, and 6 of this report.

[23] These best practices are based on a review of U.S. and foreign programs analyzed by the committee and summarized in this report.

3. A focus on supply chain and infrastructure development to build and sustain U.S.-based manufacturing.
4. Sponsorship of collaborative academic and industry R&D for new manufacturing materials, equipment, and processes.
5. Sponsorship for testing and demonstration of manufacturing advances.

RECOMMENDATIONS

While cognizant of resource limitations, our recommendations build upon and reinforce efforts that are already under way at the federal and regional levels to foster a strong domestic capability in flexible electronics manufacturing, which could help secure the jobs and the growth that can be generated by this promising industry. Major advances in recent years in materials performance and circuit designs have created additional opportunities for a viable U.S. flexible electronics manufacturing industry. Consistent with the challenges and opportunities outlined in the findings above, these recommendations seek not only to stimulate innovation, but also to keep the technology and commercialization process in the United States, which is essential for the United States to secure a leading position in flexible electronics. While interconnected, the recommendations below address different facets of this challenge.

A. ***The United States should increase the funding of basic research related to flexible electronics and augment support for university-based consortia to develop prototypes, manufacturing processes, and products in close collaboration with contributing industrial partners.***[24]

B. ***Consortia, bringing together industry, universities, and various levels of government, should be used as a means of fostering precompetitive applied research in flexible electronics.***[25] The federal government can utilize its research funding process to encourage and incentivize the private sector to collaborate with respect to the precompetitive R&D necessary to bring emerging flexible electronics technologies to market.

1. **Mechanisms for cooperation:** In the United States, federal funding for applied research in flexible electronics should support the formation of industry consortia or other cooperative research organizations to facilitate innovation in manufacturing, development of common

[24] See Charles M. Vest, *The American Research University from World War II to World Wide Web*, University of California Press, June 1, 2007, 32. Dr. Vest noted that "in this age of increasingly cooperative innovation, and fast paced change, there are many opportunities to serve through 'relevant' research and development that will complement, not distort, our core academic mission to bring new intellectual challenges to our faculty and students."

[25] See Finding H on the merits of industry consortia and Finding I, which identifies industry-led consortia as a leading best practice in national innovation programs for flexible electronics.

standards, training, and the formation of partnerships with universities and public laboratories.

2. **Need for diversity:** Flexible electronics refers to a family of technologies with many applications. Given the diversity of potential flexible electronics applications, manufacturing processes, and base materials, there is need for multiple pathways to address the challenges faced by industry.
3. **Proximity and location:** The location of individual clusters should be conditioned on the availability of university and industry partners with the ability to provide significant financial resources, as well as the requisite infrastructure, skills, and related research programs in science, engineering, and mathematics.
4. **Built-in evaluation:** The consortia should feature built-in evaluation systems and procedures to enable internal peer review and self-assessment of team and individual performance and progress toward agreed consortium objectives.

C. ***The United States should establish and support a network of user facilities dedicated to flexible electronics.***

1. **Partnership of user facilities:** An integrated networked partnership of user facilities can provide universities greater access to federal fabrication equipment and expertise and provide incubation facilities for small companies. The NSF National Nanofabrication Infrastructure Network is an example of such a nationwide network.[26]
2. **NNMI:** The National Network for Manufacturing Innovation (NNMI) initiative could also include topics relevant to flexible electronics manufacturing. The NNMI could play a positive role in developing and initially supporting institutions, research, and relevant consortia.[27]

---

[26] See remarks by Dr. Ananth Dodabalapur and Dr. Pradeep Fulay in National Research Council, *Flexible Electronics for Security, Manufacturing, and Growth*. "The National Nanotechnology Infrastructure Network (NNIN) is an integrated partnership of fourteen user facilities, supported by NSF, providing unparalleled opportunities for nanoscience and nanotechnology research. The network provides extensive support in nanoscale fabrication, synthesis, characterization, modeling, design, computation, and hands-on training in an open hands-on environment available to all qualified users." Accessed at <*http://www.nnin.org/*>.

[27] The United States lacks a well-established integrated network of centers of applied research comparable to Germany's Fraunhofers. In Germany, the network of Fraunhofer institutes possesses extensive competencies and infrastructure relevant to flexible electronics that are being deployed to facilitate the commercialization of flexible electronics products. For a description of the role of a Fraunhofer Institute in advancing flexible electronics, see the summary of the presentation by Dr. Christian May, "Organic and Flexible Electronics in Germany—A Snapshot," in National Research Council, *Flexible Electronics for Security, Manufacturing, and Growth*. For a detailed description of the Fraunhofer system, see Appendix A2, "Fraunhofer Gesellschaft: The German Model of Applied Research," in National Research Council, *21st Century Manufacturing, The Role of the MEP Program* (Washington, DC: The National Academies Press, 2013).

3. **Focus on processes:** These centers should be designated to develop manufacturing processes with applications in flexible electronics, such as roll-to-roll production and processing of deposition of electronic materials on plastic substrates at low temperatures.[28]

D. ***Where possible, federal efforts to support the growth of competitive flexible electronics industries should leverage state and regional developmental efforts with the objective of establishing co-located local supply chains and capturing the associated cluster synergies.***

1. **Leverage regional initiatives:** Any effort at the federal level to foster an indigenous manufacturing base in flexible electronics should seek to reinforce and build upon state and regional initiatives or engage state cost-sharing to build local intellectual and manufacturing infrastructure.
2. **Focus on small- and medium-sized enterprises (SMEs):** Funding to support prototyping capabilities to help new and incumbent local small- and medium-sized companies launch new products should be given a priority.[29]

E. ***Agency mission needs should help drive demand for flexible electronics technologies, while lowering costs, improving capabilities, and contributing to the development of a skilled workforce.***[30]

1. **Market pull:** The uncertainties surrounding "market pull" for emerging flexible electronics technologies remain a significant hurdle to the establishment of U.S.-based manufacturing capability for these technologies. Government procurement historically has given impetus to the commercialization of promising innovations (as the "lead user"), providing an initial revenue stream and the ability to achieve learning economies that make broader commercialization feasible.[31]

---

[28] As the NNMI is envisioned, each institute will "integrate capabilities and facilities required to address cross-cutting manufacturing challenges that have the potential to retain or expand industrial production in the U.S. on an economically rational basis." Testimony of Patrick D. Gallagher, Under Secretary of Commerce for Standards and Technology, before the House Subcommittee on Technology and Innovation, Committee on Science, Space and Technology, May 31, 2012, 6.

[29] Interview with Dr. Miko Cakmak, University of Akron, Akron, Ohio, June 4, 2013. Dr. Cakmak led the committee on a tour of the University of Akron National Polymer Innovation Center. The Department of Commerce Economic Development Agency has been providing support to "Proof of Concept Centers" to promote commercialization of green technologies. These centers feature prototyping design and development facilities and equipment. See National Research Council, *Best Practices in State and Regional Innovation Initiatives,* 95–98.

[30] See Finding C on military and civilian needs in flexible electronics.

[31] Stuart W. Leslie, "The Biggest Angel of All: The Military and the Making of Silicon Valley," in *Understanding Silicon Valley: The Anatomy of an Entrepreneurial Region,* ed. Martin Kenney (Stanford: Stanford University Press, 2000); Vernon W. Ruttan, "Will Government Programs Spur the Next Breakthrough?" *Issues in Science and Technology,* Winter, 2006; Jakob Edler and Luke Georghiou, "Public Procurement and Innovation—Resurrecting the Demand Side," *Research Policy,* 2007.

2. **Procurement opportunities:** Promising examples of flexible electronics applications in the United States are arising from the U.S. Army's need for flexible displays for use by U.S. troops.[32]
3. **Consumer applications:** The vast range of potential applications for flexible electronics technologies suggests that public procurement opportunities exist not only for the military, but also for products in fields such as health care, power generation, environmental protection, and transportation. While ultimately the competitiveness of the U.S. industry will depend on the innovative capabilities of U.S.-based firms, cooperation across industry, universities, and national laboratories can play a key role in the development of new flexible electronic technologies and applications. Such progress is unlikely to be achieved to the same degree by individual U.S. firms.

[32] See the summary description in Chapter 7 of this report of the Flexible Display Center. The FDC is a research and innovation center focusing on the commercialization of flexible displays formed through a collaboration between the U.S. Army and Arizona State University.

Bibliography

Acetech. 2011. Organic Electronics in Germany: Assessment and Recommendations for Further Development. Munich.

Advanced Display Research Center (ADRC) at Kyung Hee University. 2012. *Korea Times*. June 14.

Advances and Improvements with Manufacturing. 2012. *Flexible Substrate*. March.

Advances in Conductive Inks. 2013. *Flexible Substrate.* August.

Air Force Award to FlexTech Alliance Will Accelerate Development of New Nano-Bio Devices. 2013. *Red Orbit.* January 30.

Air Force Award to FlexTech Alliance Will Accelerate Development of New Nano-Bio Devices. 2013. *Flexible Substrate.* February.

Air Force Research Lab Awards Optemec Contract to Advance Fully Printed Transistor Technology. 2013. *Printed Electronics Now.* December 12.

AIST Develops Large Area (A4) OTFT Backplane. 2010. *OSA Direct Newsletter.* February 22.

AIST Develops Droplet Simulation Technology for Printed Electronics. 2014. *Flexible Substrate.* March.

AIST Nanotube Research Center Fabricates and Develops Ultrafine Processing Technology for CNT Wafer; One Step Closer to Mass Production and Practical Application of MEMS Devices. 2009. *Tokyo Semiconductor FDP World.* January 5.

AIXTRON Celebrates 30th Anniversary. 2013. *Printed Electronics Now.* December 5.

AIXTRON Plays Key Roles in Two-Dimensional Nanomaterial Projects. 2013. *Printed Electronics Now.* October 31.

AIXTRON and Manz AG Agree to Strategic Cooperation for OLED Manufacturing. 2014. *Flexible Substrate.* January.

Alam, Muhammad A. and Satish Kumar. 2014. Definition of Flexible Electronics, ed. Bharat Bhushan. *Encyclopedia of Nanotechnology.* Dordrecht: Springer Reference.

Alexander. D. Jane A., Acting Director, DARPA. Testimony before the Subcommittee on Emerging Threats and Capabilities, Committee on Armed Services, United States Senate. June 5, 2001.

Amazon Buys Liquavista from Samsung. *Flexible Substrate.* May 2013.

American Semiconductor Releases Industry's First Physically Flexible Microcontroller. 2013. *Flexible Substrate.* August.

Ames Laboratory Scientists Develop Iridium-Free Organic Light-Emitting Diodes. 2012. *Materials Views*. December 4.

AMOLED Production: Entering a New Era? 2013. *Information Display.* March/April.

Antibodies in Ink Dots Could Revolutionize Health Tests. 2011. *Western Mail.* August 22.

Anti-Korean Spleen in Taiwan. 2010. *Asia Sentinel.* November 25.

Application of Printed Electronics Technologies to OLED Lighting. 2013. *Flexible Substrate.* May.

Applied Research: The Fraunhofer Method. 1987. *Industries et Techniques.* October 20. JPRS-ELS-88-006.

Arizona State Researchers. Army, Build World's Largest Flexible Display. 2012. *Phoenix Examiner.* June 12.

Army Awards ASU $43.7 Million for Flexible Display Development. 2004. Arizona State University Press Release. February 10.

Army-Backed Flexible Display Effort: A Symbol of Public-Private Partnership. 2006. *IEEE Computer Society.* July-September.

Army Invests $50 Million in Flexible Displays. 2009. *CNET News.* January 29.

Army Partners With Arizona State University to Develop Flexible Displays. 2009. *Military and Aerospace Electronics.* January.

Army Researchers Creating Electronic Devices with Flexible Screens. 2013. *The Huntsville Times.* March 10.

Army Targets Flexible Electronics, Displays. 2013. *Micromanufacturing.* November-December.

Ascent Solar to Build New Manufacturing Plant in China. 2013. *Printed Electronics World.* July 16.

Ascent Solar Signs Definitive Agreement to Build New Manufacturing Plant in Suqian of Jiangsu Province, China. 2014. *Printed Electronics Now.* January 3.

Asia Major. 2012. *Money Marketing.* March 1.

Asia's New Model Company. 2011. *The Economist.* October 1.

ASU and University of Hong Kong to Work Together on Advanced Flexible Electronics. 2011. *Flexible Substrate.* December.

ASU Center. 2013. PARC Produce World's Largest Flexible X-Ray Detector. *Targeted News Service.* March 19.

ASU's Flexible Display Center Hits Milestone with New Technology. 2013. *Phoenix Business Journal.* July 3.

Atkinson, R. D. 2011. *Explaining Anemic U.S. Job Growth: The Role of Faltering U.S. Competitiveness.* Washington, DC: The Information Technology and Innovation Foundation. December.

Atkinson, Richard C. and William A. Blanpied. 2008. Research Universities: Core of the U.S. Science and Technology Systems. *Technology and Society* 30(2008):30–48.

AUO Aims for 10-fold Increase in E-Paper Shipments This Year. 2011. *Taiwan Economic News.* February 10.

AUO Former Executives Are Suspected of Selling OLED Technology to Chinese Companies. 2012. *OLED-Info.com.* October 16.

AUO Latest to Delay OLED Mass Production. 2012. *Plastic Electronics.* October 31.

AUO Signs OLED Cooperation Agreement with Idemitsu Kosan. 2012. *The Emitter: Emerging Displays Technology Monthly Report.* February.

AUO, Sony Co-develop World's 1st 4K OLED TV Panel. 2013. *Central News Agency.* January 9.

Autonomy Fails to Dent Cambridge's Tech Status. 2012. *Financial Times.* November 26.

Autor, D., D. Dorn, and G. Hansen. 2011. The China Syndrome: The Local Labor Market Effects of Import Competition. Cambridge, MA: MIT Working Paper.

Baily, Martin Neil and Barry P. Bosworth. US Manufacturing: Understanding Its Past and Its Potential Future. *The Journal of Economic Perspectives* 28(1):3-25.

Barker, Megan M. 2011. Manufacturing Employment Hard Hit During the 2007-09 Recession. *Monthly Labor Review.* April, p. 33.

BASF. 2009. Three Top European Universities Team Up on Functional Materials Research. *Printed Electronics Now.* July 28.

BASF. 2012. Heidelberg and TU Darmstadt Collaboration Shows Promise. *Printed Electronics Now.* August.

BASF. 2012. Max Planck Institute for Polymer Research Inaugurate Joint Research Laboratory for Graphene. *Printed Electronics Now.* September 24.

BASF. 2013. Top American Universities to Research New Functional Materials. *Printed Electronics Now.* March 14.

BASF to Set Up Electronic Materials R&D Center Asia Pacific at Sungkyunkwan University in South Korea. 2013. *Printed Electronics World.* November 7.

Batteries Get Flexible. 2013. *Chemical and Engineering News.* May 6.

Battershell, Caroll. 2013. Bringing Department of Energy Innovations to Market. National Research Council. *Future of Photovoltaic Manufacturing.* p. 155.

Bayer to Establish Two New R&D Centers in Taiwan. 2008. *Central News Agency.* October 9.

Bendable, Rollable, Profitable. 2006. *Taiwan Review.* July 1.

Beneq Wins Tekes Commercialization Breakthrough Award. 2013. *Printed Electronics Now.* December 19.

Berger, S. 2013. A Preview of the Production in the Innovation Economy Report. Cambridge: MIT Press.

Berger, S. 2013. *Making in America: From Innovation to Market.* Cambridge and London: MIT Press.

Bloch, Erich. 2003. Seizing U.S. Research Strength. *Issues in Science and Technology.* Summer.

BMBF Project NEMO on New OLED Materials under Merck Leadership Successfully Concluded. 2012. *Flexible Substrate.* September.

Boardman, C., and D. Gray. 2010. The New Science and Engineering Management: Cooperative Research Centers as Government Policies, Industry Strategies, and Organizations. *Journal of Technology Transfer.* February, p. 447.

BOE Loss-Making Operation Expected to Continue in 2011. 2012. *SimoCast.* February 3.

BOE Technology Group. 2011. BOE Technology Annual Report.

BOE Technology Group Teams with Applied Materials to Deliver Leading Edge Display Technologies for Next-Generation Televisions and Mobile Displays. 2013. *Printed Electronics Now.* July.

BOE Tech Group to Sell 80% of Energy Unit. 2011. *SinoCast.* November 15.

BOE Tech Sells Shares to Raise 1,96 Yuan. 2006. *South China Morning Post.* October 13.

BOE Tech. Signs Financial Cooperation Agreement with CDB. 2013. *SinoCast.* January 8.

BOE Technology Wins Government Aids. 2008. *SinoCast.* November 11.

BOE Transfers Mining Shares for OLED Investment. 2011. *WantChinaTimes.com.* October 29.

Borrus, Michael, James E. Millstein, and John Zysman. 1983. Trade and Development in the Semiconductor Industry: Japanese Challenge and American Response, in *American Industry in International Competition: Government Policies and Corporate Strategies,* eds. John Zysman and Laura Tyson (Ithaca and London: Cornell University Press), 154-156.

Bozeman, Barry. 2000. Technology Transfer and Public Policy: A Review of Research and Theory. *Research Policy.* 29(4-5):627-655.

Branstetter, Lee G., and Mariko Sakakibara. 2000. When Do Research Consortia Work Well and Why? Evidence from Japanese Panel Data. NBER Working Paper No. 7972. October.

Breakthrough for Flexible Solar Cells. 2011. *Smarthouse.* June 23.

Bright Future for UK's First Printed Lighting Panel. 2010. *Electronic Product Design & Test.* August 16.

Bringans, Ross. 2010. PARC, Challenges and Opportunities for the Flexible Electronics Industry. September 24.

Browning, Larry D. and Judy C. Shetler. 2000. *Sematech: Saving the U.S. Semiconductor Industry.* College Station, TX: Texas A&M University Press. pp. 86-99.

Building Activities Start on High Tech Campus Eindhoven. 2013. Solliance News Release. April 15.

Buntz, Brian. 2012. TEDMED Update: Why the Future of Medical Electronics Is Flexible. *Electronic Components.* April 12.

Bureau of Economic Analysis. 2011. *Survey of Current Business 2006-2009.* Washington, DC: U.S. Department of Commerce.

Burroughes, J., Bradley, D., Brown, A., Marks, R., Mackay, K., Friend, R., Burns, P., Holmes, A. 1990. Light-Emitting Diodes Based on Conjugated Polymers. *Nature.* 347:539-541.

Cai, M., Ye, Z., Xiao, T., Liu, R., Chen, Y., Mayer, R.W., Biswas, R., Ho, K., Shinar, R., Shinar, J. 2012. Extremely Efficient Indium-Tin-Oxide Free Green Phosphorescent Organic Light-Emitting Diodes. *Advanced Materials.* 24(31):9.

Cakmak, Miko. 2013. The Role of Regional Academic Institutions in Flexible Electronics Development. In National Research Council. *Building the Ohio Innovation Economy: Summary of a Symposium.* Washington, DC: The National Academies Press.

Call for More Investment in China's High Tech Sector. 2006. *Asia Pulse.* February 2.

Caloghirou, Yannis, Stavros Ioannides, Nicholas S. Vonortas. 2003. Research Joint Ventures. *Journal of Economic Surveys.* 17(4):541-570.

Cambridge at Heart of £45M Manufacturing Push. 2013. *Business Weekly.* February 27.

Cambridge University Opening Graphene Centre to Take Material to the Next Level. 2013. *Engadget.* January 25.

Cambrios Becomes Established Brand for Touch Sensor Technology. 2014. *Aznano.com.* January 7.

Cambrios Partners with Novaled to Produce OLED Lighting Tile with New Highly Transparent Electrodes. 2013. *Flexible Substrate.* May.

Can China Break Samsung's AMOLED Grasp? *China Focus.* April 8, 2014.

Carayannis, Elias C. and James Gover. The Sematech-Sandia National Laboratories Partnership: A Case Study. *Technovation.* 2002. 586. Berger. *Making in America.* pp. 57-58.

Carayannis and Gover. The Sematech-Sandia National Laboratories Partnership. 586. A Look at Printed Electronics: Printed Electronics Now Interview with Dr. Bernard Kippelen. 2011. *Printed Electronics Now.* July.

Centre for Process Innovation Ltd. (CPI) (TIC 28). 2011. Written evidence submitted to the House of Commons Science and Technology Committee. *Technology and Innovation Centres: Second Report of Session 2010-11.* February 9. p.103.

Centre for Process Innovation. 2012. CPI submitted a report to Parliament in 2012 indicating that of the total £17 million CPI had received from the ERDF since its inception, nearly all of it (£14 million) was used on projects linked to PETEC. Written Evidence Submitted by the Centre for Process Innovation, House of Commons Communities and Local Government Committee, April.

Centre for Process Innovation. 2013. CPI Works with SMEs to Develop High Performing Organic Thin Film Transistor. May 18.

CES Reveals Korea's IT Firms Are Lagging Their Rivals. 2010. *Chosun Ilbo Online.* January 12.

Chaebol Owners Tighten Their Grip. 2013. *JoongAng Daily Online.* May 31.

Chaebol's Economic Concentration at Record High. 2012. *The Korea Herald Online.* February 6.

Chakrabortty, Aditya. 2010. David Cameron's Talk About Reviving British Industry is Nonsense. *The Guardian.* November 1.

Chandler, Alfred D. 1990. *Scale and Scope: The Dynamics of Industrial Capitalism.* Cambridge and London: Harvard University Press. pp.181-193.

Cheil Industries Celebrates Shipment of Amoled Displays. 2013. *Joong Ang Daily Online.* April 3.

Cheil Is Into Electronics, Chemicals, Fashion. 2012. *Joong Ang Daily Online.* March 28.

Chemical Sciences and Society Summit. 2012. *Organic Electronics for a Better Tomorrow: Innovation, Accessibility, Sustainability.* White Paper. September.

Chen, Janglin. 2013. ITRI, National Research Council, Flexible Electronics for Security, Manufacturing and Growth in the United States: Report of a Symposium.

Chen, Janglin. 2010. Flexible Electronics Development in Taiwan. September 24.

Chen-Liang-gee in Bendable, Rollable, Profitable. 2006. *Taiwan Review.* July 1.

Chi-Lin. 2011. Announces Formation of Pervasive Displays Inc. Yahoo Finance. May 3.

Chen, Tain-Jy and Ying-Hau Ku. 2014. Indigenous Innovation vs. Ten-Long- Huan-Niao: Policy Conflicts in the Development of China's Flat Panel Industry. *Industrial and Corporate Change.* Oxford University Press.

China Aims to Be Major Player in AMOLED Industry. 2011. *Yonhap.* November 18.

China Eclipsing Colorado Makers—Cheap Loans, Subsidies Giving Chinese Companies a Star Role. 2013. *Denver Post.* August 18.

China Establishes Industrial Institute of Printed Electronics. 2013. *Flexible Substrate.* November.

China Firms Invest in AMOLED Lives Largely. 2014. *SinoCast.* March 12.

China Handset Vendors to Rely on Local AMOLED Panel Production from 2016. 2014. *Digitimes.* February 26.

China Imports Hi-Tech through Hydis Purchase: BOE Says an Acquisition Is the Only Way to Get Ahold of the Core Technology. 2003. *South China Morning Post.* February 20.

China Looking to Break Into Small Screen AMOLED Market 2015. 2014. *Plastic Electronics.* May 7.

China Panels Makers Expand Amid Fiercer Global Competition. 2011. *SinoCast.* December 12.

China Successfully Developed Its First Flexible Display. 2013. *ChinaAbout.net.* August 12.

China Sun Panels Face EU Levies. 2013. *Financial Times.* May 6.

China to Account for More than 70% of Flat Panel Display Equipment Spending. 2014. *CTimes.* February 7.

China's BOE Technology Group May Turn Around, Buoyed by Government Subsidy. 2010. *China Business News.* January 20.

China's BOE Technology Signs Agreement With Export-Import Bank of China. 2013. *China Business News.* June 20.

China's PE, RFID Markets Show Strong Growth. 2009. *Printed Electronics Now.* November.

China's Photovoltaic Industry: Exporting on the Cheap. 2009. *Energy Tribune.* September 3.

China's Rural Home Appliance Subsidy Boosts Sales. 2012. *China Daily.* November 12.

Chinese State-Owned Enterprises to Use Overseas Funds. 2011. *Asia Pulse.* January 9.

Choi-Ji-wan. 2012. NH Research Center, cited in Cheil Is Into Electronics, Chemicals, Fashion. *Joong Ang Daily Online.* March 28.

Chosun Ilbo Online. "5 Korean Universities Rank Among Global Top 20". 2010. September 8.

Chung, Chang-Chien and An- Chen. 2013. Top Industrial Research Body Praised for Helping Japanese Firms. *Central News Agency.* February 5.

Ciba and VTT Technical Research Centre of Finland Expand Collaboration in Printed Electronics. 2008. *Nanotechwire.* June 8.

Claussen, Mike, Bela Green, Martin Walkinshaw, and Simon Ogier. 2011. Introduction to the PETEC Printed Electronics Centre and Technology Challenges. CS Mantech Conference. Palm Springs, California. May 16-19.

Clayton, Byron. 2010. A State's Initiative: Advancing Flexible Electronics in Northeast Ohio. NorTech.

Clemson and Industry Backers Focus on Printed Electronics. 2009. *Printed Electronics Now.* January.

Clemson's Sonoco Institute Offers Opportunities for PE. 2010. *Printed Electronics Now.* September.

Cloudy Skies Remain for Taiwan's PV Industry. 2012. *Taiwan Economic News.* November 30.

COLAE. 2014. *Organic Electronics for Saxony OES.* Available online at http://www.colae.eu/companies/organic-electronics-saxony-oes-2/.

Colaneri, N. 2013. Manufacturing Flexible Displays: The Challenges of Handling Plastic. *Solid State Technology.* May 1.

Collaboration Produces Flexible Silver Nanowire Film for Touch Screens. 2013. *Printed Electronics World.* December 20.

Colorful Skin: Kent Displays Nears Full Production of Cholestoric LCD Products. 2009. *Printed Electronics Now.* June.

Combs, Kathryn L. and Albert N. 2003. Link, Innovation Policy in Search of an Economic Foundation: The Case of Research Partnerships in the United States. *Technology Analysis and Strategic Management* 15:177-187.

COMEDD Now a New Research Institute of the Fraunhofer. 2012. *Printed Electronics World.* July 12.

Comin, Diego, Gunnar Trumbull, and Kerry Yang. 2012. *Fraunhofer: Innovation in Germany.* Harvard Business School Monograph, 9-711-022. January 6.

Comments of Choi-Yang-hee, Dean of Graduate School of Convergence Science and Technology at Seoul National University, in Korea Needs Change for Convergence. 2009. *JoongAng Daily Online.* May 23.

Commission Proposes New European Industrial Strategy for Electronics—Better Targeted Support to Mobilise Euros 10 Billion in New Private Investments. 2013. *Targeted News Service.* May 25.

Company Profile: Semiconductor Start-up. 2011. *Chemistry World.* August.

Comparison of Laser Ablation of Transparent Conductive Materials on Flexible and Rigid Substrates. 2013. *Flexible Substrate.* February.

Congressional Budget Office. 2007. *Federal Support for Research and Development.*

Congressional Research Service. 2003. The Semiconductor Industry and South Korea's Hynix Corporation. May 27, CR5-12.

Congressional Research Service. 2014. The Obama Administration's Proposal to Establish a National Network for Manufacturing Innovation. January 29. 4.

Converting State-of-the-Art Research Results Into Significant Business. 2009. *Printed Electronics World.* June 11.

COPE Breakthrough May Simplify PE Manufacturing. 2012. *Printed Electronics Now.* April.

Corning Willow Glass Used to Make Flexible Solar Power Roofing Shingles, Could Lower the Cost of Solar Power Significantly. 2013. *ExtremeTech.* July 3.

Corporate Spotlight Interview: Jordi Lopez Launes, Investment Manager, Solvay Corporate Venturing. *Tech Connect News*, December, 2012.

Council for Science and Technology. 2007. Strategic Decision Making for Technology Policy: A Report by the Council for Science and Technology.

CPI & DuPont Teijin Films Work Together to Develop a Barrier Film as Plastic Substrate for OLED Lighting and Photovoltaic Devices. May 28, 2013.

CPI Manufacturers Flexible OLED Lighting Demonstrators. 2014. *Flexible Substrate.* March.

CPI Presents Ultra-Flexible OTFT Device Array Suitable for Foldable AMOLED Displays. 2014. *Flexible Substrate.* January.

CPI Produces Large Area Small Molecule and Polymer OLED Lighting Demonstrators. 2013. *OSA Direct.* June 3.

Dachs, B., Borowiecki, M., Kinkel, S., Schmall, T.C. 2012. The Offshoring of Production Activities in European Manufacturing. Austrian Institute of Technology and Fraunhofer Institute for Systems and Innovation Research ISI. December.

Daetwyler R&D Corp is Now Ohio Gravure Technologies Inc. 2011. *Printed Electronics Now.* October 12.

Delta. 2010. Bridgestone to Develop New E-Paper Products. *Central News Agency.* November 9.

Delta Electronics Blazes a Trail in Color Qr-Lpd Electronic Paper Applications at Ces 2011. 2011. *Central News Agency.* January 5.

Defense Science Board. 1986. *Task Force on Semiconductor Dependency.* November 30.

Defense Science Board. 2005. *Task Force on High Performance Microchip Supply.* (41):23-24.

Demiddeleer, Leopold. 2009. Plextronics Closes $14 Million Financing Round. *Printed Electronics Now.* August 20.

Department for Business Enterprise and Regulatory Reform. 2008/09. *Plastic Electronics in the UK: A Guide to UK Capability.* 9.

Department for Business Innovation & Skills (BIS). 2009. *Plastic Electronics: A UK Strategy for Success— Realizing the UK Potential.*

Department of Commerce. 1986. Europe Report, Science and Technology. JPRS-EST-86-022. September 11.

Department of Commerce. 2011. New Obama Administration Initiates to Spin Economic Growth in 20 Regions across the Country. September 22.

Developing Technology to Meet Market Needs is DuPont's Priority. 2008. *Supply Chains.* January 21.

Development of 5.2-Inch full-HD OLED Display. 2013. *Printed Electronics World.* May 27.

Development of an Inexpensive Method for Mass Synthesis of Single-Walled Carbon Nanotubes. 2007. *Nanowerk*. February 7.

Display Technology is Emerging in China. 2013. *Korea IT Times*. June 26.

DisplaySearch: AMOLED Manufacturing Capacity Forecast to Nearly Triple in 2012. 2013. *Green Street Journal.* July 29.

DNP Produces Printed Transparent Conductive Film to Replace ITO. 2009. *Flexible Substrate.* June.

Dodabalapur, Ananth. 2009. Panel on Flexible Hybrid Electronics. WTEC Workshop: International Assessment of R&D in Flexible Hybrid Electronics. June 30.

Dodabalapur, A. 2010. European Research and Development in Hybrid Flexible Electronics. Baltimore: World Technology Evaluation Center. July.

Dodabalapur, Ananth and Pradeep Fulay. 2013. Flexible Electronics for Security, Manufacturing, and Growth. Washington: National Research Council.

Do the Chaebol Choke Off Innovation? 2009. *Bloomberg Business Week.* December 3.

Droplet Simulation Technology for Printed Electronics. 2014. *Printed Electronics World.* January 21.

DuPont Displays Eye-Catching Material Innovations of SID. 2012. *Display Central.* June 8.

DuPont Expands Renewable Polymer Portfolio. 2007. *Chemical Week*. October 31.

DuPont Focus on Plant Genetics. 2010. *Zecks.com*. March 20.

DuPont MCM Advances PE Development Efforts by Employing Novacentrix Pulse Forge Tools. 2011. *Printed Electronics Now.* August 15.

DuPont Microcircuit Materials Expands Printed Electronics Research with Holst Centre Collaboration. 201. Holst Centre Press Release. February 16.

DuPont Sees Green in Cleaning Up Its Act. 2006. *The Hamilton Spectator.* October 11.

DuPont Teijin Films Develops 'Clean-on-Demand' Film for Roll-to-Roll Flexible Electronics. 2013. *Flexible Substrate.* December.

Dyson, James. 2010. *Ingenious Britain: Making the UK the Leading High Tech Exporter in Europe*. March.

E-fabric Spools Bring Bullet-Proof Watches, Paper-Thin Batteries. 2011. *The Christian Science Monitor.* April 15.

E Ink Acquires SiPix, May Dominate ePaper Universe. 2012. *Engadget.* August 4.

E Ink Files Patent Infringement Lawsuit Against Trekstor. 2013. *Flexible Substrate.* January.

E Ink. Sony Debut New Flexible Screen Technology. 2013. *Publisher's Weekly*. May 23.

E-Paper Makers Set Sight on Flexible Material Technology. 2010. *Taiwan Economic News.* December 7.

The Economist. 2011. Stretchable Electronics: A Shapely Future for Circuits. March 10.

Edler, Jakob and Luke Georghiou. 2007. Public Procurement and Innovation—Resurrecting the Demand Side. *Research Policy*. 36(7):949-963.

Edler, Jakob and Stefan Kühlmann. 2008. Coordination within Fragmentation: Governance in Knowledge Policy in the German Federal System. *Science and Public Policy*. p. 267.

Efficient OLED from Kyushu University Gets Rid of Heavy Metals. 2013. *Flexible Substrate.* February.

ENAB-SPOLED Project Targets Solution-Processed OLEDs for Lighting. 2013. *Flexible Substrate.* October.

Enfcell Brings Expertise to the Printed Battery Market. 2009. *Printed Electronics Now.* September.

Engineers Create New Transparent Electrodes for Display Screens. 2008. *Yonhap.* March 20.

Enhancing CAPE-abilities in Photonics and Electronics. 2008. *Research Horizons*. Autumn.

Entrepreneurs Lack Serious Support. 2009. *Japan Times.* November 27.

Equipment Industry is Crucial to Growth. 2009. *The Korea Herald Online.* August 17.

ERA-NET Plus to Hold European Competition for Collaborative R&D Funding for Plastic Electronics. 2011. *Flexible Substrate.* September.

Ergun, Arif S., Yaralioglu, G., Khuri-Yakub, B.T. 2003. Capacitative Micromachined Ultrasonic Transducers; Theory and Technology. Journal of Aerospace Engineering. April.

ERSO to Set UP Flexible Electronics Alliance in July. *Taiwan Economic News.* May 19, 2005.

Ettlie, John E., 2006. *Managing Innovation: New Technology, New Products, and New Services in a Global Economy.* Burlington, MA. Elsevier-Butterworth-Heinemann.

EU Cash on the Horizon for Innovators. 2014. *The Independent.* January 24.

EU Commission. 2008. Towards World-Class Clusters in the European Union: Implementing the Broad-Based Innovation Strategy. Communication from the Commission to the Council, COM 652 final/2, p. 2.

EU Commission. 2012. Capacities Part 6: Support for the Coherent Development of Research Programmes, Work Programme. C (2012) 4526 as of July 9.

EU Completes Project on Printed Transparent Electrodes for Flexible OLEDs and Solar Cells. 2014. *Flexible Substrate.* March.

The EU Launches a €18 Million Organic Large Area Electronics Funding Competition. 2011. *OLED-Info.com.* September 25.

EU Project Aims to Improve Organic Photovoltaics. 2011. *The Engineer.* October 31.

EU Project to Take OLEDs from Lab to Fab. 2013. *Hearst Electronic Products.* January 20.

EU R2R Program Creates Flexible OLED Desk Lamp. 2013. *Flexible Substrate.* October.

EURIS. 2012. Inventory of Good Practices on Open Innovation, Holst Centre, Eindhoven, The Netherlands.

European Collaboration Towards Efficient, Low-Cost Tandem Organic Solar Cells. 2011. *Printed Electronics Now.* October 28.

European Project Develops Flexible OLED Lighting Production Process. 2013. *Plastic Electronics.* February 8.

European Project Reaches Milestone Bidirectional Communication for Thin-Film RFIDs. 2012. *Flexible Substrate.* March.

A European Silicon Valley in Flanders—Europe's Largest Independent R&D Center Comes into Being Near Brussels. 1986. *Frankfurter Allgemeinc/Blick Durch die Wirtschaft.* June 12.

Evonik and AU Optronics Corp. Conclude Strategic Partnership. 2010. *Printed Electronics Now.* December 2.

Evonik, Holst Centre Partner to Extend Thin Film Electronics. 2013. *Printed Electronics Now.* February 12.

Evonik Offers FLEXOSKIN Barrier Film for Protection of Flexible PV. 2012. *Printed Electronics Now.* January 11.

Executive Office of the President. 2009. *A Framework for Revitalizing American Manufacturing.* Washington, DC: Executive Office of the President.

Expose Chips with X-Rays. Frankfurter Allgmeine Zeitung. October 24, 1985, JPRS-WST-86-018.

Ezell, S., and R. Atkinson. 2011. *International Benchmarking of Countries' Policies and Programs Supporting SME Manufacturers.* Washington, DC: Information Technology and Innovation Foundation. September.

Fair, Dedicated, Realistic, Coordinated, Innovative. 1990. *Keji Ribao.* March 25.

Fehrenbacher, K. 2013. A New Battery That Could Revolutionize Wearables. Gigaom.

Feller, I. A. Glasmeier, and M. Mark. 1996. Issues and Perspectives on Evaluating Manufacturing Modernization Programs. *Research Policy.* 25(2):309-319.

Feng, Dr. Jiyun. 2007. *Commercialization of Nanotechnology Products in Hong Kong: Strategy and Progress.* Presentation at International Nanotechnology Business Summit. Tokyo, February 21.

Fihn, M. 2013. Curved, Then Flex … *Flexible Substrate.* HSBC Global Research. *Flexible Display: Fantastic Plastic— A Shape Shifting Game Changer.* April, 4-5.

Financiers Focus on Flexible Displays. 2013. *Flexible Substrate.* November.

First Plastic Chip Plant to Open in Germany. 2007. *Chosun Ilbo.* January 4.

Flamm, Kenneth and Qifei Wang. 2003. Sematech Revisited; Assessing Consortium Impacts on Semiconductor R&D, in National Research Council. *Securing the Future: Regional and National Programs to Support the Semiconductor Industry.* Washington, DC: The National Academies Press.

Flanders Technology International: Flemish Chips on a World Level. *De Standaard.* May 8, 1987. JPS-ELS-87-004. August 12, 1987, p. 23.

Flex is Next. 2011. *San Jose Mercury News.* January 17.

Flexible AMOLEDs: At the Tipping Point? 2014. *The Emitter: Emerging Display Technologies.* March 10.

Flexible CIGS Solar Cell Submodule Achieves Record Energy Conversion of 15.9%. 2010. *Nanowerk.* April 5.

Flexible Circuits Unfold. 2009. *AFCEA International.* June.

The Flexible Display Center and AUO Enter Strategic Partnership to Accelerate Flexible AMOLED Development. 2010. *Nanowerk.* November 16.

Flexible Display Market to Reach $67.7 Billion by 202, Says HIS. 2013. *Flexible Substrate.* November.

Flexible Displays, Lighting and Solar Cells: Developing Technologies and Building the Supply Chain. 2012. *Flexible Substrate.* September.

Flexible Electrodes Using Single-Wall Carbon Nanotubes Patterned as Paper Substrate. 2012. *Flexible Substrate.* March.

Flexible Electronics Is the Winner. 2011. *Printed Electronics World.* March 2.

Flexible Electronics Pilot Lab and New Electronic Paper in Taiwan. 2007. *Printed Electronics World.* June 8.

Flexible Films for Photovoltaics. 2011. *Printed Electronics World.* May 31.

Flexible Graphene Memristors. 2010. *Printed Electronics World.* December 9.

Flexible Materials and Devices for Advanced Power. 2013. *Flexible Substrate.* May.

Flexible OLED Lighting Reaches High Energy Efficiency Thanks to Shared Research Effort. 2012. *Printed Electronics Now.* March 8.

Flexible PV: Three-Fold Growth in the Next 5 Years. 2013. *Flexible Substrate.* February.

Flexible Transparent Conductive Films and Its Application in Organic Solar Cells. 2013. *Flexible Substrate.* January.

FlexTech Alliance. 2012. *NanoMarkets, Special Report for the FlexTech Alliance: Transparent Conductor Market.* December.

FlexTech Alliance. 2013. *2014 Targeted and Open Solicitation Request for Proposals (RFP)*, Issued December 10.

FlexTech Alliance Announces 2013 RFP for Funding Opportunities. 2013. *Flexible Substrate.* January.

FlexTech Alliance Announces Completion of Project 154 for a Flexible Medical Triage Bandage. 2013. *Flexible Substrate.* October.

FlexTech Alliance Awards Clemson University with Contract to Benchmark Inks and Processes for PE Components. 2010. *Printed Electronics Now.* July 22.

FlexTech Alliance Introduces Wearable and Disposable Electronics User Groups. 2014. *Flexible Substrate.* January.

Flex-o-Fab: Pilot Scale Hybrid Roll-to-Roll/Sheet-to-Sheet Manufacturing Chain for Flexible OLEDs. 2012. EU Seventh Framework. November 25.

For Scientists in Korea, Careers of Stress, Insecurity. 2012. *JoongAng Daily Online.* September 24.

Foreign Professors Are Rare Breed. 2008. *The Korea Herald.* October 6.

Forrest, Stephen. 2004. The Path to Ubiquitous and Low-Cost Organic Electronic Appliances on Plastic. *Nature.* 428(4):911-918.

Forsythe, Eric W. Flexible Communications. 2012. *Army AL&T Magazine.* July-September.

The FP7-ICT Coordination Action OPERA and EM Commission DG INFSO Unit G5 Photonics. 2011. *An Overview of OLAE Innovation Clusters and Competence Centres*. September. p. 4.

Fraunhofer Develops Color-Changing Glove That Warns of Toxic Substances. 2013. *Flexible Substrate.* May.

Fraunhofer EMFT Becomes an Independent Institute. 2010. *DeviceMed.* July 15.

Fraunhofer Fights to Keep Solar Industry in Europe. 2013. *Renewables International.* January 11.

Fraunhofer IAP Opens Pilot Line for Organic Electronics. 2013. *Flexible Substrate.* February.

Fraunhofer ISE. 2013. Fraunhofer ISE Acquires Patent Portfolio from Schott Solar—Expertise on Crystalline Silicon Photovoltaics Remains in Europe. Press Release. January 10.

Fraunhofer Polymer Surfaces Alliances POLO. 2013. Transparent High Barrier Film for Organic Electronics: Roll-to-Roll Pilot Production.

Fraunhofer Printed Thermoelectric Generators Could Capture Energy from Waste Heat. 2013. *Flexible Substrate.* May.

Fuji Electric Systems Fully Utilizing Lightweight, Flexible Characteristics of Film Type Solar Cells to Expand Into Various Applications. 2007. *Tokyo Semiconductor FDP World.* June 26.

Fuji Electric Systems to Lift Capacity for Bendable Solar Cells. 2007. *Asia Pulse.* October 3.

Fujifilm and IMEC Develop New Technology for Organic Semiconductors Enabling Submicron Patterns. 2013. *Flexible Substrate.* October.

The Future of Stand-Alone E-Readers. 2012. *Flexible Substrate.* November.

Gallagher, Patrick. 2010. Strengthening the Connections: Research Innovation and Economic Growth. University of Pittsburgh. October 7.

Gallagher, Patrick D. 2012. Assembling the Facts: Examining the Proposed National Network for Manufacturing Innovation. May 31.

Gallagher, Patrick D., Under Secretary of Commerce for Standards and Technology. Testimony before the House Subcommittee on Technology and Innovation, Committee on Science, Space and Technology. May 31, 2012. p. 6.

Gamota, Dan. 2013. iNEMI Flexible Electronics Roadmap: From Concept to Product in Flexible Electronics for Security, Manufacturing, and Growth in the United States. Washington: National Research Council.

Gargini, Paolo. 2011. "Oral History of Paolo Gargini." Mountain View, California. Computer History Museum. July 27.

Gasman, Lawrence. Notable Developments in Flexible Glass. 2013. *Flexible Substrate.* January.

Gasorian, Lawrence. Notable Developments in Flexible Glass. 2013. *Flexible Substrate.* January.

GBP 3m Boost to Region's Chemics. 2004. *Manchester Evening News.* July 13.

Geisler, Eliezer. Technology Transfer: Toward Mapping the Field, a Review, and Research Directions. 1993. *The Journal of Technology Transfer*. (18, Summer–Fall): 88-93.

General Accountability Office. 2009. T*echnology Transfer: Clearer Priorities and Greater Use of Innovative Approaches Could Increase the Effectiveness of Technology Transfer at Department of Energy Laboratories.* June.

Geoghegan-Quinn, Marie. 2014. *Launch of Horizon 2020 in Greece.* January 10.

Georgia Institute of Technology. 2012. Stable Electrodes for Improving Printed Electronics. April 19. Available online at http://www.cope.gatech.edu/news/release.php?nid=124901.

Georgia Tech and Solvay Announce $3M Deal for OLED Research. 2006. *Eurekalert.* April 26.

Gergel-Hackett, N., Hamadani, B, Dunlap, B., Suehle, J., Richter, C., Hacker, C., Gundlach, D. 2009. A Flexible Solution-Processed Memristor. IEEE Electron Device Letters. 30(7):706-708.

German National Academy of Science and Engineering. 2011. *Organic Electronics in Germany: Assessment and Recommendations for Further Development.*

German OLED Project 'Exceeds Expectations.' 2013. *Flexible Substrate.* February.

German Researchers Synthesize World's First Supramolecular Thiophene Nanosheets. 2013. *Flexible Substrate.* May.

German Trade Office, Taipei. 2011. Bayer Material Science, Bayer Taiwan, and Chi-in Technology Launch a New Era in Taiwanese Opto-electronics Industry. March 11.

Germany–The New Mini-Superpower. 2011. *Christian Science Monitor.* January 30.

Glasmeier, A., K. Fuellhart, I. Feller, and M. Mark. 1998. The Relevance of Firm-Learning Theories to the Design and Evaluation of Manufacturing Modernization Programs. *Economic Development Quarterly.* 12(2):107-124.

The Glass Is Half Full … 2013. *Flexible Substrate.* May.

Goddard, John, Douglas Robertson, and Paul Vallance. 2012. Universities, Technology and Innovation Centres and Regional Development: The Case of the North-East of England. *Cambridge Journal of Economics.* p. 9.

Government Expected to Step in to End Display Feud Between Samsung and LG Display. 2013. *OSA Direct.* January 21.

Govt. to Invest $1 Trillion Won to Develop Premier Materials. 2009. *Arirang News.* November 16.

Graphene Appointed EU Future Emerging Technology Flagship. Press Release. January 28, 2013.

The Graphene Revolution. 2013. *The Daily Telegraph.* June 8.

Great Potential Seen in Soft Batteries. 2007. *Helsingin Sanomat.* March 20.

Green, Albert. 2013. Roll to Roll Manufacturing of Flexible Displays. in National Research Council. *Building the Ohio Innovation Economy: Summary of a Symposium.* Washington, DC: The National Academies Press.

Greene, Sarah M., Gene Hart, and Edward H. Wagner. Measuring and Improving Performance in Multicenter Research Consortia. 2005. *JNCI Monographs.* 35:26–32.

Greenough, J.J. 1854. English Views of American Manufacturers: Special Report of Mr. Joseph Whitworth. *Greenough's American Polytechnic Journal of Science, Mechanic Arts and Engineering.* (3-4)344-345.

Gye-man Kang and Hyung-deuk Jean, Samsung, LG. Hyundai Motor Join Hands for Next Generation Flexible Display Development. 2010. *MK English News Online.* July 5.

Haggard, Stephan, Wonhyuk Lim, and Euysung Kim. *Economic Crisis and Corporate Restructuring in Korea: Reforming the Chaebol.* 2010. Cambridge: Cambridge University Press.

Hagmann, Dieter. Reshoring Is an Issue for Europe Too. 2013. *Finanz & Wirtschaft.* October 23.

Hamburg, Christian. 1999. *Structure and Dynamic of the German Mittelstand* Heidelberg and New York: Physica-Verlag. 58–59.

Handful of Top Universities to Get State Research Funds. 2008. *The Korea Times Online.* December 1.

Hannah, A. 2010. The Global View of Printed Electronics and What It Could Mean to the U.S. September 24.

Hanwha Chemical Begins Mass Carbon Nanotube Production. 2008. *Yonhap.* December 22.

Hanwha Chemical Buys Into U.S. Graphene Maker. 2011. *Yonhap.* January 19.

Harding, Rebecca. 2000. Resilience in German Technology Policy: Innovation Through Institutional Symbiotic Tension. Industry and Innovation. December, p. 228.

Harrell, Sam. 2011. Oral History of Sam Harrell, November 16, 2004, Austin, Texas, Computer History Museum.

Harrop, Peter. Printed Electronics —Europe is Different. 2011. *Printed Electronics World.* March 9.

Harrup, Peter. Stretchable Electronics and Electrics for Electric Vehicles. 2013. *Flexible Substrate.* January.

Harvard Uses 3D Printing on Lithium-ion Technology. 2013. *Flexible Substrate.* December.

Hehui Optoelectronics President Visits Jobsite of SBC-MCC Group in Shanghai. 2013. *MCCChina.* Press Release. July 8.

Heilman, Sebastian, Lea Shih, and Andreas Hofem. National Planning and Local Technology Zones: Experimental Governance in China's Torch Programme. 2013. *The China Quarterly.* December.

Heliatek Achieves 12% Organic Solar Cell Efficiency. 2013. *Optics.org.* January 23.

Heliatek Opens Groundbreaking Production Facility for the Manufacture of Organic Solar Films. 2012. *M2Presswire.* March 13.

Heliatek's Organic Tandem solar Cell Verified at 10.7% Record Conversion Efficiency. 2012. *PV Tech.* April 27.

Helper, S. 2008. *Renewing U.S. Manufacturing: Promoting a High-Road Strategy.* Washington. DC: Economic Policy Institute.

Helper, S., and H. Wial. 2010. *Strengthening American Manufacturing: A New Federal Approach.* Washington, DC: The Brookings Institution.

Helper, S., T. Krueger, and H. Wial. 2012. *Why Does Manufacturing Matter? Which Manufacturing Matters? A Policy Framework.* Washington, DC: The Brookings Institution. February.

Helpman, Elhanan and Manual Trajtenberg. 1994. A Time to Sow and a Time to Reap: Growth Based on General Purpose Technologies. NBER Working Paper No. 4854. September.

Heuberger, Anton. 1988. Applied Research: The Fraunhofer Method. *Industries et Techniques*. February 23, JPRS-ELS-88-006.

Hey Japan. 2013. What's With Your Startup Culture. *Fastcolabs*. September 6.

High Performance OLEDs on Graphene Electrode and Thin C-Si TFT for Flexible Display and Lighting. 2014. *Flexible Substrate*. January.

Highlights from Printed Electronics Asia 2012. 2012. *Printed Electronics World*. October 10.

Highly Efficient Inkjet Printed Organic Photovoltaic Cells. 2010. *Japanese Journal of Applied Physics*.

HM Treasury. 2010. *Spending Review 2010* Cm 7941. October.

Holst Centre. 2010. Combines IMEC, TNO, and Industry's Capabilities. *Printed Electronics Now*. October.

Holst Centre. 2011. Philips Research Celebrate Successful Five-Year Collaboration with New Contract. *Printed Electronics Now*. October 24.

Holst Centre. 2012. *Executive Report 2012*. WTEC. *European Flexible Electronics*. Available online at http://www.hightechcampus.com.

Holst Centre. Public and Industrial Agreements Enable Further Growth of Holst Centre. 2012. News Release. April 5.

Holst Centre and IMEC Launch Research Program on Flexible OLED Displays. 2012. *IMEC News*. January 17.

Hon Hai Vows to Beat Samsung Display in 3-5 Years. 2012. *Chosun Ilbo Online*. June 20.

Hon Hai to Invest $1.2 Billion in Hitachi LCD Unit. 2010. *Reuters*. December 27.

Honey I Shrunk the Technology … NU Making a Big Difference in the World of Nanotech. 2007. *Chicago Sun-Times*. August 13.

Hong Kong Innovation and Technology Commission. 2011. LegGo Members Visit Hong Kong Science Park. Press Release. July 18.

Hopper, Darrel G. 2002. Defense Display Strategy and Roadmaps: Cockpit Displays IX: Displays for Defense Applications. Proceedings of SPIE. (4712):1-7.

Hopper, Darrel G. 2004. Display Science and Technology for Defense and Security: Organic Light-Emitting Materials and Devices VII. Proceedings of SPIE.

Hopper, Darrel G. and Daniel D. Desjardins. 2002. Analysis of the Defense Display Market. Information Display. Air Force Research Laboratory. 18(4-5):40-44.

House of Commons. 2009. Innovation, Universities, Science and Skills Committee. Engineering: Turning Ideas into Reality. March 18, p. 45.

House of Commons. 2009. Universities, Science and Skills Committee. Engineering: Turning Ideas Into Reality. March 18, p. 33.

House of Commons. 2011. Committee on Science and Technology. Second Report: Technology and Innovation Centres. February 9, p. 41.

House of Commons. 2012. Science and Technology Select Committee, Bridging the Valley of Death: Improving Commercialization of Research.

House of Commons. 2012. Written Evidence Submitted by the Centre for Process Innovation, Communities and Local Government Committee, April.

How an Abalone Shell Turned MIT's Angela Belcher into One of the World's Leading Scientists. 2013. *Bostinno*. November 6.

How Bad Universities Fill Their Seats. 2011. *JoongAng Daily Online*. June 17.

How KAIST and POSTECH Reforms Can Save the Country. 2007. *Chosun Ilbo*. October 8.

How the German Economy Became a Model. 2012. *Spiegel Online*. March 21.

Howell, Thomas, R. Alan Wm. Wolff, Brent L. Bartlett, and R. Michael Gadbaw. 1992. Conflict Among Nations: Trade Policies in the 1990s. Westview Press. p. 312-317.

HP Flexible Display Unfurled on Video. 2010. *Engadget*. March 20.

HSBC Global Research. *Flexible Display: Fantastic Plastic—A Shape-Shifting Game Changer*. April 2013, p. 22.

HTC Learns Lesson from Samsung Display Row. 2013. *Central News Agency*. May 28.

Hu, Mei Chih. 2014. Technological Catching-up in East Asia. International Economic Development: Leading Issues and Challenges, eds. Fu Lai Tony Yu, Wai Kee Yuen, and Dinna S. Kwan (New York and Abington: Rutledge, p. 86.

Hutcheson, Dan. 1993. Sematech Sets Model for Industry Consortium. *Fort Worth Star-Telegram.* November 10.

IBM Demos High-Performance CMOS on Flexible Plastic Substrates. 2012. *Flexible Substrate.* November.

ID Cards Drive RFID Demand in China. 2005. *Infowars.com.* August 30.

IDTechEx. Intense Study of Printed Electronics in Japan. 2013. September 16.

IDTechEx Names Top Printed Electronics Developments of 2013. 2013. *Flexible Substrate.* December.

IDTechEx, Printed, Flexible and Organic Electronics Sees 15.3% CAGR. 2013. *Printed Electronics World.* May 15.

IDTechEx. 2011. *Printed, Organic & Flexible Electronics: Forecasts, Players & Opportunities 2011-2021.* p. 30.

IDTechEx Releases Printed and Thin Film Transistors and Memory Report. 2013. *Flexible Substrate.* February.

IDTechEx Visits South Korea. 2012. *Printed Electronics World.* April 18.

Ijima, Sujio. Helical Micsotubules of Graphitic Carbon. 1991. *Nature.* 354: 56-58. November 7.

Illegal Trash Trade: E-waste Smuggling Contaminates Developing Countries. 2013. *Amman News.* August 6.

IMEC. Polyera and Solvay Set 8.3% Efficiency Record for Organic Solar. 2011. *EETimes Europe.* December 15.

IMEC and Holst Centre Unveil Fully Organic Images. 2013. *Flexible Substrate.* August.

IMEC Launches Project X10D for Development of OPV Cells with Increased Efficiency at Lower Costs. 2011. *PVTech.* October 28.

IMEC Paves Way for Intelligent Item-Level RFID Tagging to Replace Bar Codes. 2012. *Printed Electronics Now.* December 13.

IMEC. Solvay Announce World Record Efficiency for OPV Module. 2012. *Printed Electronics Now.* September 24.

In Taiwan, Lamenting a Lost Lead. 2013. *The New York Times.* May 12.

Industrial Institute of Printed Electronics. 2013. *Printed Electronics World.* October 29.

Industrial Technology Research Institute. 2011. Flexible Electronics Packing a Punch. March 31.

Industry Leaders Discuss Growth and Future of PE. 2010. *Printed Electronics Now.* June.

Inexpensive, Unbreakable Displays. 2010. *MIT Technology Review.* June 22.

Infinity Group Portfolio Company Power Paper and GE Collaborate. 2009. *Printed Electronics World.* December 11.

Inktec and Thin Film Electronics Win Prestigious Award. 2009. *Printed Electronics World.* April 14.

Innovate UK (formerly Technology Strategy Board). 2014. The TSB is funding ROBOLED (Robust OLED), a 30-month effort running from 2012 to 2015, with £493,792. Press Release. Available online at http://www.oled-info.com/roboled.

Innovation Can Solve China's Energy Predicaments. 2006. *Keji Ribao.* July 13.

Institute for Nano Quantum Information Electronics, University of Tokyo NanoQuine. 2014. *Nano-Quine: Quantum Information Electronics.* Available online at http://www.nanoquine.iis.u-tokyo.ac.jp/NQ-brochure.pdf.

Intel. Plastic Logic and Queen's University Reveal Bendable 'PaperTab' Display. 2013. *Flexible Substrate.* February.

Interdisciplinary Studies. 2011. *Korea Times Online.* July 8.

Interim Regulations on Supervision and Management of State-Owned Assets of Enterprises. Decree of the State Council of the People's Republic of China. No. 378, Premier Wen Jiabao. May 27, 2003.

InterNano. 2012. Accelerating Progress for Advanced Manufacturing. June 28.

Int'l Research Team Confirm Ripple Structure of Graphenes. 2011. *Yonhap.* June 30.

Interview with David Way, Director of Knowledge Exchange and Special Projects, Technology Strategy Board, London, June 12, 2012.

Interview with Dr. Janglin Chen, Director ITRI Display Technology Center, Hsinchu, Taiwan. February 14, 2012.

Interview with Dr. Miko Cakmak, University of Akron, Akron, Ohio. June 4, 2013.

Interview with DTC Director Janglin Chen, Hsinchu, Taiwan. February 14, 2012.

Interview with Fraunhofer Institute for Process Engineering and Packaging IVV, Friesing, Germany, June 13, 2012.

Interview with Lou Paniro and Saad Ahmed from Zenon Corporation. 2013. *Flexible Substrate,* November.

Interview with Tokao Someya. 2010. *Discovery News,* February 5.

Interview with Valtronic, Solon, Ohio, June 5, 2013.

Investments, Upheaval in the PV World. 2011. *Printed Electronics Now.* September.

Investments by 3M, Rusnano Show Interest in PE is Growing. 2011. *Printed Electronics Now.* June.

IRICO to Raise Funds to Establish Generation 4.5 AMOLED Production Lines. 2011. *GG-LED.* April 7.

iSuppli Sees China Targeting AMOLED Sector. 2011. *Optics.org*. November 17.

ITRI. 2011. Flexible Electronics Packing a Punch. March 31.

ITRI. Flexible Electronics Packing a Punch. Q&A: Dr. Janglin Chen of ITRI (Part I). 2013. *The Emitter.* July 3.

ITRI and Corning Collaborate on Roll-to-Roll Flexible Glass for Touch. 2012. *The Emitter: Emerging Display Technologies.* November 30.

ITRI and Kaneka Unveil New Flexible Display Technology. 2013. *ITRI Today.* April 3.

ITRI and Komori Announce New Fine-Line Printing Technology. 2013. *Flexible Substrate.* November.

ITRI Lifts Lid on Advanced Printing Tech for Lower Priced Touch Panels. 2013. *The China Post Online.* August 27.

ITRI Wins Further Prestigious R&D Awards. 2011. *Central News Agency*. June 23.

ITRI Wins Prominent Display Technology Award from Industry Group. 2011. *Central News Agency.* May 19.

It's Japan vs. Korea with Launch of Samsung Display, Japan Display. 2012. *PC Mag.* April 2.

Japan Backs OLED Display Research. 2008. *CBC News.* July 10.

Japan Display Begins Production. OLED from 2013. 2012. *FlatpanelsHD.* April 3.

Japan Display Develops Video-Rate E-paper. 2012. *Plastic Electronics.* November 6.

Japan Display Joins Kyushu University's OLED Project. 2012. *OLED.si.* August 1.

Japan Losing Out in Latest Applications for Carbon Nanotubes. 2013. *Nikkei Asian Review.* November 20.

Japan Takes a Gamble on Displays. 2012. *The Wall Street Journal.* April 10.

Japan Universities. Firms to Draw Up E-Printing Standard. 2010. *Jiji.* November 8.

Japan's Electronic Giants Struggle to Reboot. 2012. *The Wall Street Journal.* November 1.

Japan's Soken Chemical Develops New Display Material for E-Paper. 2008. *Asia Pulse.* October 22.

Japanese Electronics Sector Teetering on the Brink of Collapse. 2012. *Dong-A Ilbo Online.* May 30.

Japanese Build World's Biggest Carbon Nanotube Plant. 2011. *Jiji.* February 14.

Jobs Threatened as Plastic Logic Puts Down E-Reader. 2012. *Plastics News.* May 17.

Johnson, Bobie. Why Plastic Logic Failed—Despite the E-Book Boom. 2012. *GigaOM*. May 17.

Johnson, Kristina. 2011. The U.S. Department of Energy's Perspective In National Research Council. *Future of Photovoltaics Manufacturing*. Washington, DC: The National Academies Press.

Joint Research Project of BASF. 2012. Heidelberg and TU Darmstadt for Printed Electronics Centers the Next Phase. *Chemical Business NewsBase.* August 2.

Joseph von Fraunhofer and Max Planck Can Feel Satisfied. 1991. *Handelsblatt.* August 9, JPRS-EST-91-015.

KAIST Case Should Send a Warning to Elite Science Education in Korea. 2011. *Kyunghyang Shinmun Online.* April 12.

KAIST President Wins Int'l Mechanical Engineering Prize. 2009. *Chosun Ilbo Online.* July 3.

KAIST Tops Patent List Among Domestic Universities. 2011. *Chosun Ilbo Online*. July 5.

Kao, Y.L. 2009. Academic-Enterprise Cooperation Pact Inked to Promote Green Energy. Central News Agency. September 24.

Kateeva Expands Operations in Korea. 2014. *Printed Electronics Now.* January 20.

Kateeva Introduces YIELDjet. 2013. *Printed Electronics Now.* November 27.

Kateeva Launches New Inkjet Print Tool to Bring Flexible OLEDs to Market. 2003. *Flexible Substrate.* December.

Kateeva Proposes Printing to Make Displays. 2013. *The Wall Street Journal.* November 20.

Kelly, John E. 2011. Collaboration for Success in Semiconductors. In National Research Council *Future of Photovoltaic Manufacturing*. Washington, DC: The National Academies Press.

Kelly, M. 1997. From Mission to Commercial Orientation: Perils and Possibilities for Federal Industrial Technology Policy. *Economic Development Quarterly* 11(4):313-328. November.

Kenny, J. 2008. A New Industry Shapes the Future of Printing, *Printed Electronics Now*. December.

Kim Chang-woo. March 2010. A Technological Hermit Kingdom. JoongAng Daily Online.

King, Zella. 2007. AIM Research, Plastic Electronics: Putting the UK at the Forefront of a New Technological Revolution.

Kingsley, G., and H. Klein. 1998. Interfirm Collaboration as a Modernization Strategy: A Survey of Case Studies. *The Journal of Technology Transfer* 23(1):65-74. Spring.

Kodak Adds to Transparent Conductive Films Portfolio. 2014. *Printed Electronics World.* April 29.

Konarka Technologies Files for Chapter 7 Bankruptcy Protection. 2012. *Printed Electronics Now.* May 31.

Konica Minolta and Konarka Join Forces to Develop Organic Thin Film Photovoltaics. 2010. *Nanowerk.* March 4.

Kopola, Harri. 2012. Printed Intelligence Technology at VTT ERATO Seminar sponsored by JST ERATO Someya Bio-Harmonized Electronics Project. October 31.

Korea Hones in on Growth Engines. 2011. *JoongAng Daily Online.* March 22.

Korea, Israel Collaborate on Nanoparticles. 2011. *Korea Herald Online.* April 10.

Korea Kumho Petrochemical to Build Carbon Nanotube Plant. 2009. *Yonhap.* September 30.

Korea Needs Change for Convergence. 2009. *JoonAng Daily Online.* May 23.

Korea Ranks 11th in Papers for Scholarly Journals. 2010. *Chosun Ilbo Online.* December 15.

Korea S'pore Institutes in R&D Tie-up. 2009. *The Business Times.* March 20.

Korean Company Arrives to Compete in Extremely Tough Sector. 2013. *Dong-A Ilbo Online.* March 8.

Korean Engineering Faculties Jump in World Ranking. 2013. *Chosun Ilbo Online.* May 8.

Korean Researchers Demonstrate New Transparent, Stretchable Electrodes. 2013. *Printed Electronics World.* May 31.

Korean Students are Badly Served by Their Universities. 2011. *Chosun Ilbo Online.* June 8.

Korean Universities Hire More Foreign Faculty. 2010. *The Korea Herald Online.* August 2.

Korean, U.S. Firms to Set up Solar Panel Joint Venture. 2008. *Yonhap.* May 29.

Korea's POSTECH Explore Nanotechnology Frontiers. 2010. *Nano Science Works.* January 20.

Kulicke, C. Scott. 2004. "Oral History of C. Scott Kulicke." Computer History Museum. September 22.

Kunshan Hisense Electronics Introduces New Cost Competitive and Versatile Materials Printer. 2012. *Printed Electronics Now.* September 10.

Kunshan OLED Industry Makes a Breakthrough. 2013. *Kunshan Today.* August 5.

Kyushu University Grows Thin Films of Germanium. 2013. *Flexible Substrate.* October.

Lachance, Molly. 2008. Scientist Demonstrates Record Speed for Bendable Electronics, Advanced Materials, Manufacturing, Testing Information Analysis Center. U.S. Air Force. January 30.

Large Area Flexible OLEDs: Solution Processing and R2R Technologies. 2013. *Flexible Substrate.* May.

Large OLEDs Top Lists in China. 2011. *Solid State Technology.* April 7.

Last Word: Missed Opportunities. 2013. *Flexible Substrate.* August.

Lawrence Berkeley Engineers Use Nanotubes to Create Light-Activated 'Curtains.' 2014. *Flexible Substrate.* January.

Lawrence Berkeley National Laboratory Develops Micro-Activator That Flexes Under Laser Light. 2013. *Flexible Substrate.* January.

Lee, Changhee. 2010. Flexible and Printed Electronics—A Korean Initiative. Presentation to the The National Academies. September 24.

Lee, H., 2013. A Giant Leap to Commercialization of Polymer Solar Cell, *Printed Electronics World.* May 10.

Leslie, Stewart W. 2002. The Biggest 'Angel' of Them All: The Military and the Making of Silicon Valley, in *Understanding Silicon Valley: the Anatomy of an Entrepreneurial Region*, ed. Martin Kenney. Stanford: Stanford University Press.

LG and Samsung Face Delay with OLED Televion. 2012. *Plastic Electronics.* October 18.

LG Convinces South Korea to Fund Development of Flexible 60-Inch OLED Displays. 2012. *Extremetech.* July 17.

LG Display Chosen by Korean Government to Lead OLED Project. *Printed Electronics World.* 2012. August 8.

LG Throws the OLED Competition a 55-inch Curve. 2013. *JoongAng Daily Online*. April 30.

LG Unveils World's First Flexible OLED TV. 2014. *Flexible Substrate.* January.

Link, Albert N. and John T. Scott. 2001. Public/Private Partnerships: Stimulating Competition in a Dynamic Market. *International Journal of Industrial Organization.* (5):763-794.

Link, Albert N. and Donald S. Siegel. 2007. *Innovation, Entrepreneurship, and Technological Change.* New York: Oxford University Press.

Lisbon Council and Allianz Dresdner Economic Research. 2008. The Lisbon Review 2008.

Little Sign That Abe Can Shake Up Japan's Inbound FDI. 2013. *Reuters.* May 20.

Local Tech Darling Files for Chapter 11. 2014. *Pittsburgh Post-Gazette.* January 26.

Locke, Richard M. and Rachel L. Wellhausen. 2014. *Production in the Innovation Economy.* Cambridge: MIT Press.

Lucrative State-Run Lab Deal Called Model Sale. 2007. *JoongAng Ilbo.* June 8.

Lundvall, Bengt-Ake (ed.). 2010. National Systems of Innovation: Toward a Theory of Innovation and Interactive Learning. London: Anthem Press.

Lust for Upgrades Builds a Mountain of e-waste. 2007. *Sydney Morning Herald.* April 16.

MaDriX Sets New Standards for the PE Industry. 2009. *Printed Electronics Now.* September.

Making a Comeback—Bluffs Manufacturer's Revival in the Wake of the Recession Echoes Nationally. 2013. *Omaha World-Herald.* June 21.

Manchester Leads the Way in Graphene Membrane Research. 2013. *Printed Electronics World.* April 8.

The March of the Makers Seems to Have Got Going. *Coventry Telegraph.* February 10, 2014.

Market Intelligence & Consulting Institute. 2008. China's Visinox Begins OLED Production in Kunshan. October 17.

Martin, C., Markku Heino, and Markku Lämsä. 2013. *Research Europe.* April.

Materials for the Microencapsulated Electrophoretic Display. 2012. *Flexible Substrate.* June.

Mathews, John A. and Dong-Sung Cho. 2000. *Tiger Technology: The Creation of a Semiconductor Industry in East Asia.* Cambridge: Cambridge University Press.

Max Planck Institute Researchers Use an Inkjet Printer to Create Electrically Conductive Paper. 2013. *Flexible Substrate.* October.

Mazzucato, Mariana. 2013. *The Entrepreneurial State: Debunking Public vs. Private Sector Myths* London, New York, and Delhi: Anthem Press.

MC10 Develops Stretchable Skin to Monitor Health. 2012. *Flexible Substrate.* November.

MC10: Reshaping Electronics. 2012. *Flexible Substrate.* September.

Merck Exhibits Innovative Display Materials at IMID, FPD International. 2011. *Printed Electronics Now.* October 11.

Merck KGaA Inaugurates Liquid Crystals Center in Shanghai. 2013. *Printed Electronics Now.* December 3.

Merck KGaA and Plastic Logic Jointly Develop New Generation Organic Semiconductors. 2010. *Printed Electronics Now.* April 14.

Merck KGaA Receives Solar Industry Award 2013. 2011. *Printed Electronics Now.* October 11.

Military Projects Push Boundaries of Flexible Electronics in Imaging Technologies. 2012. *University of Wisconsin, Madison News.* December 28.

Mina, Andrew, David Connell, and Alan Hughes. 2010. Models of Technology Development in Intermediate Research Organizations. Paper prepared for DR UID Summer Conference, Opening up Innovation: Strategy, Organization, and Technology, June 16-18, Imperial College, London. p. 13.

Ministry of Science and Technology. 2005. The 1st Large-Scale OLED Production Line of Mainland China Settled in Kunshan. Press Release. November 23.

MIT and Korean Universities. 2011. *Dong-A Ilbo Online*. May 20.

MIT. 2012. U.S. Re-Shoring: A Turning Point. *MIT Forum for Supply Chain Innovation and Supply Chain Digest*. Cambridge, MA: MIT.

More Businesses are Returning to Made-in-USA Mentality, Spokane. 2013. *Spokesman-Review*. May 26.

More Japan Firms Planning to Invest in Taiwan. 2011. *China Post Online.* March 3.

Morell, Jonathan A. 2005. Why Are There Unintended Consequences of Program Action, and What Are the Implications for Doing Evaluation? *American Journal of Evaluation.* 26(4):444-463.

Morse, J. D. 2011. *Nanofabrication Technologies for Roll-to-Roll Processing: Report from the NIST-NNN Workshop.*

Morton, David and Eric Forsythe. 2007. Flexible-Display Development for Army Applications. *Information Display.* October.

Motorola Deal to Improve Airport Luggage Handling. 2009. *South China Morning Post.* May 12.

Mowery, David C. and Nathan Rosenberg. 1993. The U.S. National Innovation System, in Richard R. Nelson. *National Innovation Systems: A Comparative Analysis.* New York: Oxford University Press.

Much Finer Detail Possible with Inkjet in Japan. 2008. *Printed Electronics World.* March 28.

Mulhbauer, Novacentrix Enter Long-Term Collaboration for Developing a Flexible and Cost Effective RFID Antenna Printing Technology. 2013. *Printed Electronics Now.* April 26.

Nagoya University Researchers Develop Plastic Products That Are Moldable All-Carbon Circuits. 2013. *Flexible Substrate.* November.

Nano-based RFID Tag, You're It. 2010. *Rice University News & Media.* March 18.

NanoMarkets. 2012. *Flexible Substrate Markets: Special Report for the FlexTech Alliance.* April, p. 5.

NanoMarkets. 2012. *Silver Inks and Pastes Markets: Special Report for the FlexTech Alliance.* February, p. 810.

National Academy of Engineering. 2014. *Frontiers of Engineering: Reports on Leading-Edge Engineering from the 2013 Symposium.* Washington, DC: The National Academies Press.

National Academy of Science. 2013. Interview with Kent Displays, Kent, Ohio, Albert Green, Kent Displays. Roll-to-Roll Manufacturing of Flexible Displays in Building the Ohio Innovation Economy. Washington, DC: The National Academies Press. p. 122-124.

National Economic Council. 2011. *A Strategy for American Innovation: Driving Towards Sustainable Growth and Quality Jobs*. Washington, DC: The White House.

National Institute of Standards and Technology. 2012. Federal Laboratory Transfer Fiscal Year 2010. Department of Commerce.

National Institute of Standards and Technology. 2014. Global Standards Information. Department of Commerce. Available online at http://gsi.nist.gov/global/index.cfm/L1-5/L2-44/A-165.

National Kyushu Taking the Lead in Thin Film Solar Cells. 2010. *Asahi Shimbun.* November 13.

National Network for Manufacturing Innovation Frequently Asked Questions. Accessed June 7, 2013. http://manufacturing.gov/docs/nnmi_faa.pdf.

National Research Council. 1987. *Semiconductor Industry and the National Laboratories: Part of a National Strategy.* Washington, DC: National Academy Press.

National Research Council. 1991. *Mathematical Sciences, Technology, and Economic Competitiveness.* James G. Glimm, ed. Washington, DC: National Academy Press.

National Research Council. 1992. *The Government Role in Civilian Technology: Building a New Alliance.* Washington, DC: National Academy Press.

National Research Council. 1995. *Allocating Federal Funds for R&D.* Washington, DC: National Academy Press.

National Research Council. 1996. *Conflict and Cooperation in National Competition for High-Technology Industry.* Washington, DC: National Academy Press.

National Research Council. 1997. *Review of the Research Program of the Partnership for a New Generation of Vehicles: Third Report.* Washington, DC: National Academy Press.

National Research Council. 1999. *The Advanced Technology Program: Challenges and Opportunities.* Charles W. Wessner, ed. Washington, DC: National Academy Press.

National Research Council. 1999. *Funding a Revolution: Government Support for Computing Research.* Washington, DC: National Academy Press.

National Research Council. 1999. *Industry-Laboratory Partnerships: A Review of the Sandia Science and Technology Park Initiative.* Charles W. Wessner, ed. Washington, DC: National Academy Press.

National Research Council. 1999. *New Vistas in Transatlantic Science and Technology Cooperation.* Charles W. Wessner, ed. Washington, DC: National Academy Press.

National Research Council. 1999. *The Small Business Innovation Research Program: Challenges and Opportunities.* Charles W. Wessner, ed. Washington, DC: National Academy Press.

National Research Council. 2000. *The Small Business Innovation Research Program: An Assessment of the Department of Defense Fast Track Initiative.* Charles W. Wessner, ed. Washington, DC: National Academy Press.

National Research Council. 2000. *U.S. Industry in 2000: Studies in Competitive Performance.* Washington, DC: National Academy Press.

National Research Council. 2001. *The Advanced Technology Program: Assessing Outcomes.* Charles W. Wessner, ed. Washington, DC: National Academy Press.

National Research Council. 2001. *Attracting Science and Mathematics Ph.Ds to Secondary School Education.* Washington, DC: National Academy Press.

National Research Council. 2001. *Building a Workforce for the Information Economy.* Washington, DC: National Academy Press.

National Research Council. 2001. *Capitalizing on New Needs and New Opportunities: Government-Industry Partnerships in Biotechnology and Information Technologies.* Charles W. Wessner, ed. Washington, DC: National Academy Press.

National Research Council. 2001. *A Review of the New Initiatives at the NASA Ames Research Center.* Charles W. Wessner, ed. Washington, DC: National Academy Press.

National Research Council. 2001. *Trends in Federal Support of Research and Graduate Education.* Washington, DC: National Academy Press.

National Research Council. 2002. *Government-Industry Partnerships for the Development of New Technologies: Summary Report.* Charles W. Wessner, ed. Washington, DC: The National Academies Press.

National Research Council. 2002. *Making the Nation Safer: The Role of Science and Technology in Countering Terrorism.* Washington, DC: The National Academies Press.

National Research Council. 2002. *Measuring and Sustaining the New Economy.* Dale W. Jorgenson and Charles W. Wessner, eds. Washington, DC: National Academy Press.

National Research Council. 2002. *Partnerships for Solid-State Lighting.* Charles W. Wessner, ed. Washington, DC: The National Academies Press.

National Research Council. 2004. *An Assessment of the Small Business Innovation Research Program: Project Methodology.* Washington, DC: The National Academies Press.

National Research Council. 2004. *Productivity and Cyclicality in Semiconductors: Trends, Implications, and Questions.* Dale W. Jorgenson and Charles W. Wessner, eds. Washington, DC: The National Academies Press.

National Research Council. 2004. SBIR—Program Diversity and Assessment Challenges: Report of a Symposium, Charles W. Wessner, ed., Washington, DC: The National Academies Press.

National Research Council. 2004. *The Small Business Innovation Research Program: Program Diversity and Assessment Challenges.* Charles W. Wessner, ed. Washington, DC: The National Academies Press.

National Research Council. 2006. *Beyond Bias and Barriers: Fulfilling the Potential of Women in Academic Science and Engineering.* Washington, DC: The National Academies Press.

National Research Council. 2006. *Deconstructing the Computer.* Dale W. Jorgenson and Charles W. Wessner, eds. Washington, DC: The National Academies Press.

National Research Council. 2006. *Software, Growth, and the Future of the U.S. Economy.* Dale W. Jorgenson and Charles W. Wessner, eds. Washington, DC: The National Academies Press.

National Research Council. 2006. *The Telecommunications Challenge: Changing Technologies and Evolving Policies.* Dale W. Jorgenson and Charles W. Wessner, eds. Washington, DC: The National Academies Press.

National Research Council. 2007. *Enhancing Productivity Growth in the Information Age: Measuring and Sustaining the New Economy.* Dale W. Jorgenson and Charles W. Wessner, eds. Washington, DC: The National Academies Press.

National Research Council. 2007. *India's Changing Innovation System: Achievements, Challenges, and Opportunities for Cooperation.* Charles W. Wessner and Sujai J. Shivakumar, eds. Washington, DC: The National Academies Press.

National Research Council. 2007. *Innovation Policies for the 21st Century.* Charles W. Wessner, ed. Washington, DC: The National Academies Press.

National Research Council. 2007. *SBIR and the Phase III Challenge of Commercialization.* Charles W. Wessner, ed. Washington, DC: The National Academies Press.

National Research Council. 2008. *An Assessment of the SBIR Program at the Department of Defense.* Charles W. Wessner, ed. Washington, DC: The National Academies Press.

National Research Council. 2008. *An Assessment of the SBIR Program at the Department of Energy.* Charles W. Wessner, ed. Washington, DC: The National Academies Press.

National Research Council. 2008. *An Assessment of the SBIR Program at the National Science Foundation.* Charles W. Wessner, ed. Washington, DC: The National Academies Press.

National Research Council. 2009. *An Assessment of the SBIR Program at the Department of Defense.* Charles W. Wessner, ed. Washington, DC: The National Academies Press.

National Research Council. 2009. *An Assessment of the SBIR Program at the National Aeronautics and Space Administration.* Charles W. Wessner, ed. Washington, DC: The National Academies Press.

National Research Council. 2009. *An Assessment of the SBIR Program at the National Institutes of Health.* Charles W. Wessner, ed. Washington, DC: The National Academies Press.

National Research Council. 2009. Revisiting the Department of Defense SBIR Fast Track Initiative, Charles W. Wessner, ed., Washington, DC: The National Academies Press.

National Research Council. 2009. *Venture Capital and the NIH SBIR Program.* Charles W. Wessner, ed. Washington, DC: The National Academies Press.

National Research Council. 2010. *Managing University Intellectual Property in the Public Interest.* Stephen Merrill and A. Mazza, eds., Washington, DC: The National Academies Press.

National Research Council. 2011. *Building the 21st Century: U.S.-China Cooperation on Science, Technology, and Innovation.* C. W. Wessner, rapporteur. Washington, DC: The National Academies Press.

National Research Council. 2011. *Growing Innovation Clusters for American Prosperity.* C. W. Wessner, rapporteur, Washington, DC: The National Academies Press.

National Research Council. 2011. *The Future of Photovoltaics Manufacturing in the United States.* C. W. Wessner, rapporteur. Washington, DC: The National Academies Press.

National Research Council. 2012. *Building Hawaii's Innovation Economy*. C. W. Wessner, rapporteur. Washington, DC: The National Academies Press.

National Research Council. 2012. *Building the Arkansas Innovation Economy*. C. W. Wessner, rapporteur. Washington, DC: The National Academies Press.

National Research Council. 2012. *Building the U.S. Battery Industry for Electric-Drive Vehicles: Progress, Challenges, and Opportuntities*. C. W. Wessner, rapporteur. Washington, DC: The National Academies Press.

National Research Council. 2012. *Clustering for 21st Century Prosperity*. C. W. Wessner, rapporteur. Washington, DC: The National Academies Press.

National Research Council. 2012. *Meeting Global Challenges: German-U.S. Innovation Policy*. C. W. Wessner, rapporteur. Washington, DC: The National Academies Press.

National Research Council. 2012. *Rising to the Challenge: U.S. Innovation Policy for the Global Economy*. C. W. Wessner and Alan Wm. Wolff, editors. Washington, DC: The National Academies Press.

National Research Council. 2013. *Building the Illinois Innovation Economy: Summary of a Symposium*. C. W. Wessner, rapporteur. Washington, DC: The National Academies Press.

National Research Council. 2013. *Building the Ohio Innovation Economy: Summary of a Symposium*. C. W. Wessner, rapporteur. Washington, DC: The National Academies Press.

National Research Council. 2013. *Competing in the 21st Century: Best Practice in State and Regional Innovation Initiatives* . C. W. Wessner, ed. Washington, DC: The National Academies Press.

National Research Council. 2013. *Strengthening American Manufacturing: The Role of the Manufacturing Extension Partnership—Summary of a Symposium*. C. W. Wessner, rapporteur. Washington, DC: The National Academies Press.

National Research Council. 2014. *The Flexible Electronics Opportunity in 21st Century Manufacturing*. Washington, DC: The National Academies Press.

National Research Council. 2014. *SBIR at the Department of Defense*. Washington, DC: The National Academies Press.

National Science and Technology Council. 2013. *National Network for Manufacturing Innovation: A Preliminary Design*. Washington, DC: Executive Office of the President.

National Science Board. 2010. *Science and Engineering Indicators 2010*. Arlington, VA: National Science Foundation.

National Science Foundation, National Center for Science and Engineering Statistics. 2011. *Research and Development in Industry: 2006-07*. NSF 11-301. Arlington, VA: National Science Foundation.

National Taiwan University Develops Stacked Nanoparticle Layers. 2013. *Flexible Substrate*. January.

National Tsing Hua University Develops Direct Growth Fabrication for Paper-Based Electronics. 2010. *Flexible Substrate*. November.

National University of Singapore. 2014. BASF Embark on Joint Graphene Research. *Printed Electronics Now*. January 20.

National Universities Face Sweeping Reforms. 2010. *Yonhap*. September 28.

The Nature of Innovation Market Failure and the Design of Public Support for Private Innovation. *Research Policy* 29. no. 4-5 (2000): 437-448.

NETPark Invests in Future Success, Newcastle. 2008. *The Journal*. November.

New Developments in Highly Efficient OLEDs. 2013. *Flexible Substrate*. May.

New Funding for Promising Hasselt University Spin-offs Camargus and Lumoza. 2012. *LRM*. December 12.

Next Generation Printing Innovations with Electronic Intelligence. 2013. *Printed Electronics World*. October 14.

A New Industry Shapes the Future of Printing. 2008. *Printed Electronics Now*. December.

New Kodak Comes Into Focus After Bankruptcy. 2013. *Orlando Sentinel*. September 4.

New Korea-China Research Program Focuses on Nanomaterials for Flexible Solar Cells. 2010. MIT *Technology Review*. April 19.

New Path to More Efficient Organic Solar Cells Uncovered at Berkeley Lab's Advanced Light Source. 2013. *Flexible Substrate.* February.

New Pilot Line for Organic Electronics. 2013. *Printed Electronics World.* February 11.

New Prototype e-Paper Shown Off by Japan Display. 2012. *Good Reader.* November 6.

A New Role for Plastic. 2004. *St. Louis Post-Dispatch.* July 12.

New Technique Creates First Plastic Solar Cell. 2012. *Forbes.* April 25.

New Technology Makes E-Paper Animation Displays Possible. 2011. *Central News Agency.* January 4.

New Trends in OLED Lighting. 2014. *Flexible Substrate.* January.

Nitto Denko Invests $10 Million to Pioneer Organic Electronic Device Research in Singapore. 2008. *JCN Network.* November 30.

Non-Chaebol Firms Losing Ground to Chaebol in S. Korea. 2013. *Yonhap.* July 3.

NorTech Wins U.S. Small Business Administration Contract to Grow Flexible Electronics Cluster. 2013. *Flexible Substrate.* February.

NorTech. 2013. NorTech FlexMatters Roadmap Final Report. Available online at http://www.nortech.org/images/final-report/FlexMatters-2012-19-Strategic-Roadmap-V2-0-FINAL.pdf.

Northeast Ohio's Flex Matters Cluster Attracts $14 Million in 2010. 2011. *Printed Electronics World.* February 11.

Northern Lights. 2008. *Plastic Electronics.* April-May.

Novacentrix Wins 2010 R&D Award for Low Cost Metal on ICI Copper conductive Ink. 2010. *Printed Electronics Now.* July 20.

NREL and Solarmer Energy to Extend Lifetime of Plastic Solar Cells. 2010. *Printed Electronics World.* July 5.

NREL Uses Corning's Flexible Willow Glass to Develop Cheap, Efficient Solar Cells. 2013. *Flexible Substrate.* October.

NSF Funds Flexible Electronics and Self-Folding Structure Projects. 2012. *Flexible Substrate.* November.

Oak Ridge National Laboratory (ORNL). 2014. Manufacturing Demonstration Facility. Available online at http://web.ornl.gov/sci/manufacturing/mdf.shtml.

Obama Names Youngstown as Model for the New Tech, Youngstown. 2013. *Vindicator.* February 13.

Obama Push for 3-D Hub to Turn 'Rust Belt' City Into 'Tech City'. 2012. *Columbus Examiner.* August 17.

OE-A. *Organic and Printed Electronics.* 2014. Fraunhofer ILT, Laser Ablation for Thin Film Structuring. Available online at http://www.ilt.fraunhofer.de/en/publication-and-press/brochures/brochure_laser_ablation_for_thin_film_technology.html.

OE-A. *Organic Electronics 2nd Edition.* 2007. Fraunhofer IAP, Functional Polymer Systems. Available online at http://www.iap.fraunhofer.de/en/Forschmgsbereiche/Functionale-Polymerssysteme.html.

OE-A. 2013. Organic and Printed Electronics. p. 70.

Off, Ron. 2008. ORNL, in Oak Ridge National Laboratory Announces Granting of Exclusive Solar and Printed Electronics License to Novacentrix. Novacentrix Press Release. November 13.

Offshoring and Reshoring Trends: European Data. 2013. *The Operations Room.* September 27.

Oh Soo-hyun, O., Yoo-taeh, K. 2013. $5 Bn to Pour Into SMEs in Next 3 Yrs. *MK English Online.* May 23.

OLED Brings Out the Shine. 2013. *Printed Electronics World.* May 28.

OLED Lighting Research Backed by German Government. 2010. *Electronics Weekly.* January 13.

OLED Project Focuses on Lighting, Signage. 2009. *EE Times.* February 11.

On a Bright New Wing. 2013. *The Economist.* September 7.

Opening the Screen Door: New Consortium Hopes to Take United States into Crucial Flat-Panel Display Territory—Before it's too Late. 1993. *Austin American-Statesman.* June 28.

Optemec wins Contract to Advance Fully Printed Transistor Technology. 2013. *Flexible Substrate.* December.

Organic Electroluminescent Diodes. 1987. *Applied Physics Letters.*

Organic Electronics and Flexible Displays. 2014. *Flexible Substrate.* January.

Organic Solar Cells Reach 6% Confirmed Efficiency. 2009. *Printed Electronics World.* September 1.

Organisation for Economic Co-operation and Development (OECD). *OECD Reviews of Innovation Policy: Korea 2009.* 2009. Available online at http://www.oecd.org/sti/inno/oecdreviewsofinnovationpolicykorea.htm.

ORNL Develops Carbon Nanotube Conductive Coatings for Flexible Electronics. 2011. Flexible Substrate. September.

ORNL Wins Tech Awards. 2013. Knoxville News-Sentinel. March 21.

Orofeo, C. M., Ago, H., Yoshihara, N., Tsuji, M. 2010. Methods to Horizontally Align Single-Walled Carbon Nanotubes on Amorphous Substrate. *Journal of Novel Carbon Resource Sciences.* 2: 36-40.

OSRAM Reports Advances in Transparent OLED Development, To Start Production in 2014. 2013. *Flexible Substrate.* January.

Page, Adam and Smithers Pira. 2013. Market for Organic and Printed Electronics, in OE-A. *Organic and Printed Electronics*

Panasonic and Sony Confirm Plans to Join Forces for OLED TV Development. 2012. *Whathifi.com.* June 25.

Panel Makers Depend Much on Government Subsidies. 2012. *SinoCast.* January 4.

PARC Launches Printed Battery Project. 2013. *Flexible Substrate.* August.

PARC's Partnerships with Thinfilm, Soligie Hold Much Promise for PE. 2010. *Printed Electronics Now.* December.

Park, Jong-Woo. Samsung and the Growing Market for OLEDs. 2013. *Printed Electronics Now.* August.

Park, Seung-Rok and Ky-hyang Yuhn. 2012. Has the Korean Chaebol Model Succeeded? *Journal of Economic Studies.* 39(2):260-274.

Partners Join to Pioneer Plastic Electronics. 2004. *Financial Times.* April 6.

PELG and ESP KTN. 2012. *Capability Guide: UK Plastic Electronics.* pp. 4-5.

Peratech Works on New Ink Formulations with CPI to Enable QTC Sensors to be Printed. 2013. *Flexible Substrate.* May.

PETEC: Pushing PE into the Supply Chain. 2008. *Printed Electronics Now.* August.

Peter Pan SMEs Loath to Grow UP. 2013. *JoongAng Daily Online.* January 2.

Peter Pan Syndrome. 2013. *The Korea Herald Online.* January 5.

Phil McKinney on Device Evolution and Flexible Displays. 2010. *San Francisco Examiner.* July 17.

Photonics, European Commission. 2011. An Overview of OLAE Innovation Clusters and Competence Centres: Their role and further development for the competitiveness and growth of Europe's OLAE industry.

Pisano, G., and W. Shih. 2009. Restoring American Competitiveness. *Harvard Business Review.*

Plastic Logic, Novaled Partner to Demonstrate a World First for Displays. 2014. *Printed Electronics Now.* February 7.

Plastic Logic Abandons Pioneering E-Reader: British Firm Leapfrogs to Second Generation Device. Competition from Kindle and iPad killed off Product. 2010. *The Guardian.* August 12.

Plastic Logic Changing Strategy. Puts Off Building Plant in Russia. 2012. *Interfax.* May 17.

Plastic Logic Scraps E-Reader Development. 2012. *Business Weekly.* May 16.

Plastic Memory Firm Partners with Organic ink Startup. 2011. *EETimes Europe.* December 1.

Plastic Surgery. 2009. *The Engineer.* April 20.

Plextronics Awarded SBIR Grant for Reducing OLED Lighting Costs. 2013. *Flexible Substrate.* August.

Plextronics Lands Deal with Army Research Lab. 2007. *Pittsburgh Post-Gazette.* June 26.

Plextronics Lands $21 Million to Fund Expansion, Marketing. 2007. *Pittsburgh Post-Gazette.* August 31.

Plextronics Shorts Out. 2014. *Chemical and Engineering News.* January 27.

Plextronics Signs MOU to Supply Solar Technology to Korean Company. 2008. *Printed Electronics World.* June 9.

Polcari, Michael. 2013. The Sematech Model: Potential Applications for PV in *Future of Photovoltaics Manufacturing*. Washington: National Academy Press. p. 182.

PolyIC, Kurz. 2013. Present Touch Applications for Automotive Industry at IC 2013. *Printed Electronics Now.* November 13.

PolyMap. 2010. PolyMap Report, WP1: Survey of National and Regional Funding with OLAE Context. February 2, p. 3.

POPUP: Novel Organic Solar Cells. 2013. *Science Daily.* December 13.

POSCO Buys into U.S. Graphene Manufacturer. 2011. *Yonhap.* June 8.

POSTECH Becomes All-English Campus. 2010. *JoongAng Daily Online.* March 13.

Potential Threat to Samsung: Chinese AMOLED Panels to Pour into Market. 2014. *Business Korea.* May 2.

Powers, Charlotte Marquerite. The Changing Role of Chaebol: Multi-Conglomerates in South Korea's National Economy. 2010. *Stanford Journal of East Asian Affairs.* pp. 105–116.

PPG Expands OLED Production to Support Demand for OLED Display Products. 2013. *Flexible Substrate.* December.

PragmatIC's Imprinted Planar Nano-Devices Included in Focus of Holst Centre Program. 2013. *Flexible Substrate.* February.

Prah, Pamela M. 2013. Has U.S. Manufacturing's Comeback Stalled? *USA Today*. July 30.

President Obama Launches Advanced Manufacturing Partnership. 2011. White House Press Release. June 24.

President's Council of Advisors on Science and Technology (PCAST). 2011. *Report to the President on Ensuring American Leadership in Advanced Manufacturing.* June, pp. 19-20.

Price Waterhouse Cooper. 2011. Shale Gas: A Renaissance in U.S. Manufacturing? Accessed online April 14, 2014 at http.www.pwc.com/en_US/us/industrial-products/assets/pwc-shale-gas-as-us-manufacturing-renaissance.pdf.

Prime Minister's office, General Policy. 2003. Speech by Prime Minister Junichiro Koizumi to the 156th Session of the Diet. January 31.

Printable Transparent Conductive Films and Its Applications in Organic Solar Cells. 2013. *Flexible Substrate.* January.

Printed Battery Company in China. 2013. *Printed Electronics World.* July 22.

Printed Electronics Asia: Visits to Local Center of Excellence, Part 2. 2008. *Printed Electronics World.* October 15.

Printed Electronics at Taiyo. 2009. *Printed Electronics World.* October 14.

Printed Electronics in Korea. 2011. *Printed Electronics World.* February 15.

Printed Electronics Is the Next Big Thing. 2014. *Electronics Weekly.* January 30.

Printed Electronics—Many New Directions. 2011. *Printed Electronics World.* February 21.

Printed Electronics Research Project Moves Into Next Phase. 2012. *Package Print Worldwide.* August 2.

Printed Electronics Sector Takes a Hard Look at the Flexible Future. 2013. *Solid State Technology.* March.

Printed Electronics Suppliers Look to Learnings from Electronics World to Help Scale Volume Production. *Printed Electronics Now.* January 2010.

Printed Electronics USA 2012 Awards Recognize New Developments. 2013. *Flexible Substrate.* January.

Printed, Flexible and Organic Technology Sees 15.2% CAGR Over the Next Decade. 2013. *Solid State Technology.* May.

Printed Graphene Transistors Promise High-Speed Wireless Communications. 2013. *OSA Direct.* August 21.

Printing Batteries. 2013. *MIT Technology Review.* November 25.

Printing from Laboratory to Commercial Scale. 2013. *Printed Electronics Now.* January 7.

PrintoCent Consortium Seeks to Develop, Commercialize Printed Intelligence. 2012. *Printed Electronics Now.* December.

Product Codes Help Synchronize Flow of Goods. 2008. *South China Morning Post.* October 10.

Production Line Planned. 2012. *Akron Beacon Journal.* March 2.

Progress of Torch Plan Reviewed. 1990. *Keji Ribao.* April 11.

Project IMAGE Yields Printable Transparent and Flexible Electrodes for OLEDs and Solar Cells. 2014. *EETimes Europe.* January 15.

Public Funded OLED Project So-Light Successfully Concluded. 2013. *Printed Electronics World.* February 6.

PV Armageddon: The Rapid Market Swings Concealed Major Efficiency Gains. 2012. *Printed Circuit Design & Fab.* April.

PVI, Hyds Technologies and LG Display Launch Co-operation. 2010. *Printed Electronics World.* January 4.

Q&A, Dr. Janglin Chen of ITRI (Part 1). 2013. *The Emitter.* July 2.

Q&A: Dr. Janglin Chen of ITRI (Part 1). 2013. *The Emitter: Emerging Display Technologies.* July 3.

Qingshan L: Appointed as New Lab Director for AIXTRON China Ltd. 2013. *Printed Electronics Now.* March 21.

R2R Manufacturing of Organic PV Using Gravure and Rotary Screen Printing Technique. 2013. *Flexible Substrate.* May.

Radio Barcodes Printed on Consumer Packaging. 2005. *Printed Electronics World.* June 23.

Rand Corporation. 2002. *Federal Investment in R&D.* Science and Technology Policy Institute, study prepared for the President's Council of Advisors on Science and Technology.

REC Regrets Escalation in Solar Industry Trade War, Denies Any Wrongdoing. 2012. *Printed Electronics Now.* July 20.

Recent Progress on High-Performance OLED Technologies for Lighting Applications. 2014. *Flexible Substrate.* January.

Regulation Hindering Technological Advances. 2010. *Dong-A Ilbo Online.* March 27.

Reshaping Japan's Tech Sector a Struggle. 2012. *Financial Times.* June 4.

Reshoring Trend in the UK. 2014. *FDI—Foreign Direct Investment (UK).* February 1.

Restructured G24: Emerges Under New Ownership. 2013. *Flexible Substrate.* May.

Reuss, R.H., Chalamala, B.R., Moussessian, A., Kane, M.G., Kumar, A., Zhang, D.C., Rogers, J.A., Hatalis, M., Temple, D., Moddel, G., Eliasson, B.J., Estes, M.J., Kunze, J., Handy, E.S., Harmon, E.S., Salzman, D.B., Woodall, J.M., Alam, M.A., Murthy, J.Y., Jacobsen, S.C., Olivier, M., Markus, D., Campbell, P.M., Snow, E. 2005. Macroelectronics: Perspectives on Technology Applications. *Proceedings of the IEEE.* 93(7):1239-1256.

RFID. PE Industries Enjoy Growth in China. 2010. *Printed Electronics Now.* September.

Rho, Sungho. 2010. Technological Discontinuity and Industrial Catch-up: A Study on the Chinese Flat Panel Display Industry (presented at Globelics 2010 8th International Conference. University of Malaya, Kuala Lumpur, Malaysia, November 1-3). p. 22.

Rise of Chinese LCD Makers Threatens Local Makers. 2012. *Taipei Times.* November 12.

Rivals Furious As Hynix Set to Sell Dollars 1bn of Non-core Assets. 2003. *Financial Times.* January 31.

Rochford, K. 2011. Comments of Kent Rochford, Acting Director, Electronics and Electrical Engineering Laboratory, National Institute of Standards in Measurement and Standards: the Role of NIST in Future of Photovoltaic Manufacturing, Washington, DC: The National Academies Press.

Role of Solution Processing in the Future of OLED TVs. 2013. *Flexible Substrate.* February.

Roll-to-Roll Atomic Layer Deposition Technology for Producing Single Layer Ultra Barrier Films. 2013. *Flexible Substrate.* February.

Roper, S., J. Youtie, and A. Fernandez-Ribas. 2010. Knowledge, Capabilities and Manufacturing Innovation: A USA–Europe Comparison. *Regional Studies.* 44(3):253-279.

Rose, Doug. 2013. Sunpower, Observations on Building a PV Roadmap Model in *Future of Photovoltaic Manufacturing.* Washington: National Academy Press. pp. 198–201.

Royal Society of Chemistry. *Profile: Cambridge Display Technology.* Available online at http://www.rsc.org/about-us/.

Rural Home Appliance Subsidy Fruitful. 2013. *China Daily.* November 17.

Rusnano to Become Private Equity Fund. Remain State-Owned. 2013. *Interfax.* June 14.
Rusnano Provisions 22 bln Rubles for Failed Projects. 2013. *Interfax.* April 2.
Rusnano Raises Stake in Plastic Logic. 2011. *Plastics News.* October 18.
Ruttan, Vernon W. 2006. Is War Necessary for Economic Growth?, Clemons Lecture. Saint Johns University. Collegeville, Minnesota. October 9, p. 30.
Ruttan, Vernon W. 2006. Will Government Programs Spur the Next Breakthrough? *Issues in Science and Technology.* Winter.
Samsung Acquires Dutch Display Tech Firm. 2011. *Yonhap.* January 20.
Samsung Agrees to Buy German Screen Lighting Firm Novaled. 2013. *Bloomberg.* August 9.
Samsung Aims to Introduce Flexible Smartphone Before 2015. 2010. *MK English News Online.* July 22.
Samsung Cannot 'Kill Taiwan'—CEOs. 2013. *Taipei Times.* March 23.
Samsung Delays its Flexible Displays. 2013. *PhoneArena.com.* April 17.
Samsung Delays Flexible OLED Displays. 2012. *Plastic Electronics.* October 2.
Samsung Electro-Mechanics Launching Industrial Inkjet Business. 2009. *Printed Electronics World.* April 1.
Samsung Electronics Halts E-Paper Production. 2010. *Yonhap.* August 23.
Samsung Electronics Striving to Win Absolute Superiority in Global Market Share This Year. 2010. *MK English News Online.* January 6.
Samsung Group Unit Cheil Ind Considering Buying Novaled AG. 2013. *Financial Express.* April 2.
Samsung, LG Display to Invest in High-end Display Market. 2010. *Yonhap.* May 3.
Samsung, LG, Hyundai Motor Join Hands for Next Generation Flexible Display Development. 2010. *MK English News Online.* July 5.
Samsung. LG's Key Display Technologies Leaked by Israeli Firm. 2012. *Yonhap.* June 27.
Samsung. LG to Join Development of Thin Film Solar Cells. 2011. *Yonhap.* May 31.
Samsung Mobile Display Starts Operation of New AMOLED Plant. *Yonhap.* May 31, 2011.
Samsung to Buy Germany's Novaled, Raising Bet on Next-Generation Success. 2013. *The Wall Street Journal.* July 30.
Samsung to Invest $1.8 Billion on OLED. 2012. *The Korea Times.* November 14.
Samsung Unveils Curved OLED TV. 2013. *The Korea Times.* June 27.
Samsung's Courtship of Apple Parts Suppliers May Drive Up Costs. 2013. *The Argus.* May 16.
Savastano, D. 2011. A Look at Printed Electronics: Printed Electronics Now Interview with Dr. Bernard Kippelen, *Printed Electronics Now.* July.
SC Digest Editorial Staff. Supply Chain News: How is US Manufacturing Doing Five Years After the Great Recession? 2013. *Supply Chain Digest.* April 10.
Schaller, Robert R. 2004. Technological Innovation in the Semiconductor Industry: A Case Study of the International Technology Roadmap for Semiconductors (ITRS). Ph.D dissertation, George Mason University. p. 458.
Schaller. 2000. *Technological Innovation in the Semiconductor Industry* (citing telephone interview with Jim Daughton). Honeywell. August 4.
Schoo, Herman. 2010. *Introduction of PolyMap.* June 14.
Schrank, A., and J. Whitford. 2009. Industrial Policy in the United States: A Neo-Polanyian Interpretation. *Politics and Policy.* 37(4):521-553.
Schroder, K.A., S.C. McCool, and W.F. Furlan. 2006. Broadcast Photonic Curing of Metallic Nanoparticle Films. *NSTI-Nanotech.* (3):198-201.
Science Park Looks Forward to a Bright Future, Newcastle. 2009. *The Journal.* January 29.
Scrambling for Raw Materials: EU Plans Measures to Tackle Resource Crunch. 2010. *Spiegel Online.* November 20.
SDK and Nova Centrix to Cooperate in Printed Electronics. 2011. *JCN Network.* April 11.
Sealing Up ALD Moisture Barrier Technology. 2013. *Flexible Substrate.* May.
Seiko Epson Announces OLED TV Breakthrough. 2009. *Flexible Substrate.* June.
SEL Develops Flexible OLED Displays with New Materials. 2014. *Flexible Substrate.* March.

SEL Shows a Flexible Rechargeable Li-Ion Battery Prototype. 2013. *Flexible Substrate.* November.
Semanza, Paul. OLEDs in Transition. 2011. *Information Display.* October.
Seoul National University Develops Bandage that Senses Tremors and Delivers Drugs. 2014. *Flexible Substrate.* April.
Seoul's Commerce Ministry to Change Name with Added Role. 2013. *Yonhap.* March 22.
Seung-byung, Chae. Manufacturing Sector at its Worst in Three Years. 2012. *The Korea Herald Online*. December 9.
Shanghai Tianma to Invest in GEN 4.5 AMOLED Pilot Line. 2010. *OSA Direct.* August 30.
Shapira, P. 2009. Innovation and Small and Midsize Eenterprises: Innovation Dynamics and Policy Strategies. In R. Smits, S. Kuhlmann and P. Shapira, eds., *Innovation Policy: Theory and Practice. An International Handbook.* Cheltanham, United Kingdom: Edward Elgar.
Shapira, P., and J. Youtie. 1995. *Assessing GMEA's Economic Impacts: Towards a Benefit-Cost Methodology*. Atlanta, GA: Georgia Institute of Technology. (Revised October 1996).
Shapira, P., J. Roessner, and R. Barke. 1995. New Public Infrastructures for Small Firm Industrial Modernization in the USA. *Entrepreneurship and Regional Development*. 7:63-84.
Shapira, P., J. Youtie, and J. D. Roessner. 1996. Current Practices in the Evaluation of U.S. Industrial Modernization Programs. *Research Policy.* 25(2):185-214.
Shapira, P., J. Youtie, J. Wang, D. Hegde, D. Cheney, Q. Franco, and S. Mohapatra. 2004. *Re-assessing the Value of Information and its Impact on Productivity in Small and Midsize Manufacturers*. Atlanta, GA, and Arlington, VA: Georgia Tech Policy Project on Industrial Modernization and SRI International.
Shapira, P., J. Youtie, and L. Kay. 2011. Building Capabilities for Innovation in SMEs: A Cross-Country Comparison of Technology Extension Policies and Programs. *International Journal of Innovation and Regional Development* 3-4:254-272.
Shimpo, Hiroko. 2011. *Business Alliances Between Japanese and Taiwanese Companies.* Osaka Sang Yo University, (http://www.nikkeibp.co.jp/article/NEWS/20110614/1925531).
Showa Denko Develops Printable Silver Nanowire Ink. 2013. *Flexible Substrate.* January.
Shunpei Yamazaki Holds Almost 2000 Patents. 2007. *Wired.* October 26.
Six Technium Innovation Centers to Be Cut. 2010. *Western Mail.* November 19.
Sleeckx, Eric. 2008. Implementing and Monitoring the Flemish Innovation System in *Innovative Flanders: Innovation Policies for the 21st Century: Report of a Symposium*. Washington, DC: The National Academies Press. pp. 38–41.
Slim Leap Forward. 2012. *South China Morning Post.* September 28.
S. Korea Develops World's 1st Imprintable, Flexible Battery. 2013. *Yonhap*. January 15.
S. Korea Launches Project to Develop New Materials. 2010. *Yonhap.* September 30.
S. Korea to Spend 40 Mln USD on Graphene Development. 2013. *Xinhua.* May 21.
SmartKem Joint Development Agreement with Asian Display Manufacturer. 2013. *Printed Electronics World.* December 10.
Smits, Ruud E., Stefan Kuhlmann and Philip Shapira (eds.). 2012. *The Theory and Practice of Innovation Policy: An International Research Handbook* (Cheltenham and Northampton, MA: Edward Elgar.
Solar Market Continues to Evolve. 2012. *Printed Electronics Now.* July.
Solar Power from Plastic Foils. 2012. *Printed Electronics World.* July 5.
Solar Shakeout: Konarka Technologies Files for Bankruptcy. 2012. *PVTech.* June 2.
Soligie Enhances Printed Electronics Manufacturing Capabilities with Sheetfed, Flatbed Screen Printing Line. 2011. *Printed Electronics Now.* April 7.
Soligie Receives Two R&D Awards from FlexTech Alliance. 2013. *Flexible Substrate.* December.
Soligie (Taylor Corp). 2008. Progresses Co-deposition of Printed Electronics Components. *PI World.* September.
Solliance Develops World's First All Inkjet Printed OPVs. 2014. *Flexible Substrate.* January.
Solvay Acquires Plextronics to Accelerate its OLED Display Development. 2014. *Flexible Substrate.* April.

Solvay and the Holst Centre Present an Efficient (30 lm/w) Large OLED Lighting Panel. 2012. *Flexible Substrate.* March.

Solvay Commits $15 Million to Support Plextronics' Innovative Technology Development in OLED and OPV. 2011. *Printed Electronics World.* July 27.

Solvay Expands its Printed Electronics Development with Investment in Minority Stake in Polyera. 2010. *Hugin.* September 7.

Solvay Has Made a $10M Investment in Plastics. 2007. *Plastics Engineering.* October.

Sony. 2013. Sony Develops the World's First and Largest '56-inch 4K OLED TV' Prototype to be Exhibited at 2012 International CES. Press Release. January 7.

Sony More Open to TV Panel Sourcing. 2011. *Central News Agency.* December 28.

Sony, Panasonic OLED Link is Official. 2012. *Advanced Television.* June 26.

Sony to Buy All of AUO's OLED Capacity for Its Mobile Phones? 2012. *OLED-Info.com.* April 6.

Sony, Toshiba, Hitachi, Unload LCD Units to Japan government-Backed Fund. 2011. *Bloomberg.* August 31.

South Korea to Speed Up Overseas Development of Indium. 2011. *Yonhap.* May 11.

South Korean Electronics Companies Are Beating Japanese Competitors to the Punch in Getting Cutting Edge Technologies Developed by Japanese Researchers into Their Product Ranges. 2010. *Asahi Shimbun.* July 9.

Special Supplement: KSND, Rising Star in the Yangtze River Delta. 2007. *China Daily.* September 7.

SQW Ltd., The Evaluation of the Northern Way 2008-11: Final Report —Full Report, April 2011. 74.

Stanford and UNL Engineers Make World's Fastest Organic Transistor. 2014. *Flexible Substrate.* January.

Stanford Engineers Monitor Heart Health Using Paper-Thin Flexible Skin. 2013. *Flexible Substrate.* August.

Start-up Spirit Emerges in Japan. 2013. *The New York Times.* December 25.

Startups in Materials: An Interview with Antonio Facchetti. 2013. *Materials Views.* February 5.

State-Run Laboratory Export Advanced Nanoimprinting Technology. 2010. *Yonhap.* March 15.

The Status and Outlook of R2R Atomic Layer Deposition Technology. 2014. *Flexible Substrate.* January.

Stewart, S.L. and James A. St. Pierre. 1997. NIST, Experiences with a Manufacturing Framework. *Business Object Design and Implementation.* p. 138.

Stretchable Electronics: A Shapely Future for Circuits. 2011. *The Economist.* March 10.

Stretchable Electronics Enable Minimally-Invasive Cardiac Electrophysiological Sensing and Actuations. 2014. *Flexible Substrate.* January.

Stretchable Electronics Maker Raises $19.8M. 2013. *Boston Business Journal.* December 27.

Stowsky, Jay. 2003. Secrets to Shield or Share? New Dilemmas for Dual Use Technology Development and the Quest for Military and Commercial Advantage in the Digital Age. BRIE Working Paper 151.

Sumitumo Chemical Joins Holst Centre OLED Research Program. 2012. Holst Centre Press Release. September 18.

Sumney, Larry. 2012. The Semiconductor Research Corporation (SRC): A Proven Means to Fund Relevant Research in *Future of Photovoltaic Manufacturing.* Washington, DC: The National Academies Press. p. 186.

Sun Burn 2: Global Changes Slow Solar Growth. 2012. *Toledo Free Press.* July 26.

Swamidass, P. 1994. *Technology on the Factory Floor II: Benchmarking Manufacturing Technology Use in the United States.* Washington, DC: The Manufacturing Institute.

Swansea Unveiled as UK Centre of Excellence for Plastic Electronics. 2009. *Tendersinfo.* December 10.

Szondy, D. February 2013. Acrobatic Polymer Film Developed at MIT Harvests Energy from Water Vapor," Flexible Substrate.

Ta-Lin Hsu, founder of private equity firm H&Q Asia Pacific, in Japan's Once Mighty Tech Industry Has Fallen Far Behind Silicon Valley. 2012. *San Jose Mercury News.* October 12.

Taedok to Become Mecca for Venture Firms. 1998. *Chonja Sinmun.* April 10.

Taheri, Bahman. 1998. Manufacturing of Curved Liquid Crystal Devices, in National Research Council. *Building the Ohio Innovation Economy.* p. 125.

Taiwan Enters Japan's Semiconductor Supply Chain. 2012. *Taiwan Insights.* April 4.

Taiwan Research Team Turns Silk into E-Display Material. 2011. *Asia Pulse.* March 2.

Taiwan Tries to Shore Up its Defenses Against Samsung. 2013. *The New York Times.* April 21.

Taiwan's Industrial Technology Research Institute. 2013. Taiwan's Industrial Technology Research Institute: A Cradle of Future Industries in *21st Century Manufacturing, The Role of the MEP Program.* Washington DC: The National Academies Press.

Taiyo America Adopts Knowledge from PCB Market to Printed Electronics. 2014. *Printed Electronics Now.* January.

Tamtik, Merli. 2012. Rethinking the Open Method of Coordination: Mutual Learning Initiatives Shaping the European Research Enterprise. *Review of European and Russian Affairs.* 7(2):2.

Tassey, G. 2009. *The Technology Imperative.* Northampton, MA: Edward Elgar.

Tassey, G. 2010. Rationales and Mechanisms for Revitalizing U.S. R&D Manufacturing Strategies. *Journal of Technology Transfer* 35:283-333.

Tassey, Gregory. Modeling and Measuring the Economic Roles of Technology Infrastructure *Economics of Innovation and New Technology* (January 2008). p. 624.

Tassey, Gregory. 1995. Technological Infrastructure and the Role of Government. Working Paper Series. December 19.

TD Economics. 2012. Onshoring, and the Rebirth of American Manufacturing. October 15.

Technical Difficulties Foil Sony-Panasonic OLED Effort. 2014. *Flexible Substrate.* January.

Technium Centers a Waste, Says Price. 2012. *Western Mail.* March 26.

Technology Strategy Board. 2010. Minister Announces £8.4m Investment in Plastic Electronics Technologies. Press Release. July 1.

Teijin Acquires Nano Gram Corporation. 2010. *Printed Electronics World.* August 11.

Teijin Chemicals Enters Electronic Paper Market. 2010. *Printed Electronics World.* July 6.

Teijin Acquires NanoGram Corporation—Will Accelerate Nanoparticle R&D for Printable Electronics. 2010. *Nanowerk.* August 9.

Teijin's World First Silicon-on-Plastic Integration Technology. 2009. *Printed Electronics World.* November 28.

Tekes, Beneq. 2013. Coating Manufacturer Ready to Grow. 2013. Available online at http://www.tekes.fi/en/tekes/results-and-impact/cases1/2013/beneq-coating-manufacturer-ready-to-grow.

Tera-Barrier Investment & Exclusive Distribution Agreement with Kisco. 2011. *Printed Electronics World.* January 26.

Testing Nanomaterial Smart Plastics in Real Time at Fraunhofer ICT. 2013. *Flexible Substrate.* November.

Thierbach, Dieter. Deutsche Enheit in Forschung und Technologie (BMFT 1991). *Materialforschung* (BMFT, 1986), JPRS-EST-86-036; Forschung und Technik zum Whole der Menschen: Jahresbericht, 1984, JPRS-WST-86-015.

Thompson, C. 1998. Local Politics, National Policy, and the Taxpayer-Payback of Manufacturing Extension. *The Journal of Technology Transfer.* 23(1):37-42.

Thompson, Malcolm J. A Consortium in Flexible Electronics for Security, Manufacturing and Growth in the US, in National Research Council. *Flexible Electronics for Security, Manufacturing, and Growth.*

TMR. 2013. *Flexible Electronics Market—Global Industry Size, Share, Trends, Analysis and Forecasts 2012-2018.*

Top Flexible Electronics Developments Win 2013 FLEXI Awards. 2013. *Flexible Substrate.* February.

Toppan Printing and Taiwan's Chi Lin Collaborates to Launch Electronic Paper Business Targeting Use in Industry. 2011. *Printed Electronics Now.* May 12.

Toppan Printing, Plastic Logic Collaborate on OTFT-Based Large Area Flexible Display. 2013. *Printed Electronics Now.* March.

Toxic Waste Shipped by UK Firm Still Lying at Port. 2013. *Kenya Daily Nation Online*. August 5.

Transparent OLEDs for Signage and Decorative Lighting Applications. 2013. *Flexible Substrate.* January.

Transistor Could Lead to Wearable Computers. 2007. *JoongAng Ilbo.* October 9.

Truly International Holdings Limited. 2013. Shareholders Agreement on Formation of a Joint Venture Company. December 17, p. 5.

Twilight of an Industry: Bankruptcies Have German Solar on the Ropes. 2012. *Spiegel Online.* April 3.

UC Berkeley E-Skin Responds to Touch with Promise for Sensory Robotics and Interactive Environments. 2013. *Flexible Substrate.* August.

UK Opportunity: Korean Project 2012-17. 2012. *Plastic and Printed Electronics.* May 1.

Ultra-High Basics for Encapsulation of Flexible Organic Electronic Devices. 2014. *Flexible Substrate.* March.

United States: Defense Money Goes to Idaho Projects. 2009. *TendersInfo.* December 18.

Universal Display and Fraunhofer Agreement for White OLED Lighting. 2012. *Printed Electronics World.* May 3.

Universal Display Corporation. 2013. UDC 2012 10-K risk disclosures. February 27. p. 3-6.

Universal Display's State-of-the-Art Chemistry Laboratory in Hong Kong. 2011. *Printed Electronics World.* April 29.

Universities Get a Failing Grade. 2011. *JoongAng Daily Online.* June 17.

Universities: Spin-Outs Put New Life into the Sector. 2007. *Financial Times.* September 18.

University of Cambridge. 2014. Cambridge Innovation and Knowledge Center. Accessed February 11 at http://www-cikc.eng.cam.ac.uk/category/networks/.

University of Cambridge Graphene Spin-out Funded. 2013. *Plastic Electronics.* May 1. Available online at http://www-cikc.eng.cam.ac.uk.

University of Tokyo Forms Industry-Academic Cooperation Group. 2007. *Tokyo Semiconductor FDP World.* July 31.

University Plays a Role in SmartKem's Organics Success Story. 2014. *Electronics Weekly.* March 19.

Updates on Visionox's OLED Lighting Program: New Panels and Lamps Unveiled. 2012. *OLED-Info. com.* April 22.

Upgrading Self-Aligned Imprint Lithography (SAIL) in Preparation for Roll-to-Roll Manufacturing. 2013. *Flexible Substrate.* November.

U.S. Census Bureau. 2010. *Trade in Advanced Technology Products*. Washington, DC: U.S. Department of Commerce.

U.S. Department of Commerce. 2012. *The Competitiveness and Innovative Capacity of the United States*. Washington, DC: U.S. Department of Commerce. January.

U.S. Department of Commerce. 2012. *U.S. Competitiveness and Innovation Policy*. Washington, DC: U.S. Department of Commerce. January.

U.S. Manufacturers 'Relocating' from China. 2013. *Financial Times*. September 23.

U.S. Sets Antidumping Duties on China Solar Imports. 2012. *Bloomberg.* October 10.

The User Innovation Revolution. 2011. *MIT Sloan Management Review.* Fall.

Utterback, J.M. and Murray, A.E. 1977. The Influence of Defense Procurement and Sponsorship of Research and Development on the Development of the Civilian Electronics Industry. National Institute of Standards and Technology, Massachusetts Institute of Technology Energy Laboratory.

Varma, Roli. 2000. Changing Research Cultures in U.S. Industry. *Science, Technology, and Human Values.* Autumn.

Ver-Bruggen, Sara. 2011. Nanomaterials Firms Turn to Asia for Commercial Opportunities. *Plastic Electronics.*

Vest, Charles M. 2007. *The American Research University from World War II to World Wide Web*. University of California Press. June 1, p. 32.

Visionox Announces Its Commitment to Gen 4.5 AMOLED Production. 2011. *OSADirect.* August 8. Available online at http://www.visionox.com/en/about.aspx.

Visionox Developed Full-Color 3.5 Flexible AMOLED. 2014. *OLED-Info.com.* March 27.

Visionox Starts Constructing a 5.5-Gen AMOLED Line, Develops a Flexible 3.5' Monochrome AMOLED. 2013. *OLED-Info.* August 2.

Volume Production Necessary for Flexible Electronics. 2007. *Solid State Technology.* February 26.

von Hippel, Eric. 1988. *The Sources of Innovation.* New York and Oxford: Oxford University Press.

von Hippel, Eric. 2005. *Democratizing Innovation.* Cambridge, MA: The MIT Press.

VTT. 2011. *Promoting Entrepreneurship in Organic and Large Area Electronics in Europe*. p. 4.

VTT Technical Research Centre of Finland. 2014. Introducing Printed Electronics Into Mass-Produced Articles. Available online at http://www.vtt.fi/references/introducing_printed_electronics_into_mass_produced_articles.jsp?lang=en.

VTT Upgrades PrintoCent Pilot Factory. 2012. *Printed Circuit Design & Fab.* September 20.

WAG Supports a Smart Move for Electronics. 2011. *Liverpool Daily Post.* January 19.

Wang, Lisa. 2012. AUO Signs Accord with Idemitsu to Manufacture OLEDs. *Taipei Times Online.* February 3.

Washington Post. 2010. Made in Germany, Sold in China. September 17.

Waterproof Film for Solar Panels. 2009. *Japan Market Information.* July 25.

Welch, D., E. Oldsman, P. Shapira, J. Youtie, and J. Lee. 1997. *Net Benefits: An Assessment of Manufacturing Business Networks and Their Impacts on Member Companies*. Chapel Hill, North Carolina: USNet and Regional Technology Strategies.

Western Mail. £100m and 10 Years On—Lessons from the Technium Experience. 2013. June 1.

What Happened to Japan's Electronics Giants? 2013. *BBC News Asia*. April 1.

What Lenovo's deal could mean for American manufacturing. 2014. *Engadget*. January.

What's Behind the Success of German Manufacturing Industry? 2012. *Xinhua.* February 23.

What's Next: Batteries Push Paper Into Electronics Age. 2001. *The New York Times.* May 24.

White, T. K., J. Reiter, and A. Petrin. 2012. *Plant-level Productivity and Imputation of Missing Data in U.S. Census Manufacturing Data*. National Bureau of Economic Research Working Paper No. 17816.

Whither Our National Universities? 2011. *JoongAng Daily Online,* June 22.

Why Plastic Logic Failed—Despite the E-book Boom. 2012. *GigaOM.* May 17.

Williams, Chris. Last Word: Who? Where? When? 2013. *Flexible Substrate.* December.

Winnes, Markus and Uwe Schimack. 1999. *National Report: Federal Republic of Germany.* Institute for the Study of Societies. TSER Project No. SOE1-CT96-1036. May.

Wistron Folds Up Polymer Vision—Collapsible eReaders Are Once Again Science Fiction. 2012. *The Digital Reader.* December 2.

WMU, Daetwyler R&D Announce Partnership in PE. 2009. *Printed Electronics Now.* November 25.

WMU's CAPE Plays Key Role in Integrating PE, Printing. 2009. *Printed Electronics Now.* December 22. Available online at http://www.wmich.edu/engineer/cape.

World Technology Evaluation Center. 2010. WTEC Panel Report on European Research and Development in Hybrid Flexible Electronics. July, p. xvi and 7.

World's First Pilot Factory for Printed Intelligence Industrialization Opens at VTT. 2012. *EurekAlert.* March 13.

Xenon Launches Worldwide PE Test Center Network to Help Drive Commercialization of Printed Electronics Industry. 2012. *Printed Electronics Now.* May 1.

Yase, Kiyoshi. 2008. Organic Transistor Frontline, Joint Research Between AIST and the Japan Chemical Innovation Institute, Technology for Fabrication of Organic TFT Arrays Using Full Printing Method, Step Closer to Realization of Printable Fabrication. *Tokyo Semiconductor FDP World.* December 2.

Yissum and Vaxan Collaborate and Printed Electronics Ink. 2011. *Printed Electronics World.* February 2.

Yong-min, K. 2012. A Call for Scientific Reinvention. *JoongAng Daily Online*. April 21.

Youtie, J., and P. Shapira. 1997. *GMEA 1997: Review of Results*. Evaluation Working Paper E9701. Atlanta, GA. March.

Youtie, J., and P. Shapira. 2005. *The Challenge of Manufacturing Innovation: Industry, Rurality, and Competitiveness in the State of Georgia*. Evaluation Working Paper E2501. Atlanta, GA, June.

Youtie, J., P. Shapira, L. Kay, D. Dodonova, D. Sabbarese, and C. Morales. 2010. *Innovation in Manufacturing: Needs, Practices, and Performance in Georgia 2010-2012*. Atlanta, GA: Georgia Tech Program in Science, Technology, and Innovation Policy.

YSU Expands Research Role with S. Korean Institute. 2013. *Vindicator.* April 3.

YungKai Yang. 2006. The Taiwanese Notebook Computer Production Network in China: Implication for Upgrading of the Chinese Electronics Industry (University of California at Irvine, Personal Computing Industry Center, February 2006). p. 19.